T0192528

Risk Management in Engineering and Construction

Today's businesses are driven by customer 'pull' and technological 'push'. To remain competitive in this dynamic business world, engineering and construction organizations are constantly innovating with new technology tools and techniques to improve process performance in their projects. Their management challenge is to save time, reduce cost and increase quality and operational efficiency. Risk management has evolved as an effective method of managing both projects and operations. Risk is inherent in any project, as managers need to plan projects with minimal knowledge and information, but its management helps managers to become proactive rather than reactive. Hence, it not only increases the chance of project achievement, but also helps ensure better performance throughout its operations phase.

Various qualitative and quantitative tools are researched extensively by academics and routinely deployed by practitioners for managing risk. These have tremendous potential for wider applications. Yet the current literature on both the theory and practice of risk management is widely scattered. Most of the books emphasize risk management theory but lack practical demonstrations and give little guidance on the application of those theories. This book showcases a number of effective applications of risk management tools and techniques across product and service life in a way useful for practitioners, graduate students and researchers. It also provides an in-depth understanding of the principles of risk management in engineering and construction.

Stephen O. Ogunlana is currently the Chair of Construction Project Management at the School of Energy, Geoscience, Infrastructure and Society at Heriot-Watt University, Scotland. He has an international reputation for research in the application of system dynamics to construction projects and organizations. He is the author of over 200 scholarly publications in journals and conferences. His research has been funded by several governments. His works on leadership were awarded the Emerald Literati Award for two consecutive years for the most outstanding paper in the *Engineering, Construction and Architectural Management* journal. He has a continuing interest in technology innovation for project management and in public-private partnerships for infrastructure delivery.

Prasanta Kumar Dey is a Professor of Operations Management at Aston Business School. He has been honoured as 50th Anniversary Chair of Aston University in 2017. Prior to joining Aston University in 2004, he worked for five years at the University of the West Indies in Barbados and for 14 years at Indian Oil Corporation Limited. He specializes in supply chain management and project management, and circular economy. He has published more than 150 research papers in leading international refereed journals. He has accomplished several impactful interdisciplinary research projects on a sustainable supply chain of small- and medium-sized enterprises (SMEs), which were funded by leading bodies such as the Research Council UK, British Council, Royal Society, Royal Academy of Engineering, EU Horizon 2020 and ERDF. He is the founder editor-in-chief of the *International Journal of Energy Sector Management*.

Spon Research

Publishes a stream of advanced books for built environment researchers and professionals from one of the world's leading publishers. The ISSN for the Spon Research programme is ISSN 1940-7653 and the ISSN for the Spon Research E-book programme is ISSN 1940-8005

Risk Management in Engineering and Construction

Tools and Techniques

Edited by Stephen O. Ogunlana
and Prasanta Kumar Dey

Routledge
Taylor & Francis Group

LONDON AND NEW YORK

First published 2020 by Routledge

2 Park Square, Milton Park, Abingdon, Oxon OX14 4RN
605 Third Avenue, New York, NY 10017

Routledge is an imprint of the Taylor & Francis Group, an informa business

First issued in paperback 2021

Publisher's Note

The publisher has gone to great lengths to ensure the quality of this reprint but points out that some imperfections in the original copies may be apparent.

British Library Cataloguing-in-Publication Data
A catalogue record for this book is available from the British Library

Library of Congress Cataloging-in-Publication Data
A catalog record has been requested for this book

ISBN: 978-0-415-48017-8 (hbk)
ISBN: 978-1-03-217724-3 (pbk)
DOI: 10.4324/9780203887059

Typeset in Baskerville
by Swales & Willis Ltd, Exeter, Devon, UK

Contents

Contributors

Editors

Stephen O. Ogunlana is currently the Chair of Construction Project Management at the School of Energy, Geoscience, Infrastructure and Society at Heriot-Watt University, Scotland. He has an international reputation for research in the application of system dynamics to construction projects and organizations. He is the author of over 200 scholarly publications in journals and conferences. His research has been funded by several governments. His works on leadership were awarded the Emerald Literati Award for two consecutive years for the most outstanding paper in the *Engineering, Construction and Architectural Management* journal. He has continuing interest in technology innovation for project management and in public-private partnerships for infrastructure delivery.

Prasanta Kumar Dey is a Professor of Operations Management at Aston Business School. He has been honoured as 50th Anniversary Chair of Aston University in 2017. Prior to joining Aston University in 2004, he worked for five years in the University of the West Indies in Barbados and 14 years in Indian Oil Corporation Limited. He specializes in supply chain management and project management, and circular economy. He has published more than 150 research papers in leading international refereed journals. He has accomplished several impactful interdisciplinary research projects on sustainable supply chain of small and medium sized enterprises (SMEs), which are funded by leading funding bodies such as the Research Council UK, British Council, Royal Society, Royal Academy of Engineering, EU Horizon 2020 and ERDF. He is the founder editor in chief of the *International Journal of Energy Sector Management*.

Contributors

Martinus Abednego is currently Head of Engineering Department at an oil and gas field asset in Algeria owned by Pertamina International EP, a subsidiary of PT Pertamina, which is Indonesia's National Oil Company. He has more than 20 years' experience in the application of construction engineering and project management, including in the oil and gas industry. He has been involved in various projects in Indonesia, Thailand, Qatar and Algeria.

Saheed Ajayi, BSc, MSc, PhD, PGcertAP, FHEA, MCIOB, is currently a Senior Lecturer at Leeds Beckett University. He has expertise in construction management, BIM and construction informatics. Through these areas of expertise, Dr. Saheed has led and participated in various applied and collaborative projects involving investigation, development and deployment of digital technologies for solving productivity issues affecting AEC firms.

Oluwaseyi A. Awodele is an Associate Professor in the Department of Quantity Surveying, is currently the Sub-Dean for Master Degree Programmes in the School of Postgraduate Studies, Federal University of Technology, Akure. He has supervised 15 masters and three PhD Students and has authored over 60 scholarly publications in referred journals and conferences locally and internationally.

Prince Boateng is a Senior Lecturer in the Faculty of Built and Natural Environment of Koforidua Technical University. He was a Lecturer in Construction and Project Management and served as the Course Leader for BSc (Hons) Construction Management Program at Robert Gordon University, Scotland. He gained his PhD from Heriot-Watt University and has published in several leading scientific journals. His current research projects involve STEEP risk assessment with SDANP plus, project portfolio management and infrastructure investment justification.

Atanu Chaudhuri is an Associate Professor in Operations and Supply Chain Management at Aalborg University, Copenhagen. His research interests are in digital supply chains with focus on additive manufacturing, supply chain risk management and supply chain integration. He has published more than 30 papers in reputed international journals.

Walid Cheffi is an Associate Professor of Accounting at CBE-UAE University. He earned his PhD in accounting from Paris Dauphine University. He has lectured at Paris Dauphine University, University of Dubai and Neoma Business School, France. His research interests focus on accounting for managers, performance and risk measurement, and corporate governance. He has published extensively in international refereed journals.

Ben Clegg is a professor of operations management. His research, teaching and consulting activities focus on operations and process improvement. He draws upon theoretical ideas, such as systems thinking and multi-organization enterprise management, to implement practical solutions for organizations. He got his PhD from De Montfort University. He has published extensively in international refereed journals.

John S. Edwards is Professor of Knowledge Management at Aston Business School, Birmingham, UK. His research interests include how knowledge affects risk management; knowledge management strategy and implementation; and its synergy with analytics and big data. He has published over 75 research articles and three books. He is consulting editor of the journal *Knowledge Management Research & Practice.*

Konstantinos I. Evangelinos is a senior researcher in Corporate Social Responsibility and Environmental Management with over 60 papers in academic journals in the field. He is editor in chief for *Progress in Industrial Ecology: An International Journal.*

Brian Fildes is a Professor of Research at the Monash University Accident Research Centre. He has over 250 published papers in journals and international conferences and several books and book chapters in injury prevention. He is a Visiting Professor at Loughborough University, has served as President of the AAAM USA, is advisor to governments in Australia and has won numerous awards.

Farid Ezanee Mohamed Ghazali is a senior lecturer at the School of Civil Engineering, Universiti Sains Malaysia (USM) where his expert areas are in construction management and project management. He acquired his PhD in risk management from the University of Leeds, United Kingdom. He is currently involved in research studies on developing a rating tool for sustainable infrastructures in Malaysia.

Yaser E. Hawas is a Professor at the Department of Civil and Environmental Engineering at the UAE University. He also serves as the Director of the Roadway, Transportation, and Traffic Safety Research Center (RTTSRC) at the UAE University. His primary areas of expertise include the development of intelligent transportation systems, advanced control systems, traffic network modeling and safety analysis.

Mastura Jaafar is currently attached to the quantity surveying program at the School of Housing, Building and Planning, Universiti Sains Malaysia. She has numerous years of experience in the construction industry dealing with project estimation and costing, project management and development proposals. Her areas of research, publication and supervision interests include strategic management in the construction and housing development.

Soo-Yong Kim is currently the coordinator of the Interdisciplinary Program of Construction Engineering and Management at the Department of Civil Engineering at Pukyong National University, Korea. He has a Korean reputation for research in Built-Transfer-Lease (BTL), Value Engineering (VE) and project management. He is the author of over 50 scholarly publications in top-tier journals and refereed conferences.

Foteini Konstandakopoulou is an Adjunct Professor at the Engineering Project Management MSc Program at Hellenic Open University, Greece. She has over 15 papers in peer-reviewed journals, book chapters and conference proceedings. Her main research interests focus on the area of construction law and safety and environmental impacts of construction projects.

Chrisovalantis Malesios is a mathematician-statistician, and is currently occupied as a Marie-Curie research fellow at the Aston Business School, Aston University. He has published papers in international journals such as: *Business Strategy and the Environment, Ecological Economics, Annals of Operations Research, Ecological Indicators* and *South European Society and Politics.*

Ibrahim Motawa is an Associate Professor (Senior Lecturer) in the Department of Architecture, University of Strathclyde, Glasgow, UK. His research work covers the areas of: digital construction, BIM, construction simulation and optimisation, dynamic modelling in built environment, risk and knowledge management. He obtained his PhD in construction, MSc in structural engineering and BSc in civil engineering.

Moza T. Al Nahyan completed her PhD in management at the Monash Business School (Department of Management), Monash University Australia. She has published a number of journal and conference papers relating to the management of infrastructure projects. She is currently teaching in the College of Business Administration, Abu Dhabi University, UAE.

Ioannis E. Nikolaou is an Associate Professor of Corporate Environmental Management at the Department of Environmental Engineering at Democritus University of Thrace, Greece. He has over 60 papers in peer-reviewed journals, 15 book chapters and three books (in Greek and English).

Jonathan Nixon is a Senior Lecturer in the School of Mechanical, Aerospace and Automotive Engineering, Coventry University. He was a Senior Lecturer in Renewable Energy Engineering at Kingston University London and gained his PhD from Aston University. He has been published in several leading scientific journals. His current research projects involve the provision of energy services in the UK, India, China and Pakistan.

Isaac A. Odesola is currently a Lecturer in the Department of Building, Faculty of Environmental Studies, University of Uyo, Nigeria. He has authored over 60 scholarly publications in refereed learned journals, book chapters and conferences, most of which are in construction management. His area of research interests is in construction resource management and labour productivity studies.

Adekunle S. Oyegoke, DSc, MCIOB, MNIQS, is a senior lecturer at the School of Built Environment and Engineering. A distinguished academic, quantity surveyor and an expert in the theory and practice of construction procurement, Adekunle has published extensively in journals, peer reviewed conferences and has supervised many doctoral works to successful completion. He is a member of many professional bodies.

Eduardo Rodriguez is Sentry Insurance Endowed Chair in Business Analytics at the University of Wisconsin – Stevens Point, USA. He is Principal at IQAnalytics Inc. and Adjunct Professor at the University of Ottawa, Canada. He holds a PhD from Aston Business School, UK, in the field of knowledge management, an MSc in mathematics from Concordia University in Montreal, Canada, and an MBA and BSc in mathematics from Universidad de Los Andes, Colombia.

Krish Saha is a Senior Lecturer in International Business at the Birmingham City Business School. His research interests include institutional influence on business performance, country governance, supranational institutions and trade. Krish has published and advised businesses on political risk, trade regulation, country governance and their impacts on business performance. Krish holds a PhD in trade economics and an MBA.

Djoen San Santoso is currently working as an Associate Professor in the Department of Civil and Infrastructure Engineering, School of Engineering and Technology at the Asian Institute of Technology, Thailand. He has published international journal papers related to risk, PPP, culture and performance evaluation in construction and infrastructure sectors.

Anthony Sinclair until recently was Director of Pharmacy at a specialist children's hospital and currently works as a management consultant advising hospitals in a number of areas including efficiency, effectiveness, quality management and cost savings (outpatient pharmacies, homecare and special medicines) and business development. He has published a number of papers, book chapters and has a research interest in safety.

Amrik S. Sohal is a Professor in the Department of Management, Monash Business School, Monash University Australia. His expertise is in operations and supply chain management. He has published over 240 journal papers and a number of books and chapters. He is a past president and now a Life Fellow of the Australia and New Zealand Academy of Management.

Nguyen Van Thuyet is currently working for Vietnam Oil and Gas Group, the biggest state-owned group in Vietnam. He has deep expertise and experience on risk management for oil and gas projects from upstream to downstream fields. He has international publications on risk management.

Thomas A. Tsalis is an environmental engineer and a PhD holder in corporate social responsibility. He has published 13 papers in international peer-reviewed journals and he has participated in various research projects in the field of corporate sustainability.

Nguyen Van Tuan is currently the project manager of a large construction project in Vietnam. He obtained his master's degree in construction engineering and management, HCMC Polytechnic University, Vietnam in 2005. He worked for many years as a project scheduler and project manager in building construction projects in Ho Chi Minh City, Vietnam.

Luu Truong Van is currently the property manager at Maxim Realty Group LLC, USA. He received his PhD degree in construction management, Pukyong National University, Korea. He has published about 100 papers in Vietnamese journals and 15 papers in top-tier journals and refereed conferences. His H-index is of 6. His research interests focus on system dynamics simulation, performance measurement, project management and risk management.

Daniel Wright is one of the founders of Grid Edge company that changes the way that people use energy by empowering a new generation of energy consumers to become active players in the UK's energy system. Prior to this he worked as post doc fellow at European Bio Research Institute at Aston University and was awarded a PhD degree from Aston Business School.

Introduction

Prasanta Kumar Dey

The engineering and construction industry is often considered a risky business due to the complexity of its demand, supply and internal operations management. Unfortunately, it has a poor reputation of managing these risks. However, there is no project and operation that has no risk. Risk can be minimised, shared, transferred, and accepted. It can't be ignored (Latham 1994). Risk can be simply defined as any possible event that can negatively affect the viability of a project or action. Risk management includes the processes concerned with identifying, analyzing, and responding to risk. It includes maximizing the results of positive events and minimizing the consequences of adverse events. Risk and uncertainty are inherent in all projects no matter the size. Although size can be one of the major causes of risk, other factors carrying risk with them include complexity, speed of implementation, location of the project, and familiarity with the work.

Risk management was adopted formally in the 1990s (Renuka et al. 2014). The sources of risk were categorized based on controllable and uncontrollable factors, which lead to cost and time overrun in a project (Akincl and Fischer 1998). Accordingly, different risk assessment models were developed using multiple criteria decision-making frameworks (e.g. Wirba et al. 1996, Dawood, 1998). In the early 2000s, a more robust approach was undertaken to identify risk and as a result risks were better classified and their management became more sophisticated using tools like influence diagrams, Monte Carlo Simulation, the Analytic Hierarchy Process and decision support tools (Dey 2001). Post 2010, risk assessment and decision support tools are integrated so as to make major project management decisions. Researchers suggested a hierarchical structure for risk management (Goh et al. 2013 and Dey 2010). Researchers also have related project risks with complexity of the surrounding environment (Lazzerini and Mkrtchyan 2011) and revealed that implementation of risk management results in improvement in project quality, cost and schedule performance of projects. Through a review of the literature of three decades Renuka et al. 2014 came up with a list of risk factors in construction. They are scope and design changes; technology implementation; site conditions and unknown geological conditions;

inflation, country economic conditions and rules and regulations; unavailability of funds and financial failure; inadequate managerial skills and improper coordination between teams; lack of availability of resources; weather and climatic conditions; statutory clearance and approvals; poor safety procedures; and construction delays. Recent research (Gupta et al. 2019) reveals through a literature survey of the last five decades that the common causes for project failure are top management's commitment and involvement/support; allocation of scarce resources; communications among various stakeholders; team configuration and structure; and social cohesion in the team and complexity of the project and organizational culture.

Risk identification is studying a situation to realize what could go wrong in the product design and development project at any given point of time during the project. Sources of risk and potential consequences need to be derived, before they can be acted upon to mitigate them (Ahmed at al. 2007). The most serious effects of risks are failure to keep within the cost estimate, failure to achieve the required completion date, and failure to achieve the required quality and operational requirements. The common tools that are used for risk identification are brainstorming, the Delphi technique, checklists, influence diagrams, cause and effect diagrams, failure mode and effect analysis, hazard and operability studies, fault trees, event trees, SWOT analyses, and risk registers.

The function of risk analysis is to determine the influence of risk factors on the system as a whole. Several techniques have been adopted to pursue risk analysis in the engineering and construction industry. They are probability and impact grids, estimation of system reliability, sensitivity analysis and simulation, decision tree analysis, portfolio management, and the multi-criteria decision-making method.

Risk management attempts using all the information as derived in the previous two steps – risk identification and analysis – to suggest risk mitigating measures. A reactive approach to risk mitigation measures often uses contingency plans, while a proactive approach makes use of risk analysis outcome and undertakes action accordingly (e.g. insurance). A combination of these two approaches is generally followed to mitigate, avoid, and reduce the likelihood and impact of risk.

Additionally, one must understand project and project management success factors in the engineering and construction industry. According to Shokri-Ghasabeh and Kavousi-Chabok (2009), the factors are project schedule management, project budget management, quality/specification management, effective project control mechanisms, scope management, effective project change management, stakeholders' satisfaction management, project team management, management commitment, resource management, project contracts management, and project risk management. It is important to note that project risk management is an important factor for making a successful project in the engineering and construction industry.

Today's businesses are driven by customer 'pull', technological 'push' and 'pressure' from policymakers and customers. To remain sustainable in this dynamic business environment, the engineering and construction industry constantly innovates new technology, and tools and techniques to transform processes in order to enhance performance of projects and operations. Their management's challenge is to save time, reduce cost and increase quality and operational efficiency. Risk management has recently evolved as an effective method of managing both project and operations management. Risk is inherent in any project and operation, as managers need to plan projects with minimal knowledge and information, but its management helps managers to become proactive rather than reactive. Hence, it not only increases the chance of project success, but also helps ensure better performance throughout its operations phase.

Various qualitative and quantitative tools are routinely deployed by practitioners and researchers for managing risk and have been reported extensively over the past decade. These have tremendous potential for wider applications. Yet the current literature on both the theory and practice of risk management is widely scattered. Most of the books emphasize risk management theory but lack practical demonstrations and give little guidance on the application of those theories. This book showcases a number of effective applications of risk management tools and techniques across product and service life in a way useful for practitioners, graduate students and researchers. And it also provides an in-depth understanding of the principles of risk management in engineering and construction. Additionally, it demonstrates the application risk management tools and technique across continents. Through 20 chapters, this book presents a few state of the art risk management tools and techniques and demonstrates their applications. This will help the researchers to further develop the tools and techniques through appropriate analysis of pros and cons, and facilitate the industry practitioners to adopt them in their system to achieve benefit with their usage.

References

Ahmed, A., Kayis, B. and Amornsawadwatana, S. (2007), A review of techniques for risk management in projects, *Benchmarking: An International Journal*, 14 (1), 22–36.

Akincl, B. and Fischer, M. (1998), Factors affecting contractors' risk of cost overburden, *ASCE Journal of Management in Engineering*, 4 (1), 67–76.

Dawood, N. (1998), Estimating project and activity duration: A risk management approach using network analysis, *Construction Management and Economics*, 16, 41–48.

Dey, P. K. (2001), Decision support system for risk management: a case study, *Management Decision*, 39 (8), 634–649.

Dey, P. K. (2010), Managing project risk using combined analytic hierarchy process and risk map, *Applied Soft Computing*, 10 (4), 990–1000.

Goh, C. S., Abdul-Rahman, H. and Samad, Z. A. (2013), Applying Risk Management workshop for a public construction project: Case Study, *ASCE Journal of Construction Engineering and Management*, 139 (5), 572–580.

Gupta, S. K., Gunasekaran, A., Antony, J., Gupta, S., Bag, S. and Roubaud, D. (2019), Systematic literature review of project failures: Current trends and scope for future research, *Computer & Industrial Engineering*, 127, 274–285.

Latham, M. (1994), *Constructing the team*. London: HMSO.

Lazzerini, B. and Mkrtchyan, L. (2011), Analyzing Risk Impact Factors Using Extended Fuzzy Cognitive Maps, *IEEE Systems Journal*, 5 (2), 288–297.

Renuka, S. M., Umarani, C. and Kamal, S. (2014), A review on critical risk factors in the life cycle of construction projects, *Journal of Civil Engineering Research*, 4 (2A): 31–36.

Shokri-Ghasabeh, M. and Kavousi-Chabok, K. (2009), Generic project success and project management success criteria and factors, *Literature Review and Survey*, 8 (6), 456–468.

Wirba, E., Tah, J. and Howes, R. (1996), Risk interdependencies and natural language computations, *Engineering Construction and Architectural Management*, 3, 251–269.

1 Monte Carlo Simulation as a risk management tool in construction and engineering projects

Isaac A. Odesola

Highlights

- Tracing the advent of Monte Carlo Simulation method
- An overview of the principle and procedure for Monte Carlo Simulation
- General applications of the Monte Carlo Simulation method
- Application of Monte Carlo Simulation method to construction and engineering projects
- Demonstrating Monte Carlo simulation method for project cost and schedule risk management.

Introduction

Construction and engineering projects often entail risks and uncertainties either from the complexity of the project or from its internal and external environments which could pose serious challenges in decision making. According to the Project Management Institute (PMI) (2008), project management involves the application of knowledge, skills, tools and techniques to project activities which invariably are directed at achieving project requirements. Apart from identifying and addressing project requirements and the various needs, concerns and expectations of the stakeholders, project management will also strive towards balancing project constraints of scope, quality, schedule, budget, resources, and risk (PMI 2008). It is widely acknowledged that construction projects are faced with enormous complexities and uncertainties primarily because the production processes take place in-situ where the projects are subject to various influencing factors or conditions that could alter project scope, schedule, cost, quality and other project requirements. In addition, the unique nature of all construction projects underscores the importance of risks and uncertainties in the management of construction projects which will require advanced modelling techniques capable of handling such situations.

Robinson (as cited in Martijn 2017, pg. 4) considered simulation as an "experimentation with a simplified limitation of an operations system as it progresses through time, for the purpose of better understanding and/or improving that system". Hence, simulation is seen as an excellent tool for addressing complex real-life problems for which mathematical modelling approaches are infeasible and modelling behaviour of complex systems for which conducting experiments may be expensive, time consuming, or dangerous (Martijn 2017). Construction project managers have in recent times explored simulation techniques to model a project's internal and external environments in a bid to optimize the system and address arrays of risks and uncertainties challenging the management of construction projects. One of such simulation techniques is the Monte Carlo Simulation (MCS) otherwise referred to as probability simulation. It has found application in almost all spheres of life and construction management is not an exception.

Consequently, this chapter is aimed at discussing the MCS technique and how it has been applied to the management of risks in construction and engineering projects. The chapter begins with an overview of the origin of the technique, continues with the underlying principle of the MCS, MCS applications and application of MCS in project management, and concludes with an illustration of schedule risk analysis using Primavera Risk Analysis.

Origin of the MCS method

The response to solving problems has always necessitated research, discovery of new theories and invention of various devices, tools, equipment and machines. The origin of MCS is not unconnected to this paradigm. MCS evolved in the process of proffering a solution to complicated mathematical integrals associated with the theory of nuclear chain reactions involving prediction of neutron histories (neutronics), hydrodynamics, neutron diffusion in fissionable material and thermonuclear detonation. Ideas for solving identified problems may be birthed but may not be utilized due to lack of appropriate technology. The nature of the problem that led to the discovery of the Monte Carlo method which has its root in the theory of probability and statistical sampling was first conceived in 1777 by Georges-Louis Leclerc, Comte de Buffon and was called the Buffon Needle problem.

However, lack of appropriate technology at that time could not allow such ideas to come to the fore because of the length and tediousness of the calculations involved. Invention of devices such as the Fermiac by Enrico Fermi and Percy King and ENIAC (Electronic Numerical Integrator and Computer) – the first electronic computer – by a team of scientists, engineers, and technicians led by John Mauchly and Presper Eckert (Metropolis, 1987), provided the impetus for the development of the modern day MCS. This method emerged in the mid twentieth century through the works of Stanislaw Ulam, John von Neumann and Nicholas Metropolis who were succinctly referred to as the pioneers of the modern day MCS. Nevertheless, the credit for the discovery of this novel approach to solving complex problems that ordinarily would have been tedious and practically impossible with numerical and analytical approaches is given to Stanislaw Ulam, a

Polish-born mathematician who suggested to John von Neumann the statistical sampling approach to evaluating complicated mathematical integrals that arise in the theory of nuclear chain reactions.

The idea for the invention of the Monte Carlo method by Stanislaw Ulam in 1946 was birthed while considering the probabilities of winning a card game of solitaire. Interestingly, the name Monte Carlo has no link with the inventor as normally would be expected, it was suggested by Nicholas Metropolis, who headed the team that developed the first algorithm and carried out the first actual Monte Carlo calculations on the ENIAC computer, from the premise that gave rise to the idea. According to Metropolis (1987, p. 127), "I suggested an obvious name for the statistical method – a suggestion not unrelated to the fact that Stan had an uncle who would borrow money from relatives because he 'just had to go to Monte Carlo'". The term Monte Carlo refers to the city of Monte Carlo in Monaco where lots of gambling takes place.

There are different variants of the MCS methods. Santra (2015) categorized them into two major groups, namely, classical Monte Carlo in which samples are drawn from a probability distribution, often the classical Boltzmann distribution to obtain thermodynamic properties or minimum-energy structures, and quantum Monte Carlo in which random walks are used to compute quantum-mechanical energies and waver functions, often to solve electronic structure problems, using Schrodinger's equation as a formal starting point.

Underlying principle and procedure for MCS

The MCS is essentially sets of computational algorithms involving statistical sampling which could be applied to solving vast ranges of problems especially those that are complex and tedious for the usual numerical or analytical methodologies. The underlying principle to the MCS is the generation and processing of random input variables based on their identified statistical properties to produce output or a series of output through successive runs. It provides approximate solutions by approximating the expectations or considering the mean of the outputs emanating from the several simulation runs performed on the basis of the law of large numbers. Hence, Monte Carlo is a form of stochastic process because it utilizes random variables but differs in the sense that it aims at obtaining a non-random output. Likewise, it is not a deterministic approach because all the data is not known at first, it is only drawn through repeated statistical sampling from an identified distribution based on the statistical property of the input variable. The basic steps involved in utilizing the Monte Carlo principle are:

1   Design a deterministic model that proffers a solution to the problem in terms of the input variables involved and the attendant mathematical relationships or formula that will give the desired output. The model is run using the base case for all the input variables without considering factors that could influence them to produce an output.

2   Assess the various input parameters involved in the problem and select appropriate statistical distributions that fit their properties through a process referred to as distribution fitting. This process will require historical data of the various input variables. Methods for distribution fitting include method of maximum likelihood (ML), method of moments (ME) and nonlinear optimization. The probability distributions that are available for consideration could be discrete, continuous or multivariate. Examples of discrete distributions are Bernoulli distribution, Binomial distribution, Geometric distribution, Poisson distribution and Uniform distribution. Continuous distributions involve Beta distribution, Cauchy distribution, Exponential distribution, Gamma distribution, Normal distribution and Uniform distribution. Multivariate distributions could be Dirichlet distribution, Multivariate Normal distribution or Multivariate Student's t distribution

3   Generate a set of independent and identically distributed random numbers for each of the input variables from its probability distribution. The set of random numbers so generated are inputted into the deterministic model to obtain a set of outputs. Available methods for sampling or drawing variates from a distribution include the inverse transformation method, composition method, convolution method and acceptance-rejection method. This procedure is repeated as many times as would be necessary for the application of the principle of large numbers in the approximation of the expectations. This is why the Monte Carlo method is referred to as a method of repeated statistical sampling. Several random number generators are available for use and include linear congruential generators, multiple-recursive generators, matrix congruential generators, modulo 2 linear generators, and combined generators.

4   Obtain the Monte Carlo estimate(s) of the expected value(s) for the many sets of outputs generated from the simulation runs. This estimate is simply the average or mean of the sets of outputs. The MCS process could be illustrated mathematically by considering a random variable X with probability density function $f_X(x)$ greater than zero over sets of values χ. The expected value of a function g of X will be

$$\mathbb{E}\big(g(\mathrm{X})\big) = \sum_{x \in \chi} g(x) f_X(x)$$

if X is discrete, and

$$\mathbb{E}\big(g(X)\big) = \int_{x \in \chi} g(x) f_X(x) dx$$

if X is continuous. Taking n samples of X's $(x_1,\ldots,x_n)$, then the mean of $g(x)$ over the sample being the Monte Carlo estimate of $\mathbb{E}(g(X))$ could be computed as

$$\tilde{g}_n(x) = \frac{1}{n}\sum_{i=1}^{n} g(x_i)$$

However, computing the mean in terms of the random variable which is called the Monte Carlo estimator of $\mathbb{E}(g(X))$ is

$$\tilde{g}_n(X) = \frac{1}{n}\sum_{i=1}^{n} g(X)$$

Consider the weak law of large numbers for any arbitrary ϵ

$$\lim_{n\to\infty} P(|\tilde{g}_n(X) - \mathbb{E}(g(X))| \geq \epsilon = 0)$$

This implies that as n gets larger, there is a small probability that $\tilde{g}_n(X)$ will deviate much from $\mathbb{E}(g(X))$. In the same vein, the strong law of large numbers equally suggests that provided n is large enough, the Monte Carlo estimate $\tilde{g}_n(X)$ resulting from the MCS runs will be close to the expected value $\mathbb{E}(g(X))$. Therefore, for $\tilde{g}_n(x)$ being unbiased for $\mathbb{E}(g(X))$ we have

$$\mathbb{E}(\tilde{g}_n(X)) = \mathbb{E}\left(\frac{1}{n}\sum_{i=1}^{n} g(X_i)\right) = \frac{1}{n}\sum_{i=1}^{n} \mathbb{E}(g(X_i)) = \mathbb{E}(g(X))$$

A number of statistical analyses could also be performed on the stored sets of outputs resulting from the simulation runs to make inferences.

The accuracy of the solution provided by the Monte Carlo method is a function of the number of runs made and also on the accuracy of the distribution fitting process for the input variables. Convergence rate of the process is considered slow which is estimated to be $\frac{1}{\sqrt{N}}$ from the Central Limit Theorem. This is an indication that the larger the value of N the smaller the statistical error or approximation error, which is random. There are however, error or variance reduction methods that could be employed to limit the errors in the Monte Carlo method, namely antithetic random variables, control variables, conditional Monte Carlo and importance sampling.

Monte Carlo applications

The MCS method is amenable to a vast range of problems in virtually all disciplines including agriculture, basic medical sciences, business management, engineering, environmental sciences, medicine, natural and applied sciences, pharmaceutical studies and social sciences. In mathematics it is used to solve multi-dimensional integrals with complex boundary conditions, and to solve multi-dimensional optimization problems; in statistics it is used to address difficult problems like the traveling salesman problem; in computer graphics it is used in the ray tracing rendering technique; in engineering and business fields it is used to assess the overall risk due to a variety of ill-defined factors, when taking a decision e.g. the program SAROPS used by the US Coast Guard to predict the likely position of debris and survivors of a maritime accident; in astronomy it is used in modelling light scattering by dust grains, and generation of initial conditions for N-body simulations (Aguilar, 2012).

Construction and engineering projects involve real-life situations and problems that are essentially characterized by uncertainties and probabilities which are technically referred to as risks. The involvement of risks in real-life problems is what makes it complex and difficult for ordinary numerical and analytical approaches to handle, hence, the application of MCS in construction and engineering projects.

Application of Monte Carlo method in project management

There could be quite a number of applications of MCS in construction and engineering projects. Ng, Xie and Kumaraswamy (2010) established the probability distribution of two indicators (namely amount of equity and return on equity) under the influence of risks for a concession-based public-private partnership (PPP) through MCS. Bock and Truck (2011) assessed uncertainty and risk in public sector investment projects using the MCS method. Risk analysis of bridge construction projects in Pakistan using MCS was undertaken by Choudhry and Aslam (2011). Valderrama and Guadallupe (2013) applied MCS to compare tenders in construction projects. Rui-mei (2015) discussed the properties of Monte Carlo and its application to risk management with emphasis on predicting and controlling management activities with uncertainties. Ganame and Chaudhari (2015) and Kong, Zhang, Li, Zheng, and Guan (2015) analysed schedule risks in construction projects using MCS. Peleskei, Dorca, Munteanu, and Munteanu, (2015) considered the influence of risks in the cost estimation of construction projects using MCS. Joubert and Pretorius (2017) used MCS to develop a ranked checklist of risks in a portfolio of railway construction projects for use by stakeholders in similar projects.

Figure 1.1 is a typical framework model for conducting MCS in construction project risk management. It is presented by Ikediashi and Ogwueleka (2018) as an adaptation from previous studies. The processes involved in the utilization of MCS for construction project risk management are structured into a flow chart and are illustrated in the Figure.

Nevertheless, there has been very limited use of this method in project management probably because of the lack of a good understanding of the method, an erroneous misconception that it is complex and tedious, and ignorance of the enormous benefits inherent in its judicious applications. In the past, prior to the advent of computers, it was quite cumbersome and impracticable to use this method. In addition, software that could perform this simulation were scarce. However, with the recent pace in information

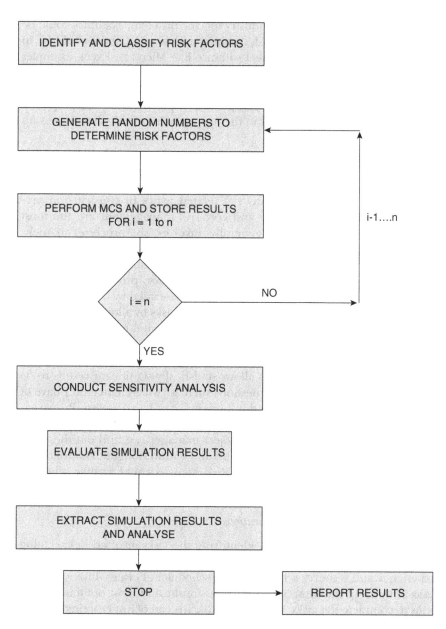

Figure 1.1 Framework for conducting MCS in construction project risk management

Source: Ikediashi and Ogwueleka (2018)

and communication technology there are a number of project management software available to model an industry-specific problem, examples could be Microsoft Project, Primavera Risk Analysis, and so on. Others could be in the form of adds-in to spreadsheet software like Microsoft Excel, examples include RiskAMP, Crystal Ball, @RISK, Solver add-in and so on. Examining the various applications of the MCS method in the different disciplines is outside the scope of this book. Consequently, emphasis will be on the application of the Monte Carlo method in the management of risks associated with construction projects.

Schedule risk analysis using Primavera Risk Analysis

As noted earlier, there are many risk-related issues in construction project management but schedule and cost risk management are the most explored. Risk management generally involves six processes namely; Risk Management Planning, Risk Identification, Risk Qualification, Risk Quantification, Risk Response Planning, and Risk Monitoring and Control. MCS is essentially utilised in the risk quantification process to effectively quantify, model and statistically analyse risks associated with project schedule and budget or cost. Risk analysis quantifies risks by allowing the Project Manager to assign durations and costs as a distribution rather than a single value. Having entered risk data, the software simulates the project many times. Each iteration of the analysis is one way the project could run. The combination of many iterations allows statistically significant results to be generated. From these results questions such as "what chance do I have of finishing the project on time and in budget?" can be answered. Primavera Risk Analysis software could help to analyse the time and cost risk within a project. By using risk analysis, the project manager can find out the probability of completing a project on time and within budget. The following basic steps illustrate how Primavera Risk Analysis could analyse schedule risks associated with a simple construction project.

1. Inputting and modelling task durations

The exercise begins by carefully identifying the tasks involved in executing the project and from available productivity norms, historical records, past experience and resource scheduling, the scheduler allocates duration to the tasks in terms of the most likely, most pessimistic and most optimistic durations to complete the tasks. When using risk, instead of just entering the best guess or estimate of the task duration (i.e. the most likely), the minimum and maximum durations are also entered as well. By entering minimum and maximum durations and applying risk analysis the user is modelling the project more accurately. The duration probability distribution from which independent identically distributed variables will be drawn to perform the simulation runs is selected for each of the activity. These information serves as the input variables that are inputted into the software as indicated in Figure 1.2.

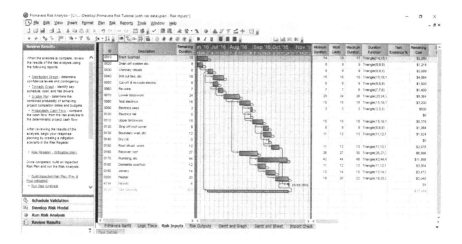

Figure 1.2 Tasks scheduling for a simple maintenance project

2. Entering uncertainty

Using risk analysis within Primavera Risk Analysis allows the user to easily enter extra information about the uncertainty of the duration and cost of the tasks. This extra information can be used to produce more accurate and realistic plans. It can also help the user manage the project more effectively by answering questions such as:

1 What is the chance of finishing the project on time?
2 What chance do I have of finishing the project on a particular date?
3 What date can I be 80% confident of finishing the project?
4 What tasks are most likely to cause project delay?

By using the Risk Register feature in Primavera Risk Analysis, users can integrate pre-developed risk registers as well as define new ones. Users can also employ this feature to produce both qualitative and quantitative models of positive and negative risk events (threats and opportunities) and their associated response plans (such as mitigation). Risk Register also automatically integrates identified risk events into the schedule, by creating a risk event plan, which users can then analyse to determine both key risk drivers and the cost-effectiveness of the identified mitigation strategies.

3. Running the analysis

During the analysis, Primavera Risk Analysis looks at each task that has a distribution and sets its duration to a value between the minimum and maximum. Once Primavera Risk Analysis has changed all the task durations

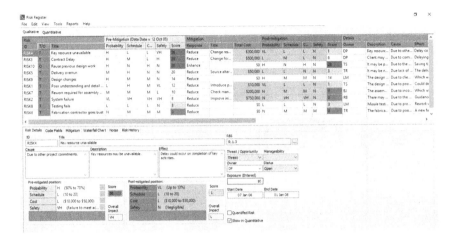

Figure 1.3 Risk register to accommodate the effects of identified risks on project schedule and cost

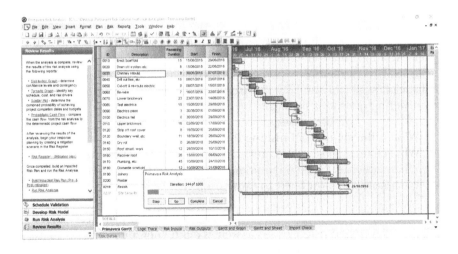

Figure 1.4 Running the risk analysis

it calculates and records the finish date. This process is repeated for the required number of iterations and the finish date of the project is stored each time. What Primavera Risk Analysis is doing is simulating the project repeatedly and seeing how the finish date varies. Once the analysis is complete the first thing the user can do is look at the chance of finishing by a certain date. Two questions that are often asked about a project are:

"What finish date can I be 80% confident of finishing by?" (the P80 finish date).

'What chance do I have of finishing by, say, the 12th of August?" (or any date that is relevant).

Both of these questions can be answered by running a risk analysis and simply reading the results from the Distribution Graph report. So let's run the risk analysis.

4. Simulation outputs

As mentioned earlier, several statistical tests could be performed on the stored data set of outputs obtained from each successive simulation runs. One of the important output results is the distribution graph. From this graph the user can:

- Determine confidence levels, P schedules, and schedule and cost contingency
- Report confidence levels with regard to finish dates.

Now we will look at reading and generating useful information from the analysis that Primavera Risk Analysis has just performed. Look at Figure 1.5

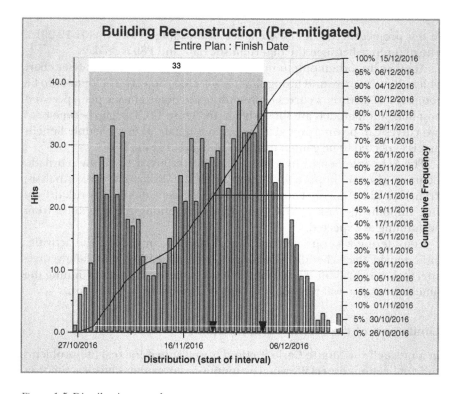

Figure 1.5 Distribution graph

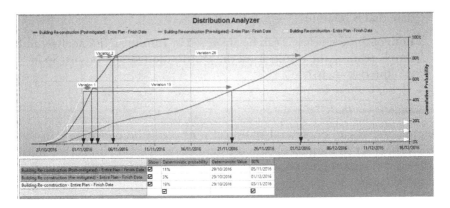

Figure 1.6 S-curves from the distribution analyser

that has been generated. The bars on the graph represent how often during the analysis the plan finished on a certain date or between a certain date range. Let's look at answering one of our questions:

What date can I be 80% confident of finishing by?

In our project there is an 80% chance of finishing before the 01-12-2016. The difference between the deterministic date and P80 is 33 days.

Another important output information is the distribution analyser chart which allows s-curves and histograms for selected projects and/or tasks to be compared. The three s-curves representing the deterministic plan, pre- and post-mitigation models are shown below in Figure 1.6. The gap between the two curves can be interpreted as being an estimate of the schedule benefit that is expected to be gained from the mitigation at the P80.

The Schedule Sensitivity Chart otherwise referred to as the Tornado Chart is also an important output from the Primavera Risk Analysis. Primavera Risk Analysis Tornado graphs help users identify key risk drivers and pinpoint the task or risk event that's preventing the schedules from performing as expected.

Considering appropriate risk mitigation plans for each of the activities indicated in the Schedule Sensitivity Chart, indicating accordingly their effects on the qualitative assessments of the associated risk and running the simulation again, will remodel the project finish date and duration.

Limitations of the Monte Carlo Method

In a nutshell the Monte Carlo method is well suited for real-life problems and applications especially when assumptions governing statistical theories cannot be sustained. However, being a forecasting tool certain assumptions

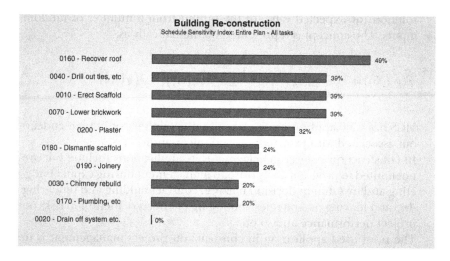

Figure 1.7 Schedule Sensitivity Chart

have to be made with respect to the input variables, for example, it could be the duration of tasks involved in a construction project in terms of the most likely, optimistic and pessimistic duration for the completion of the tasks under given resources and conditions. These assumptions result in estimates of the expected values based on experience, historical data and expert opinions. It therefore follows that the simulation can only be as good and precise as the estimates made. Hence, the Monte Carlo method simply presents well guided and evaluated probabilities and not certainties. This is the major limitation of the application of this method. The process also lends itself to several iterations, though much easier now with advancement in technology which has produced faster software, it could still take some time to arrive at an optimum solution compared to analytical procedures that are deterministic.

Key Points

- The Monte Carlo method has its root in the theory of probability and statistical sampling.
- MCS is a technique for solving complex problems by approximating expectations of outcomes from a probability distribution function of randomly drawn input variables through repeated sampling.
- It is a form of stochastic process because it uses random numbers but differs because its output is non-random.
- It is also non-deterministic because its inputs are not known initially and are random variables but it involves a deterministic model.

- It relies on the principle of large numbers to provide an approximate solution for expected value of the output from a number of random inputs. The concept is represented mathematically as:

$$\mathbb{E}(\tilde{g}_n(\mathbf{X})) = \mathbb{E}\left(\frac{1}{n}\sum_{i=1}^{n}g(\mathbf{X}_i)\right) = \frac{1}{n}\sum_{i=1}^{n}\mathbb{E}\left(g(\mathbf{X}_i)\right) = \mathbb{E}\left(g(\mathbf{X})\right)$$

- MCS has wide applications in virtually all disciplines of human endeavour associated with probabilities and uncertainties.
- In construction project management, its applications include but are not limited to program and portfolio management during capital budgeting and investment decision making, contract bidding and tendering decision making or strategies, modelling of effects of identified risks on project performance and so on.
- The most used application in construction project management is in schedule and cost risk analysis.
- The major limitation is that its accuracy is largely dependent on the reliability of the estimates and probability distribution provided for the input variables.

References

Aguilar, L. A. (2012). Principles of Monte Carlo simulation [On-line lecture note]. Retrieved from www.utinam.cnrs.fr/spip.php?action=acceder_document&arg=1419&cle=9d2179954957b5e3414a025c74ac62f4761cecd3&file=pdf%2FMonte Carlo_Besancon2012-L1.pdf.

Bock, K. & Truck, S. (2011). Assessing uncertainty and risk in public sector investment projects. *Technology and Investment*, 2, 105–123 doi:10.4236/ti.2011.22011.

Choudhry, R.M. & Aslam, M.A. (2011). Risk analysis of bridge construction projects in Pakistan. In: Proceedings of International Council for Research and Innovation in Building and Construction (CIB) Working Commission W99, International Conference on Prevention: Means to the End of Injuries, Illnesses, and Fatalities, Purdue University, Virginia Tech, University of Kentucky, 24–26 August 2011, Omni Shoreham Hotel, Washington, DC, USA, 13–28.

Enrique, F. & Richtmyer, R. D. (1948). Note on census-taking in Monte Carlo calculations. LAM 805 (A), Declassified report Los Alamos Archive.

Ganame, P. & Chaudhari, P. (2015). Construction building schedule risk analysis using Monte-Carlo simulation. *International Research Journal of Engineering and Technology*, 02(04), 1402–1406.

Ikediashi, D. & Ogwueleka, A. (2018). Exploring Monte Carlo simulation technique for construction project risk management. Conference Proceedings of International Sustainable Ecological Engineering Design for Society (SEEDS), 6–7 September, Dublin, 168–178.

Joubert, F.J. & Pretorius, L. (2017). Using Monte Carlo simulation to create a ranked check list of risks in a portfolio of railway construction projects. *South African Journal of Industrial Engineering*, 28(2), 133–148.

Kong, Z., Zhang, J., Li, C., Zheng, X. & Guan, Q. (2015). Risk assessment of plan schedule by Monte Carlo simulation. International Conference on Information Technology and Management Innovation (ICTMI 2015). Retrieved from www. google.com/url?q=https://download.atlantis-press.com/article/25839948.pdf &sa=U&ved=2ahUKEwiB3LKo5JLhAhVUtHEKHbyxCTUQFjACegQIBxAB&usg =AOvVaw3IBSOwEE6J0OKqR_CGZE-o.

Martijn, R. K. M. (2017). Simulation modelling using practical examples: A plant simulation tutorial [e-book]. Retrieved from www.utwente.nl/en/bms/iebis/staff/ mes/plantsimulation/tutorialplantsimulation13-v20170911.pdf&sa=U&ved= 2ahUKEwiJ44Ch3JLhAhWDUhUIHf0fAL04ChAWMAF6BAgKEAE&usg=AOvVa w3xuLEqdh9qtkqukOCan7wE.

McKean, H. P. (1967). Propagation of chaos for a class of non-linear parabolic equations. Lecture Series in Differential Equations, Catholic Univ. 7, 41–57.

Metropolis, N. (1987). The beginning of the Monte Carlo method. *Los Alamos Science Special Issue.* Retrieved fromhttp://jackman.stanford.edu/mcmc/metropolis1.pdf.

Ng, S. T., Xie, J. & Kumaraswamy, M. M. (2010). Simulating the effect of risks on equity return for concession-based public-private partnership projects. *Engineering, Construction and Architectural Management*, 17(4), 352–368.

Peleskei, C. A., Dorca, V., Munteanu, R. A. & Munteanu, R. (2015). Risk consideration and cost estimation in construction projects using Mont Carlo simulation. *Management*, 10(2), 163–176. Retrieved from www.fm-kp.si/zalozba/ISSN/1854-4231/10_163-176.pdf&sa=U&ved=2ahUKEwiB3LKo5JLhAhVUtHEKHbyxCTU QFjAIegQIBBAB&usg=AOvVaw08QvKk6p5lxqFYvWOY1fP4.

Project Management Institute (2008). *A guide to the project management body of knowledge (4th ed.).* Pennsylvania, USA: Author.

Rui-mei, L. (2015). Properties of Monte Carlo and its application to risk management. *International Journal of U – and e – Service, Science and Technology*, 8(9), 381–390.

Santra, S. B. (2015). Monte Carlo simulation technique [On-line lecture note]. Retrieved from www.iitg.ac.in/nwast2015/slides/sbs.pdf&sa=U&ved=2ahUK EwixrdLTnZThAhVBWxUIHdPGBNoQFjAAegQIBBAB&usg=AOvVaw1G_ Ar2AVkB4qdaY2aSaaqR.

Valderrama, P. and Guadallupe, R. (2013). Monte Carlo method applied to comparison of tenders in construction projects. 17th International Congress on Project Management and Engineering Universidad de La Rioja, 17–19 July, Logroño, Spain. Retrieved from www.rib-software.es/pdf/Art%C3%ADculos/ AEIPRO-Montecarlo.pdf.

2 Modelling risk effect using Monte Carlo Technique

An application for innovative projects

Ibrahim Motawa

Chapter summary

The need to gain competitive advantage stimulates many construction organisations to exploit innovative products and processes. However, the high level of uncertainty associated with innovative construction leads many organisations to focus on the application of traditional construction processes and products. Implementing construction innovation often involves experimentation, iteration and refinement of activities that are reliant on volatile information. The acceptance of any innovation in construction often only comes after very significant advantages of this innovation on several projects. Therefore, construction organisations should exhibit specific characteristics to promote new technology and to overcome the expected barriers to innovation in order to achieve the desired competitive advantage, pursue new markets and improve productivity. Because of the considerable unknown risks, implementing innovative technologies creates a greater need for co-operation among businesses to address the planning, development and implementation of technological capabilities to shape and accomplish the strategic and operational objectives of an organisation. This chapter introduces a model to simulate the risk effect on the process of implementing technological innovations in construction. It includes the model hypothesis, techniques and its application on a case study. The model identifies sources of risks within the innovation process and schedules activities taking into account stochastic analysis of the information influencing implementation. The proposed model is a decision support tool that simulates different scenarios to control the implementation phase.

Introduction

Implementing construction innovation is the process through which new ideas turn into new components that have economic, functional or technological value. The new components may revolutionise the process itself and result in traditional and accepted processes being replaced by new approaches that require deliberate and informed action and control.

Effective implementation of technological innovations requires an understanding of the complexity underpinning the process and consideration of many factors. Cost/benefit factors have to be evaluated along with risks and uncertainties. Innovative ideas have little historical data to aid implementation and this increases the degree of uncertainty. Most innovative projects do not fulfil their time and cost objectives and it is essential for construction organisations to plan and control their implementation.

Although several models have been developed to assess the performance of new technology, little attention has been paid to implementation. The implementation phase includes several steps at which decisions need to be analysed. According to Wakeman (1997), project development moves from the debate to action level where decision-makers face four potential barriers to success; namely, technical, financial, institutional and public/perceptual. To successfully implement construction innovation, managers should control innovative activities and overcome these barriers. This chapter presents a generic conceptual model that deals with the implementation phase of innovation. The innovation process and the associated risks will be defined first. A model is then proposed to simulate the effect of risk on the process. The model uses the Monte Carlo technique to show the impact of risk on different implementation scenarios. The technique will be first explained then a case study is presented to show how the model works.

Methodology

This chapter introduces a generic conceptual model developed to simulate the implementation phase of innovative construction projects. The model was developed by reviewing case studies models and from interview results. A comprehensive review for relevant literature has been conducted to identify the stages of the innovation process and the potential risks. A number of interviews with construction professionals have been conducted to verify the identified process and risk factors. A study on the available modelling techniques has helped in identifying the suitability of Monte Carlo technique to simulate the effect of risk on the innovation process. Monte Carlo simulation is used to model influence information on the innovation implementation phase. A planning tool was also developed to schedule and analyse costs of the implementation tasks. The proposed model was developed using Visual Basic environment, which has been validated by a number of case studies one of which is shown at the end of the chapter. In addition, specific semi-structured interviews were conducted with industry professionals associated with innovative construction projects to test and validate the proposed model. The case study was about the installation of a location control system using satellite facilities for a company's vehicle fleets which were used in highway maintenance within the UK. Validation of the model resulted in refinements and suggestions being incorporated into the final version. The simulation model is used to investigate different scenarios of typical events that occur during the implementation of innovations. It also addresses the impact of changes on other activities and on project durations and costs. The following sections review the relevant literature and present the proposed model, its technique and application.

Literature review

To '*innovate*' means to 'bring in new methods, ideas, etc.' (Oxford Dictionary 2000). Derived from this definition, an *innovative project* can be defined as 'a project that has a new process or product that the project team has not previously dealt with'. Practically, 'innovation' is seeking, recognising and implementing new technology to improve the functions a company performs. What may be considered as new technology to one company, may not be considered as new technology to another. Simply, the author considers that innovation takes place when a new approach replaces a set of traditional and accepted processes or products. Innovations can be driven by problems that cannot be solved by current technology prompt innovations. Also, owner demands are not only for safe and economic products, but for more functional facilities and aesthetic criteria. The high standard of regulatory demands may cause design and construction teams to innovate to fulfil these regulations. The driving forces to innovations also include changes in the construction environment, any related science, engineering, industry and society that may have a significant effect on the construction industry. It has been argued that competitive performance depends not simply on success with a single innovation, but success with a sequence of innovations and post-innovation improvements. This approach involves a shift in perspective – from treating innovations as isolated, discrete events – to treating them as an evolving flow of developments in a technological agenda (Arditi et al. 1997). While the risk effect may decrease due to lessons learnt from various trials of implementation, other risks may also be created. The following sections illustrate how innovation is taking place within construction organisations and introduces a model for managing its implementation.

Innovative organisations

For innovation to take place, an environment that stimulates new ideas must be created, and this remains the responsibility of management. Models reviewed for the innovation process helped in identifying the dynamic framework of innovative organisations, as shown in Figure 2.1. Factors such as company size, type of innovation and breadth of innovation affect the implementation of innovations. Large companies are more able to afford new investment for innovation and tolerate the risks associated with adopting them, whereas smaller companies are more likely to value technology and have less complex decision-making processes. A description of the framework is presented in the following sections.

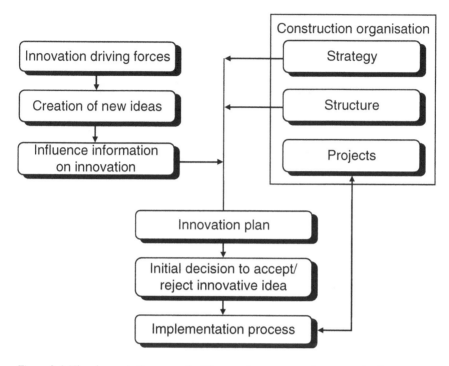

Figure 2.1 The dynamic framework of innovative construction organisation

Organisation strategy

Competition based on advanced technology and an ability to innovate requires long-term strategic planning. Figure 2.2 classifies these strategy goals into two areas. These goals are directed to innovation in all aspects. As the construction industry is highly competitive, volatile, attracts low margins and is subject to increasingly stringent standards, success becomes difficult to achieve without innovation.

Organisation structure

Organisational structure and environment have an effect on the success of innovation. To create a climate conducive to innovation, short-lines of communication between senior management and project teams, with effective information flows to identify and resolve problems resulting from new technologies, should be present. This climate should expect some failures and accept the risks inherent in innovative processes. An innovative structure and culture should establish supportive policies and priorities, including long-term viewpoints, implicit vertical integration,

Figure 2.2 Main organisation strategy

emphasis on planning mechanisms, broad views on risk, flexibility through open project teams and use of slack resources. Integration among project partners, owners, designers, contractors and suppliers also motivates innovation. Decentralisation and informal decision-making enhance the team culture necessary for innovation. As innovative improvements may be generated by any employee as 'technical innovator', Winistorfer (1996) stated that empowerment-allowing decisions to be made at the most appropriate level in an organisation have an important role.

An organisation's ability to enhance competitiveness can foster innovation. Strong and unbiased management commitment to select technologies that best support project goals is one of these abilities. This can be fostered by establishing designers' knowledge of technology. As this knowledge is dynamic and fragmented, the rate and scope of innovation depends on how this knowledge is managed. Continuous learning is also essential if innovation and adoption of new technologies are to be accepted (Lansley 1996). While stability of employees for a period of time reduces training costs and focuses experience, a lack of varied experience also produces lack of creativity, flexibility and innovations (Winistorfer 1996).

The success of innovative processes often requires the creation of an innovation team. The role of this team is to keep the organisation in tune with technological advancement, expend energy and take risks necessary to make innovations happen (Tatum 1989). An individual will have to be identified to champion or manage the process to completion. Winistorfer (1996) described four key categories of individuals: 'Technical Innovator', 'Business Innovator', 'Chief Executive' and 'Product Champion'. These champions do not exist in all construction firms. However, line managers may assume the role of champion, but this often takes second place in the face of other problems and opportunities. Nam and Tatum (1992) suggested an *integration champion*, who facilitates inter-organisational co-operation and learning, to ease this function. Technology gatekeepers, who link between organisations and sources of technology, identify, monitor and evaluate any improved or new technologies used by other companies that may also be effective (De La Garza and Mitropoulos 1991).

Influence information on innovation

The interviews conducted for this research identified the implementation stages and the main information that influences the implementation of innovation. Table 2.1 summarises the identified barriers to innovation and potential risks. Some of the key points emphasised by the interviewees were also addressed by relevant literature as discussed in the next sections.

Risks increase as more resources are committed to innovation. Capital intensiveness makes risk-aware decision-makers invest in structures built through mainstream, well-tested designs, materials and methods, rather than innovative ways (Skibniewski and Chao 1992). The extensive, unstable, highly fragmented and geographically dispersed construction market's characteristics create an uncertain climate for investment in innovation, especially for small companies (Technology Foresight Panel on Construction 1995). Nam and Tatum (1988) described construction as a system locked to any attempt to change the status quo. The perception of a locked system explains why construction innovation that is technologically superior does not often follow the route that diffusion theorists, economists or engineers may anticipate. The system players include various owners, craft unions, subcontractors, local governments that enforce obsolete building codes and interest groups and coalitions that have stakes in construction technology development. The dynamics and friction among these parties that slow the rate of innovation are too complex to measure in quantitative terms.

Innovation plan

Managers should identify the groups involved within proposed innovations and use the performance indicators to measure the progress. A list of performance indicators has been identified for the innovation process, further details can be found in Motawa et al. (2003). The availability of the facilities required, expected changes for the management system and the achievement of the overall strategic objectives should be evaluated. Replacing existing technology with new or using both concurrently, as well as the negative effects or termination of existing technology created by innovation should be considered. Providing detailed descriptions of the development process to those involved and receiving feedback from them are important. This phase ends with the initial decision to accept or reject the innovative idea, as shown in Figure 2.1.

Table 2.1 Influence information

Barriers	Risks and Uncertainties
Codes	**Economic sources**
Reaction of other construction partners	Yield (financial returns)
Labour relations issues	Costs (financial estimates)
Organisation culture	Time (how long it takes)
Individual roles in the organisation	Training requirements
Level of design/construction integration	Availability of human resources
Safety considerations	Contractual claims
Economic and political conditions	Market changes
Capital intensiveness	
Resistance to change	**Physical sources**
Fragmented nature of the industry	Substructure conditions
Workforce skills	Weather conditions
Company size (capability of implementation)	
Governmental regulations	**Capability sources**
Environmental and social constraints	Damage to existing utility construction lines
Procurement procedures	Safety risks
The priority attached to the project**	Productivity decline (learning curve)
Functional requirements due to the type of building	Practicality of design and buildability
Funding and resources made available	Technological function risk
Owner's view**	
Operational requirements	**Political and social sources**
Project aesthetics	Contractual and tendering methods
Market circumstances	Environmental risks
Level of complexity of the project	Government rules and regulatory bodies

** It has been argued that some of these barriers can act as a motivator rather than a barrier, such as these examples.

Implementation phase

For a traditional (non-innovative) project that has clearly defined end-objectives and traditional processes, the project activities can be easily directed towards achieving the project's objectives, as illustrated in Figure 2.3. Innovative projects always have incomplete knowledge, a high level of uncertainty and often result in work being iterated to achieve satisfactory performance. Considerable refinement or experimentation is required before acceptance of the final innovation. The path to achieving the end objective, as shown in Figure 2.5, does not follow the same sequence as traditional projects. A project may have one start event but may often have

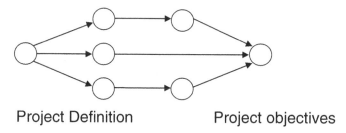

Project Definition   Project objectives

Figure 2.3 Sequential process of a traditional project

several targets or combinations of final targets (A or B or AB) which have great uncertainties in their features. The whole process may require several scenarios to be considered before the final outcomes are achieved, as indicated by decision nodes in Figure 2.4. This makes traditional planning techniques unsuitable when planning the implementation of innovation.

The main elements of analysis for the implementation phase include: influence information, process conditions, and performance indicators as shown in Figure 2.5. This phase includes technological and economical risk analysis. Feedback, iteration and process documentation are shown throughout this phase. The analysis results may change any new method completely, refine the present idea or require more experimentation.

Modelling the implementation phase of innovation involves deciding whether or not to accept the product of a new construction process. If the process is faulty, it should be rejected and vice versa. Reliable performance indicators should be used to help ascertain the condition of the process, as shown in Figure 2.5. The direction of the arrow indicates the direction of influence. The probability of process perfection is dependent on the actual

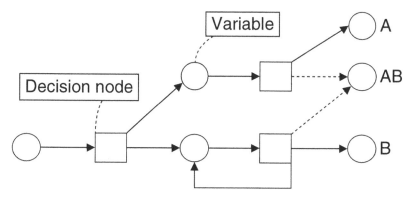

Figure 2.4 Process of an innovative project

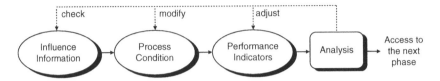

Figure 2.5 Phase of an innovation implementation

status of the process. Even though the process condition influences the performance indicators, the measurement of the performance indicators is known before the true process condition is determined. In this situation, the timing of the nodes should be opposite to the direction of the arrow. The decision node 'Analysis' is therefore added and is based on the measurement results. The arrow from 'Performance Indicators' to 'Analysis' indicates that the measurement result is known before the decision to access or iterate is made, then the true condition of the process is learned. According to the results of this 'Analysis', the decision-maker can select any of the shown alternatives; check the 'Influence Information', modify the 'Process Conditions', adjust the 'Performance Indicators' or move to the next implementation phase. As time progresses, transitions take place from one phase to the next. The following section will illustrate how the risk effect on this process can be modelled.

Proposed model

A fundamental challenge to the construction industry regarding innovation is the planning and control of work during the implementation phase. The implementation of innovation highlighted the influence information that affects the innovation process. Developing a simulation tool to plan the implementation requires testing this process against information that influences its implementation. As this information may alter the innovative project objectives (i.e. time or cost) from the initial ones, there is a need to simulate these deviations using a sensible and reliable tool to help managers monitor their plans. The proposed simulation will change the planning type from deterministic to stochastic. A planning tool has been designed for this purpose.

The proposed tool uses the Monte Carlo technique to analyse the effect of risk on the innovation process. The use of this technique will allow managers to investigate:

- the sensitivity of the implementation tasks to the subjective estimates used within the influence information;
- different durations/costs of the interdependent tasks for each stage;
- information that most critically influences implementation.

The principles of the Monte Carlo technique will be illustrated first. Then an example will be presented to demonstrate the use of the technique.

Monte Carlo technique

For the purpose of simulating construction processes, such as innovation in construction, two main considerations have to be taken into account, namely deterministic or stochastic simulation and discrete or continuous change.

The deterministic simulation is for systems whose behaviour is completely predictable. An example of this system is the traditional planning of projects. A stochastic simulation is for a system whose behaviour cannot be completely predictable which fits well with the characteristics of innovation process information, as described above. The stochastic simulation refers to using mathematical models to study systems that are characterised by the occurrence of random events.

The main difference between discrete and continuous change is that the former deals only with variables that are not changed during the simulation process while the latter allows continuous change for the variables' values during the simulation run. These changes could be represented by differential equations that, theoretically, allow variables to be computed at any period of time.

To simplify the modelling approach for the innovation process, an innovative task is assumed to be completed at a discrete point of time so the discrete event simulation is the one that will be considered in this research. Many discrete event stochastic simulation models have been developed in the field of construction such as those used to schedule construction activities and simulate repetitive cyclic construction operations.

Apart from the fact that the Monte Carlo technique is the most popular technique of the stochastic process simulation, it has a considerable edge in computational efficiency over other methods of approximation as the size of the problem (the number of the studied factors) increases (Fishman 1996). Therefore, the Monte Carlo technique will be adopted in this research. It provides approximate solutions to a variety of mathematical problems by performing statistical sampling experiments. The basic elements of Monte Carlo technique to deal with discrete event simulation for non-repetitive events, as assumed for the implementation of innovations in construction are: probability and subjectivity.

Probability and subjectivity

To emphasise the distinction between the frequentist and subjective approaches, consider the probability 'P'. To a frequentist, 'P' is the long-run relative frequency with which the person being observed chooses object 'A' when repeatedly offered the choice between 'A' and 'B'. To a subjectivist, 'P' represents the observer's degree of belief that the person will select

'A' in a choice between 'A' and 'B'. Note that a frequentist must conceive of a sequence of choices, whereas a subjectivist need only imagine the person being offered the choice once. The frequentist approach cannot be used to encode the uncertainty present in the majority of decision problems such as implementing innovations. Decisions are made almost invariably in unique circumstances that may not arise again. Thus, the frequentist approach is quite inappropriate to the decision analysis needs of this research. The subjective view of probability does fulfil these needs in decision analysis. The application of subjective probabilities within decision analysis gets its importance because an uncertainty that cannot be resolved cannot affect the consequence of a decision.

Probability is then taken as representing the observer's degree of belief that a system will adopt a certain state. In decision theory terms, $P(Oj)$ represents the decision-maker's degree of belief an Observation (Oj) will occur; the stronger the belief, the greater the Probability $P(Oj)$ is. Different people have different beliefs, thus, different observers and different decision-makers may assign different probabilities to the same event. Probability is, therefore, personal; it belongs to the observer. Subjective probability has a personal, non-objective meaning. Although we may interpret $P(Oj)$ as quantifying a personal degree of belief, we are not at liberty to call it a probability, at least among mathematicians, unless we have shown that it combines with other subjective probabilities according to Kolmogorov's laws (French 1988). Subjective probability is a discipline to measure uncertainties about an event considering the knowledge base at the measurement time (Lindley 1994). In other words, it reflects the decision-maker's belief about an uncertain event. Changing knowledge might change the uncertainty measure. Ferrell (1994) emphasised that subjective probability provides a normative framework for the representation and updating of beliefs. The probability of a hypothesis is conditional on one or more items required to identify information relevant to the problem at hand. The identification of an item of evidence influences the degree of belief in a hypothesis.

Decision analysis quality depends upon the process being comprehensive, having a sound theoretical basis and being carefully and systematically applied. Concerning the theoretical basis, probability, as a mathematical principle, is well grounded, but there is considerable debate about the philosophical and psychological status of subjective probability. However, because of its ambiguity there should be an especially strong emphasis in decision analysis on the careful and systematic application of a comprehensive subjective probability elicitation process. Subjective probability changes with time, where the probability of a correct action increases by more trials of the action due to learned experience and corrections.

During the course of an analysis, the decision-maker may gather information that causes him to revise his beliefs, and consequently, his/her subjective estimate. Due to insufficient data for such type of modelling, decision analysts may tend to use the simplest functions to express his/her beliefs about uncertainties such as, using linear probability functions than sophisticated functions (normal, beta, . . . etc.).

Monte Carlo technique (main steps)

This section illustrates the main steps of the Monte Carlo technique that are used to conduct experiments of the stochastic process simulation. These steps are as follows:

1. Estimate a range of values for each variable of the influence information affecting the system behaviour. This range consists mainly of two values. The minimum value expresses the minimum impact of this variable on the project time or cost (as the main performance indicators to be assessed for implementing innovations). This impact is considered as a percentage of the original estimate of the task time/cost. For example, a +/- 10 per cent means the time or cost of the tasks affected by this variable will be increased/reduced at most by 10 per cent.

2. Identify the most suitable Probability Distribution Function (PDF) for each variable that describe the pattern of the variable variation), see below for further details, see section below for further details on PDF.

3. Generate the cumulative frequency function for the distribution function of the variable, which can be obtained by the inverse probability method.

4. Select a value for each variable from its cumulative distribution using a random number (RN), see section below for further details on RN. During simulation, each variable will have a random estimate from its identified range and then the duration/cost of each project task will be changed according to this estimate.

5. Using these random values, compute the desired objective function for the system variables (which is the project time or cost). A model has been developed for this research to compute these objective functions.

6. Repeat steps 1 to 5 for N-times, using successive and independent streams of uniform random numbers, to get N-realisations of the desired function. The 'N-times' is determined where steady results are achieved (i.e. where more iterations do not affect the results).

7. The output of the simulation runs gives the cumulative distribution function for the project objectives (time or cost). Using this output, decision-makers can determine the probability of a project time or cost and estimate the mean and standard deviation of the objective function.

Figure 2.6 illustrates an example of using the sampling technique in the proposed simulation tool using a triangular PDF for the 'x' variable. The following sections will discuss a few technical terms used for the Monte Carlo technique.

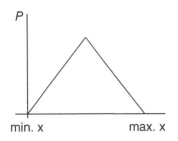

Probability distribution function

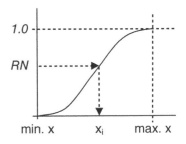

Cumulative function

Figure 2.6 Monte Carlo Simulation sampling technique

Technical terms used in Monte Carlo Simulation

Probability Distribution Functions (PDF)

A random variable is one whose values have more than one possible value that cannot be predicted with certainty at the time of decision-making. For each possible value of the random variable there is an associated likelihood of occurrence.

Random variables are sometimes called stochastic variables to denote the fact that the likelihood of the values occurring is stochastic or probabilistic in nature. On the other hand, if the value of a variable is known or can be predicted with certainty at the time of decision-making the variable is called a deterministic variable.

The probability distribution is the graphical representation of the range and the likelihood of occurrence of a random variable. It can be discrete or continuous, depending on the nature of the random variable. If the distribution can be represented by a function, this function is called the probability density function. Every distribution can be expressed in an equivalent graphical form called the Cumulative Frequency Distribution. A cumulative frequency point expresses the summation of all the previous probability values of the variable to this point. There are a great number of distributions that are in common use. Each distribution has some features, and is used to describe a variable according to the data available for that variable and with some tests for these data to fit the suitable distribution.

Decision makers specify a type of distribution for each variable by their subjective judgement. PDF can be selected from a number of functions such as Uniform, Triangular, Normal, Beta, etc. The selection is mainly based on the observer's degree of belief on the way the values of each variable behave. Simple profiles are always advocated in the absence of statistical data. For example, triangular distribution can be approximated to a normal distribution. Trapezoidal or rectangular distributions are useful in representing situations where there is no evidence that one particular estimate value is any more likely than another within the prescribed range.

A difficult situation occurs when the analyst wishes to define a distribution for a random variable but has no data available and has no idea what the shape of the distribution is, or should be. In these cases the analyst needs first to try at least the range of values – a minimum value and a maximum value. Then if there is any value or a range of values within the limits that might be more likely to occur than other values, triangular or trapezoidal distribution can be selected. If not, a uniform distribution may be suitable. If the most likely estimate for the random variable does not exceed a certain probability value (P), then the weighted triangular distribution can be used.

Random Numbers (RN)

Random Numbers can be generated using a computer source of pseudo-random numbers. Many off-shelf applications (such as MS Excel) have built-in functions to generate RNs. Pseudo-random numbers generation is an algorithm to produce a fixed and deterministic sequence numbers that can at best be called 'Pseudo-Random' which behaves, according to statistical tests, like a truly random sequence. Pseudo-Random numbers are uniformly distributed within the unit interval (0,1) with equal likelihood. Uniformly distribution random number provides a basis for generating the random varieties required in a wide variety of realistic simulation problems. It is not correct to use the same random number to sample all distributions on a specific pass. The reason for this is that using the same random number would automatically imply fixed values for all variables (all values will be near their upper or lower limits).

Number of iterations

Crandall (1977) selected five networks for empirical testing to determine the impact of varying the number of iterations during simulation upon the generated time distribution, the criticality of individual activities and the most likely critical paths. A simulation was performed on each of the five networks varying the number of iterations from 250 to 8000. Results obtained for each network based on varying the number of iterations during the individual simulations are useful in determining the number of iterations to utilise when processing networks by the Monte Carlo technique. Statistical comparisons of the simulated distributions as a function of the number of iterations indicate that the data generated at 500 iterations were adequate to forecast the desired probabilities of project completion even though the densities were not sufficient to clearly define the actual time distribution. But, it was desired to test the ability to forecast the probability that a given activity would be critical and which paths would be critical. Therefore, Crandall concluded that the 1000 iteration simulation is adequate to determine the relative degree of individual activity criticality.

Dealing with correlation

In practice, numerous interrelationships and dependencies exist among a system's variables. These dependencies may be included in the simulation by means of explicit equations linking the relevant variables.

For any deterministic analysis, each estimate is made with complete knowledge of the values attributed to all other variables in the model. For stochastic simulation based on multiple runs, however, it is possible that the expert may allow (consciously or subconsciously) for relationships between the probability distributions that have been selected for the system variables.

It may be, for example, that high values of one variable will tend to be associated with high values of another. Then independent sampling from the prescribed distributions will not fully reflect management's expectations, and consequently, sets of conditional probability distributions are required. A simulation procedure that takes account of such relationships must be based on conditional sampling. Many approaches have been developed to deal with this problem. A popular one was reported by Van Gelder (1967). This approach is based on random numbers correlation and is called 'Markovian Correlation'. The approach assumes that if two RNs are correlated, the two sample values of variables estimated by these random numbers will be correlated. RNs are chosen independently using RN generators. However, to correlate two random numbers, modification of the second RN somewhat in relation to the first is required. An acceptable formula to achieve this is found by 'Markovian Correlation' in the following expression:

$$RN_2 = RN_2 + a \ (RN_1 - RN_2)$$

where: RN_1 = first random number drawn
RN_2 = second random number drawn (independently)
RN_2 = corrected second random number drawn (correlated)
a = weighting factor

This formula corrects the second (independent) random number RN with a proportion (a) of the difference between the first random number and the second one. The effect of the factor (a) is easily recognised. If $a = 0$ then the formula maintains the original independent second random number RN$_2$ which means no correlation. If a =1 then the second random number is replaced by the first which means full correlation. If the factor (a) has a value between 0<a<1, it obviously obtains partial correlation. Therefore, if the amount of correlation (r) is known or predefined, and after relating (r) to a value of (a), the correlated values for the variables could be determined.

Application

This section introduces a case study that has been used to test and validate the proposed simulation tool of innovation implementation in construction. The case study was launched in a UK construction company. The project is about 'Highway Maintenance Satellite Support System' (HMSSS) that aims to install a location control system using satellite facilities for the company vehicle fleets.

The data of the case study was collected in interviews held with the project designers and managers. This resulted in some refinements and suggestions have been incorporated into the simulation tool. The simulation tool was used to investigate different scenarios of typical events that occur during the implementation and the impact of changes on other activities and on the project duration and costs. The proposed tool requires users to input the planning data of the implementation tasks that include tasks' durations, resources, costs and the required dependencies among them; and the stochastic data of the influence information. The output of the tool includes the deterministic and stochastic results of each stage's tasks.

Background to the project

The installation of a location control system was seen to be beneficial to the company from two major perspectives: the improvement of company efficiency and the extra service supplied to the client. Clients were likely to be concerned with the Vehicle Location System (VLS) for the following reasons:

- it enhances the supervision of the operatives and gives the client a better service;
- the client's GIS systems and inventories can be augmented; and
- it provides proof of work.

A trial period has been undertaken first to test the new system's validity for cash flow and planning reasons. The decision-making process for choosing suppliers of the proposed VLS considered that most of the vehicle fleet used by the company was on contract hire and as such the installation of the system might pose problems for the following reasons:

- Approval to modify the vehicles has to be gained from the hiring company. Some of them have already approved the modifications. However, vehicles owned by the company may be a more realistic target for the trial. This situation could mean an agreement for lower hire-rates if the system's installation suggests that the hire will be longer term.
- Hired vehicles may need to be changed. While attempts will be made to ensure the hardware is easily interchangeable between vehicles, this may cause the hire companies some concern.
- Many vehicles in the fleet are very old. The use of the vehicle may be improved but its remaining lifetime may not be long.
- The initial installation would be performed by the supplier's operatives, but the possibility of training company workshops to install equipment to large numbers of vehicles might save time and money.
- The issue of operator acceptance was very important. Systems run by the company have seen vehicle mistreatment as drivers do not wish to be tracked. The operators must be involved from the start and made aware of the implications of the new system. If the equipment is not used properly the gains will not be fully realised.
- It is important to consider that, for full use of the system, there must be a base station at each set of sites. This is not a problem as such but will have to be examined for cost implications.

Currently the 'scout' vehicle travels along set patrol routes and records 'outages' on a dictaphone. These are written down by office staff from the dictaphone tape on the next day, then passed to work gangs who determine their route to complete the work for the day. If the scout can record the position of the item and its current status on the proposed control system, the database will hold this information so that it can be passed directly to the work gangs via their vehicle console, cutting out the middle person, which was the main aim of this innovative application.

The data flow diagram, as shown in Figure 2.7, explains the data sent when problems are spotted by the scout, or during the course of a work crew's day. Data will also be fed to the database after an emergency call. This information is available to the vehicle on site to facilitate rectification of identified problems. Once the work has been completed, the system allows the work to be recorded in terms of the labour activity and the materials used, by feeding data back to the database.

The data capture process should be more 'automatic' and less reliant on the operators making decisions. The ideal solution for recording the material used within the process is by bar-coding. This record could be attached to the data regarding the unit position and the labour activity carried out. It may also be possible to create a similar tool to record the labour activities. A sheet of paper with all relevant bar-codes for each activity may be produced. A hybrid of the picklist and bar-code ideas could perhaps work well.

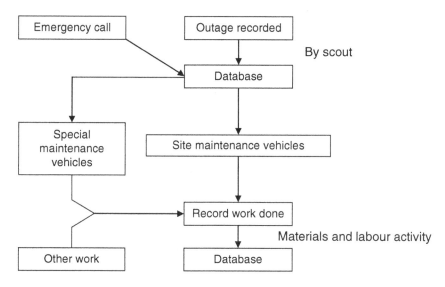

Figure 2.7 Data flow of the case study

The initial programme for the installation procedure indicated two milestones for assessing the required preparation. The first milestone is 21 days after the initial plan while the second milestone is 40 days after it.

The first milestone assessment

At this stage the current system will continue to run, with the new service running alongside it. This enables verification of the records made by the HMSSS. Details of the emergencies will have to be placed into the database after extraction from the daily log. The current work procedure includes ten steps, and implementing the new procedure will change five of them.

The second milestone assessment

The full capacity of the system is available at this stage. Interaction between the base station and the vehicles is two way; the completion of many of the daily log sheets will be internal to the system and completed electronically partially by the office and partially on site. At this stage of the implementation, time and actions according to each type of work can be recorded, without the need for individual log sheets.

The installation of HMSSS required a coding system and programming models to be set up. These models have been refined and tested several times before being accepted for the proposed installation.

Application of the proposed simulation tool to the project

Applying the proposed tool needs the project team to review the pre-defined project phases and the influence information. The project team will allocate durations, resources and costs for the project tasks. The Monte Carlo technique will assist in simulating the high level of uncertainty inherent in the implementation of this project by simulating the influence of information on the tasks of each phase. Running the tool will produce a stochastic estimation for the project time and cost. If these estimations do not fit the completion time and cost objectives of the project, the project team would attempt different scenarios on different durations and resources or would manipulate the influence information' impact values by advanced actions towards achieving the project objectives.

Input Data

The available data for this case study were detailed mainly for pre-installation and installation phases. Figure 2.8 shows how these data could be entered on the model interface. The planning tool should be run first to get the deterministic duration and cost of the implementation stage tasks, eliminating the information that causes uncertainty. This will mean that satisfactory

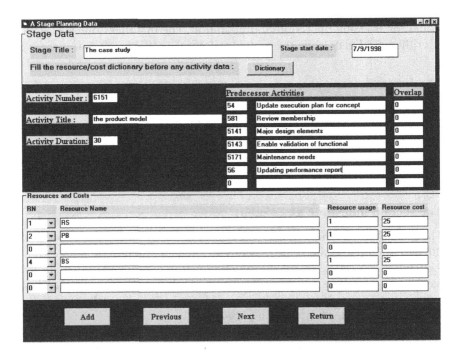

Figure 2.8 The input of the phase tasks

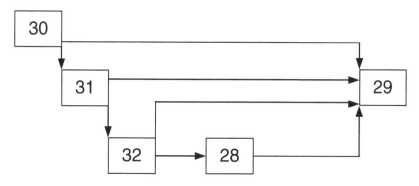

Figure 2.9 A block of implementation tasks

performance is gained by implementing this group of tasks under ideal conditions (i.e. no uncertainties). A block of tasks was chosen from the design stage of the innovative project to show the output of this run, as shown in Figure 2.9, which are as follows.

Task 28: The project execution should be firmly set to enable construction works and facilitate the measurement of the performance criteria.

Task 29: Evaluation of the performance criteria for the project stage.

Task 30: Co-ordinated, structural, mechanical and electrical elements should be prepared to a high level of technical detail with corresponding specifications.

Task 31: Include phasing of construction works.

Task 32: Detail description of work packages and interfaces between them to enable 'trouble free' construction work.

Information that influences the implementation phases changes the duration/costs of certain tasks. Each information item will be simulated by a range of values of their effect on the deterministic estimate of the time/cost of the implementation tasks. The stochastic range of the information influence on the project implementation, as shown in Table 2.2 (column 'Impact at the first iteration'), was estimated according to subjective judgement of the project managers. Figures 2.10, 2.11 and 2.12 show how a user can enter these stochastic data. The tool has a pre-defined list of information that can help entering data as a check-list of risk sources. Users can add/remove any other risks if they are not on the pre-defined list.

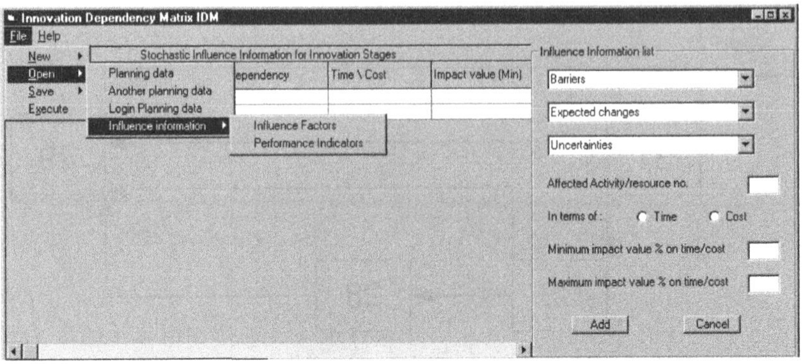

Figure 2.10 The stochastic data input

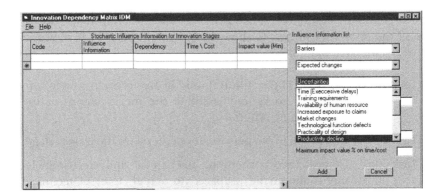

Figure 2.11 Model checklist for the stochastic information

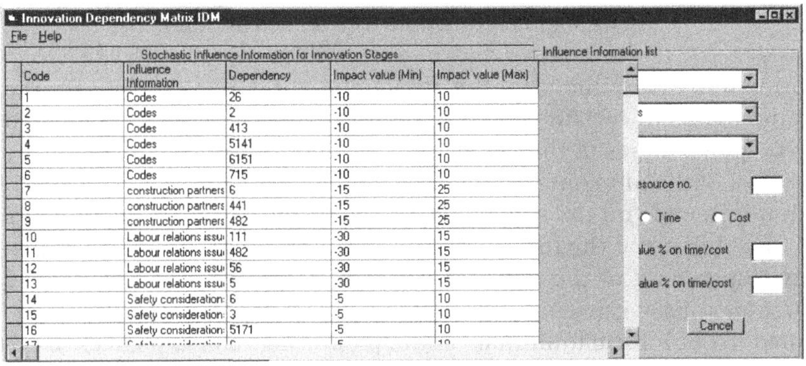

Figure 2.12 The stochastic data of the case study project

Table 2.2 The influence information on the innovative tasks of the case study

ID	Information title	Impact at the first iteration		Impact after first iteration	
1	Codes	−10 :	+10	−5 :	+5
2	Reaction of other construction partners	−15 :	+25	−10 :	+10
3	Labour relations issues	−30 :	+15	−10 :	+10
4	Safety considerations	−5 :	+10	−5 :	+10
5	Economic and political conditions	0 :	+25	0 :	+15
6	Capital intensiveness	0 :	+20	0 :	+10
7	Resistance to change	0 :	+20	0 :	+10
8	Workforce skills	0 :	+15	0 :	+10
9	Company size (capability of implementation)	−15 :	+10	−10 :	+10
10	The priority attached to the project	0 :	+10	0 :	+10
11	Functional requirements due to the type of facility	−5 :	+5	−5 :	+5
12	Funding and resources made available	0 :	+20	0 :	+10
13	Owner's view	−5 :	+5	−5 :	+5
14	Operational requirements	0 :	+15	0 :	+15
15	Market circumstances	0 :	+5	0 :	+5
16	Level of complexity of the project	−10 :	+10	−10 :	+10
17	Damage to existing utility construction lines	0 :	+10	0 :	+10
18	Productivity decline (learning curve)	0 :	+15	0 :	+10
19	Technological function risk	0 :	+25	0 :	+5
20	Contractual and tendering methods	−5 :	+5	−5 :	+5
21	Rules and regulatory bodies	−10 :	+10	−5 :	+5

Discussion of the model results

The results for implementomg these tasks under the ideal conditions (no uncertainty) were 59 time units for the schedule and 5450 cost units, as shown in Figure 2.13 (left and central parts). Figure 2.13 (right part) shows the model results of the stochastic analysis of the tasks duration. The stochastic results for duration and cost are shown in Table 2.3.

These results mean that the chance of having the duration of this group of tasks under the ideal conditions is about 2 per cent, which is very low. Also, Table 2.3 shows that the minimum cost is 5641, which is more than the cost of the tasks if no uncertainty is considered (5450). This means the target cost under ideal conditions cannot not be met. If the project team is looking for a higher chance to finish the tasks within the deterministic estimation, a risk mitigation strategy should be adopted to minimise the effect of uncertainties. Risk mitigation and response is

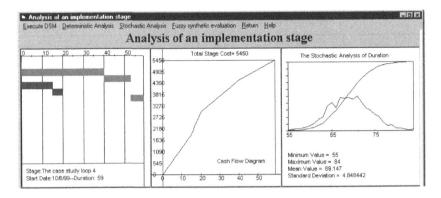

Figure 2.13 The model results according to the initial information

Table 2.3 Model results for the stochastic analysis

	Minimum	*Maximum*	*Mean*	*Standard Deviation*
Duration	55	84	69	4.84
Cost	5641	8768	7045	464.7

Table 2.4 Model results for the stochastic analysis – second iteration

	Minimum	*Maximum*	*Mean*	*Standard Deviation*
Duration	48	64	56	2.84
Cost	4816	6387	5520	259.1

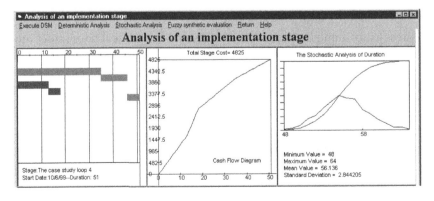

Figure 2.14 The model results according to the adjusted information

one of the key areas of risk management, but it is beyond the scope of this study. Another scenario has been made to show the effect of reducing the effect of uncertainty on the selected group of tasks. The data used for this trial is shown in Table 2.2 under the column 'Impact after first iteration'. The results obtained by running the model after adjustment of information are shown in Table 2.4 and Figure 2.14.

The adjusted information resulted in an increase of the chance of having the duration of this group of tasks under the ideal conditions to 85 per cent. Risk mitigation strategies do not usually eliminate the effect of risk on projects. They are mainly to control and reduce the effect of risks, as shown in this example.

Conclusion

This chapter discussed an approach to model the risk effect on the implementation phase of innovation in construction using the Monte Carlo technique. While risks should be considered for traditional projects as well as innovative ones, the characteristics exhibited for innovative projects are affected by higher levels of uncertainty. The chapter first introduced the innovation process in construction and identified its planning and monitoring stages. Information affecting the process was identified and discussed. The chapter also highlighted the principles of subjective estimates of these information impacts that should be considered. The idea of modelling these impacts on project time/ cost was discussed. Then, the modelling technique of Monte Carlo was explained in addition to a few technical terms always used for the development of a Monte Carlo model. A case study was then used to illustrate how the technique can be applied to simulate the effect of risks on a project time and cost.

The main conclusions drawn from this application prove that although current planning techniques such as network analysis and bar-charts are suitable for planning deterministic activities, they are not suitable for planning activities affected by risks and uncertainty. More advanced modelling techniques should deal with the various uncertain outcomes inherent in innovative projects, define all situations of a particular innovation and plan the innovation activities.

Implementing innovations in construction requires many scenarios to be analysed where decision-makers may use subjective judgements for the likelihood of particular scenarios. This analysis is complicated by uncertainties because, invariably, the decision-maker may lack control over the consequences of one or more of the scenarios under consideration. A structured methodology, that puts uncertainties into perspective and then takes them into account, is an effective way to deal with these problems. The developed model has transformed the simulation of implementation from a static state to a dynamic state through allocating durations, resources and costs to the implementation tasks. In addition, the model simulates uncertainties inherent in the implementation phase using the Monte Carlo technique. The tool has the capability of identifying influence variables on either the time or the cost of a specific task. The developed tool allows managers to investigate:

- the sensitivity of the implementation tasks to the subjective estimates used within the influence information;
- different durations/costs of the interdependent tasks for each stage;
- information that most critically influences implementation.

References

Arditi, D., Kale, S. and Tangkar, M. (1997). Innovation in construction equipment and its flow into the construction industry. *J. Constr. Engrg. and Mgmt., ASCE*, 123 (4), 371–378.

Crandall, K. C. (1977). Analysis of schedule simulation. *Journal of Construction Division, ASCE*, Sep. 77, 387–394.

De La Garza, J. M. and Mitropoulos, P. (1991). Technology Transfer (T2) Model for Expert Systems. *J. Constr. Engrg. and Mgmt., ASCE*, 117(4), 736–755.

Ferrell, W. R. (1994). Discrete subjective probabilities and decision analysis: elicitation, calibration and combination. In G. Wright and P. Ayton (eds), *Subjective Probability*. London: John Wiley & Sons Ltd.

Fishman, G. S. (1996). Monte Carlo – concepts, algorithms and applications. *Springer series in operations research, Springer – Verlag*, New York.

French, S. (1988). *Decision Theory: an introduction to the mathematics of rationality*. Hemel Hempsted: Ellis Horwood.

Lansley, P. (1996). Innovation: The role of research, education and practice. Construction Papers, No. 59. Editor; Peter Harlow. Published by the Chartered Institute of Building (CIOB).

Lindley, D. (1994). *Subjective probability – foundation*. London: John Wiley & Sons Ltd.

Motawa, I. A., Price, A. D. F. and Sher, W. (1999). Scenario planning for implementing construction innovation. *Proceeding of COBRA 1999 Conference*, Salford University, Manchester, UK, 172–181.

Motawa, I. A., Price, A. D. F., and Sher, W. (2000). Simulating innovation performance in construction technology – a fuzzy logic approach. *Proceeding of 2nd international conference on decision making in urban and civil engineering*, Lyon, France, 1087–1098.

Motawa, I. A., Price, A. D. F. and Sher, W. (2003). A fuzzy approach for evaluating the iterated implementation of innovations in construction, *International Journal of IT in Architecture, Engineering and Construction (IT-AEC)*, 1(2), 105–118.

Nam, C. H. and Tatum, C. B. (1988). Major characteristics of constructed products and resulting limitations of construction technology. *J. Constr. Mgmt. and Econ.*, 6, 133–148.

Nam, C. H. and Tatum, C. B. (1992). Government-industry cooperation: fast-track concrete innovation. *J. Constr. Engrg. and Mgmt. ASCE*, 118(3), 454–471.

Oxford Advanced Learner's Dictionary (2000). www.oxfordlearnersdictionaries. com/definition/english/innovation (accessed on 15 July 2019).

Skibniewski, M. J. and Chao, L. (1992). Evaluation of advanced construction technology with AHP method. *J. Constr. Engrg. and Mgmt. ASCE*, 118(3), 577–593.

Tatum, C. B. (1989). Organising to increase innovation in construction firms. *J. Constr. Engrg. and Mgmt. ASCE*, 115(4), 602–617.

Technology Foresight Panel on Construction (1995). Progress through partnership, 2, *Office of Science and Technology*, London: HMSO.

Van Gelder, A. (1967). Some new results in Pseudo Random Number generations. *Journal of the association for computing Machinery*, 14, 785–792.

Wakeman, T. H. (1997). Engineering leadership in public policy resolution. *J. Mgmt. in Engrg. ASCE*, 13(4), 57–60.

Winistorfer, S. G. (1996). Product Champions in Government Agencies. *Journal of Management in Engineering*, 12(6), 54–58.

3 A structural equation modeling approach for assessing external and internal risk factors in projects

Chrisovalantis Malesios and
Prasanta Kumar Dey

Highlights

- An effective technique for analyzing risk in projects is demonstrated
- Structural equation modeling is applied to simulated data for illustration of the method
- Applications of structural equation modeling in analyzing risk are scant
- The proposed approach may assist in measuring impact of projects' risks.

Introduction

The success parameters for any project are on-time completion, within specific budget and with requisite performance (technical requirement) (Dey et al. 2011). Today's project managers believe that a conventional approach to project management is not sufficient, as it does not enable the project management team to establish an adequate relationship among all phases of project, to forecast project achievement for building confidence of the project team, to make decisions objectively with the help of an available database, to provide adequate information for effective project management and to establish close co-operation among project teams (Dey 2010).

As a market entity, the enterprise should be able to face various risks resulting from external market environment and internal operation management (Zhang et al. 2012). Managing risk (MR) is the systematic process of identifying, analyzing and responding to risk. It includes maximizing the probability and consequences of positive events and minimizing the probability and consequences of adverse events in projects. MR is an expanding field, growing beyond the rich work done in finance and insurance (Wu and Olson 2009; Lloyd-Jones et al. 2019). Research has revealed that risk management can be used as philosophy for greater rewards and not just to control against loss (Wu and Olson 2008). Managing risks is one of the most important tasks for the construction industry as it affects project outcomes (Dey 2010; Fan and Stevenson 2018).

Although risk management standards are helpful to manage risk, they do not help to choose the right tools for risk identification and analysis. The current literature on project risk management consists of empirical research on risk management practices of the construction industry, and conceptual and applied frameworks of risk management using various mathematical models (Kwak and Anbari 2009). Moreover, both business and operational risks constitute project risks. Actually, business risks affect projects as a whole and operational risks affect specific work packages and/or specific activities (Dey 2010). It is thus necessary for the organizations to deploy risk management approaches throughout the project's life-cycle (Dey et al. 2013).

Although today's organizations appreciate the benefits of managing risks in various projects (e.g. constructions), formal risk analysis and management techniques are rarely used due to lack of knowledge and to doubts on the suitability and effectiveness of these techniques for construction industry activities (Akintoye and MacLeod 1997).

The structural equation modeling (SEM) approach has been applied across management science disciplines in order to examine relationships among the variables within a system. However, application of the SEM approach for risk management is scant. This chapter demonstrates how the SEM approach could be adopted to study the impact (combination of likelihood and severity) of risk factors on project outcomes. This will enable project managers to develop strategies for risk responses objectively. This chapter uses a hypothetical case with simulated data from a project organization to: (a) describe and develop a SEM model to analyze the impact of internal/external risk factors with project goals; (b) illustrate fit of this theoretical construct through the use of standard user-friendly statistical software; and (c) provide interpretation of the derived results in a manner that can be practically used by non-expert practitioners for risk evaluation.

In particular, this chapter contributes to the field of risk management by analyzing the impact of both internal and external factors on industry-based project goals. We follow a statistical modeling approach for risk analysis through deriving association among the variables and their latent constructs for MR. More specifically, in order to test the influence of the various latent variables of MR, we fit structural equation models (Bollen 1989; Jöreskog and Sörbom 1979), testing the conceptual model that we have hypothesized.

The next section demonstrates the proposed method with a step by step approach to identification, analysis and response development for internal and external risk factors. The third section demonstrates the application of the proposed method in a hypothetical case. The final sections provide a discussion and conclusion with pros and cons of the proposed method along with elaboration on scope for further research.

Method

This section first demonstrates structural equation modeling and then presents the proposed method for risk analysis through SEM using a step by step approach.

Structure equation modeling

A structural equation model is a system where causal relationships are modeled between variables. The distinguishing feature is that variables here – in contrast to typical regression analysis techniques – can be either directly observed or latent or a mixture of both of these.

This chapter illustrates the utilization of SEM in analyzing the combined effect of different risk factors on management risk and deriving an estimate for the total effect of these factors on risk management. Whereas one may use regression analysis techniques to test the connection of each one of these factors separately, SEM allows for simultaneously analyzing the relationship of different proxies on the dependent measure. Another distinctive feature is that here the dependent can be either observed or latent (i.e. not directly measurable item), a feature that cannot be addressed by typical regression analyses. Hence, SEM possesses a distinctive characteristic of latent variables being regressed on other latent variables, such as those selected for our illustrative example.

The SEM setup allows including several risk measures in parallel, hence deriving a more complete picture of relationships between management risk and the potential factors that may affect it. SEM models essentially consist of multiple regression equations for both observed and latent items that can be visually illustrated by graphical structures usually known as "SEM diagrams" or "path diagrams". In these graphs, latent constructs are depicted as circles or ellipses, observed items as squares and single-headed arrows are used to imply a direction of assumed causal influence between the variables.

Despite the obvious advantages of SEM methodology, the latter has also some disadvantages (Tomarken and Waller 2005). The large flexibility of the SEM approach makes difficult the identification of the "best" model for the available data since there are numerous ways to identify relationships, especially when dealing with datasets consisting of many variables, both of latent and observed type. Other disadvantages are the requirement of non-linear approximation methods for parameter estimation, leading frequently to instability problems as regards to model convergence, and the inaccuracy of some commonly used rules of thump for assessment of model fit.

To briefly illustrate the methodology in rough lines let us assume the correlation between a single latent factor (say ξ) with n observed items denoted by x_i $(i = 1,2,..,n)$ as shown in Figure 3.1. The model of Figure 3.1 is a simple special case of the SEM model, known as a Confirmatory Factor Analysis (CFA) model.

In order to fit the model and derive parameter estimates, we assume the following relation between the latent factor ξ and the observed items x_i that can be expressed as:

$$x_{ij}=\lambda_j \, \xi+ \delta_{ij} \ (i=1,2,...,n; j=1,2,...,k) \tag{1}$$

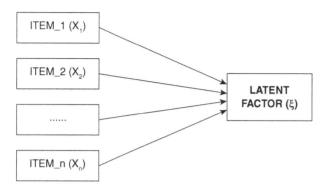

Figure 3.1 A typical SEM model

where x_{ij} denotes the jth collected measurement on the ith observed item x_i, ξ is the (1×1) scalar of the (unknown) single common factor, the λj's terms are the factor loadings to be estimated connecting ξ to the the x_{ij}'s, and δ_{ij} is the measurement error in x_{ij} (i.e. the part of x_{ij} that cannot be accounted by the underlying factor ξ). It is further assumed that the error terms δ_{ij} and the common factor ξ have a zero mean and that the common and unique factors are uncorrelated, i.e. $E(\xi-E\xi)(\delta_{ij}-E\delta_{ij})=0$, for every i, j. In vector notation, to each measurement i corresponds a n×1 vector of items $\mathbf{X}_i$:

$$\mathbf{X}_i = \wedge\xi + \delta_i \tag{2}$$

where $\Lambda=(\lambda_1, \lambda_2,...,\lambda_n)^t$, and $\delta=(\delta_{i1}, \delta_{i2},...,\lambda_{in})^t$.

In turn, the SEM model based on the complete data set can be written as:

$$\mathbf{X}=\Xi\Lambda^t+\Delta \tag{3}$$

where $\mathbf{X}$ is the (k×n) matrix of observed variables for the k measurements, $\Xi = \xi\mathbf{1}$, where boldface $\mathbf{1}$ is a (k×1) vector of 1's, Λ^t is the transpose of the (n×1) vector of factor loadings, and finally, Δ denotes the (k×n) matrix of measurement errors.

Such a model is usually fit by the maximum likelihood method (Jöreskog 1970; Maydeu-Olivares 2017). If we denote by $\mathbf{S}$ the empirical covariance matrix of the matrix of the observed variables $\mathbf{X}$ (i.e. the sample variance-covariance matrix), then to obtain the ML estimates of Λ, σ^2 and Θ, one needs to maximize the following log-likelihood function logL given by:

$$\log L\left(\wedge,\sigma^2,\Theta\right)=-\frac{1}{2}n\left[\log|\Sigma|+tr\left(\mathbf{S}\Sigma^{-1}\right)\right]. \tag{4}$$

Fitting an SEM model with maximum likelihood assumes multivariate normal data and a reasonable sample size (e.g. n>=200 measurements). However, with non-normal data, for instance to apply structural equation modeling with ordinal variables, there exist alternative methods such as the method of Weighted Least Squares (WLS) (Jöreskog, 1994).

As regards assessing the fit of an SEM model, there exist a large variety of goodness-of-fit measures that are mostly functions of the model's chi-square. Typical examples of such indices are the GFI (goodness-of-fit index) and the AGFI (adjusted goodness-of-fit index) devised by Jöreskog and Sörbom (1989), with AGFI adjusting the GFI for the complexity of the fitted model. Another popular measure is the Root Mean Square Error of Approximation (RMSEA). If the fit of the model is good, GFI and AGFI should approach one, whereas RMSEA should be small (typically less than 0.05).

Framework for risk analysis using structural equation modeling

The general outline of the proposed methodology for the assessment of project risks is described by the following few steps. First, the practitioner (e.g., the analyst in the company) designs a risk management model as per specific need. The model includes all the potential risk factors that have been identified by experts using their prior experience. Second, based on this model framework, relevant data will be gathered from a specific project either through objective data sources or through a perceptional survey involving experienced project people. Third, the MR model is fitted to the collected data through structural equation modeling and the utilization of suitable statistical software. We have to note that there exists a large variety of similar SEM-specific software that can be relatively easily utilized for conducting SEM analyses, for instance AMOS (Arbuckle 2014),[1] LISREL (Jöreskog and Sörbom 1989) or Mplus (Muthén and Muthén 2007). The SEM model utilizes all gathered data and tests the theoretical MR model and its relationships among the variables, by estimating structure coefficients through suitable estimation methods. Through this, the project's risks can be identified and assessed based on measurable items, as well as the main moderators for these risks. Several estimation techniques are available for this, including maximum likelihood, generalized least squares etc. or methods based on the Bayesian paradigm. The results of model estimation are typically illustrated through a path diagram (i.e., a graphical representation of the estimated model). Fourth, based on the derived SEM results, e.g. in the form of estimated coefficients of the associations between the variables of the MR theoretical model, managers can assess the (relative) impact of the various risk factors on the project's overall management risk and correspondingly apply suitable mitigation plans to reduce or eliminate these risks.

Applications

Application using simulated data

A hypothetical example of project management risk scenario is considered in order to illustrate the application of structural equation modeling on the risk analysis framework. This illustration example is based on a hypothetical firm's effort to examine, evaluate and take actions on a project's risks related to various aspects, for example based upon social, environmental and operational risk factors. Based on an evaluation of certain variables measuring these risk factors, the company wants to identify and manage these risks – separately and/or as an overall risk outcome – for the successful implementation and completion of the project.

Therefore, let us hypothesize that the general latent construct of MR is operationalized as three-dimensional constructs of "social risks", "environmental risks" and "operational risks". Each one of the three risk factors is supposed to comprise of three observed items, as were ranked for example by a certain sample (e.g. the company's managerial personnel). We further assume that the observed items are measured using a 5-point likert scale, where the hypothetical respondents are ranking the importance of each one of the items by asking for their opinion as concerns their agreement on the importance of each one of the items. According to this framework, respondents are assigning one of the following values: (1) totally disagree; (2) disagree; (3) neither agree nor disagree; (4) agree; (5) totally agree. Under this set up, the hypothetical dataset consists of the following observed items:

> **Social risks:** risks that exist when there is a social unrest (hypothetical selected indicator items: strikes (SOCIAL_1), religious turmoil (SOCIAL_2), and terrorism (SOCIAL_3)).

> **Operational risks:** The risk that a firm's internal practices, policies and systems are not rigorous or sophisticated enough to cope with unexpected market conditions or human or technological errors (Karam and Planchet 2012) (hypothetical selected indicator items: damage to physical assets (OPER_1), employment practices and workplace safety (OPER_2) and business disruption and systems failures (OPER_3)).

> **Environmental risks:** risks related to various environmental risks facing a business (hypothetical selected indicator items: global warming (ENV_1), air pollution (ENV_2) and industrial waste (ENV_3)).

To acquire the previously described hypothetical sample we simulate the data required for running the example using an algorithm described in Johnson, Kotz and Balakrishnan (1997). The devised algorithm generates multinomial variables. If you draw observations with replacement from a

Table 3.1 Chosen weights for the 5-category response choices for each observed item

Factor	Item	Weights				
		Totally disagree	Disagree	Neither agree nor disagree	Agree	Totally agree
environmental	1	0.5	0.2	0.1	0.1	0.1
	2	0.4	0.3	0.1	0.1	0.1
	3	0.3	0.3	0.2	0.1	0.1
operational	1	0.1	0.1	0.5	0.2	0.1
	2	0.1	0.2	0.3	0.3	0.1
	3	0.1	0.1	0.4	0.2	0.2
social	1	0.1	0.1	0.2	0.3	0.3
	2	0.1	0.2	0.4	0.2	0.1
	3	0.1	0.3	0.2	0.1	0.3

population with k classes of objects, where k>2, the k numbers of objects sampled from the respective classes have a multinomial distribution. We implement the algorithm with the assist of SPSS statistical package (IBM Corp. Released 2017). Specifically we use SPSS to generate a total number of nine variables (i.e. three variables for each one of the three latent factors) with a multinomial distribution for a specified number of cases.

We chose to generate a sample of size n = 200 for all nine items. For the simulation scheme arbitrary weights for each one of the five category answers were attributed according to Table 3.1, which resulted in the corresponding sample sizes shown in Table 3.2. The selected weights are chosen in a way to attribute more importance on the operational and social observed

Table 3.2 Corresponding sample sizes for the 5-category response choices for each observed item

Factor	Item	Sample size (n)				
		Totally disagree	Disagree	Neither agree nor disagree	Agree	Totally agree
environmental	1	105	43	19	13	17
	2	79	67	23	15	16
	3	69	64	31	14	22
operational	1	26	13	107	32	22
	2	15	44	47	81	13
	3	18	24	76	37	45
social	1	24	19	38	59	60
	2	17	33	88	39	23
	3	26	57	33	22	62

items and less importance on the environmental items. Subsequently, we will test if the statistical analysis can adequately recognize these attributes and supply with all necessary information the firm's analysts on how to take the right decisions in order to cope with those management risks.

Model description

Our main hypothesis is that the environmental, social and operational risks are all important factors that directly influence a firm's risk taking. We thus, utilizing the simulated observed data on potential environmental, operational and social observed variables, test for a relationship between the four latent constructs.

To achieve the stated objectives, the incorporation of the three risk factors requires the SEM to be set up on a three-level framework. The first level of this three-level SEM includes the observed items constituting the risk factors. Under the SEM framework, social, environmental and operational risk factors are defined as latent variables in the second level and the final level comprises of the general MR.

In order to test our research hypothesis we implement a SEM analysis with latent structures using the IBM SPSS AMOS software. In particular, we tested two types of structural equation models for exploring inter-individual variation using latent variables (Bollen 1989). In the first model (MODEL A) we connect directly the nine observed items to the latent construct of risk management (see the following Figure 3.2).

Accordingly we will test a second SEM model. Model B (see Figure 3.3) is an extended and more complex version of Model A, in which we add another level to the path analysis by including the three latent factors of "social", "environmental" and "operational risks". Hence, in the second SEM model we associate the observed items with the three factors and we then link these "social", "environmental" and "operational risks" to "management risk".

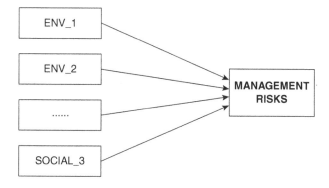

Figure 3.2 Model A (direct association of observed items with risk management)

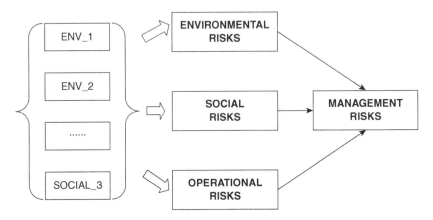

Figure 3.3 Model B (association of the three latent factors with risk management)

Hence, in the second model, we introduced internal/external risk factors in the model, investigating their impacts on MR and at the same time the relative impact of each one of the three observed items on these latent factors.

In order to perform the fit of the two previously described models with AMOS, for each factor we need to fix one loading to the value of one in order to give the latent factor an interpretable scale. In addition to the factor loadings, it is usual for statistical software to provide also standardized factor loadings that assist in better interpretation of results. For our demonstration the original unstandardized loadings will be reported.

Results

Two statistical analyses were performed to empirically test the two hypothesized models through the AMOS software. The windows environment of the specific software is shown in the following Figure 3.4.

The SEM (path) diagrams that the user has to construct in the interface of the software (see path diagram pane in Figure 3.4) are shown in the following Figures 3.5 and 3.6, for Model A and Model B respectively. Observe the set-up of one factor loading for each factor to the value of one for identifiability of the resulted factor loadings.

As one observes from the constructed path diagrams, rectangles represent the observed items whereas circles or eclipses represent the latent factors. Due to the departure of our data from normality we fit the two models through weighted least squares model estimation method.

The results obtained by the fit of the two models are shown in the following figures 3.7 and 3.8, as were obtained by the AMOS program. In these

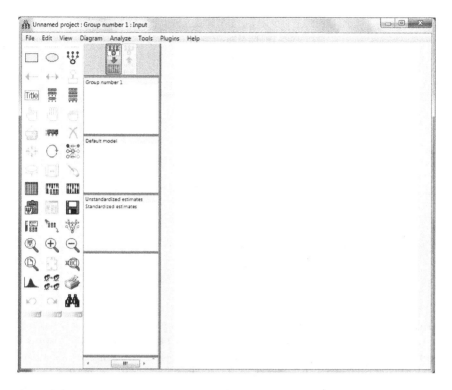

Figure 3.4 Main screen of AMOS software

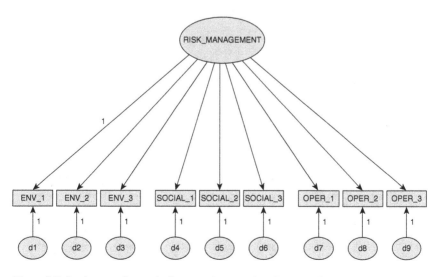

Figure 3.5 Setting-up the path diagram for running SEM analysis (MODEL A)

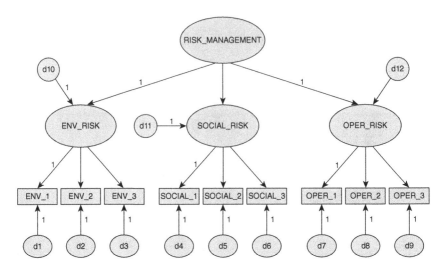

Figure 3.6 Setting-up the path diagram for running SEM analysis (MODEL B)

graphs, the single-headed arrows are used to imply a direction of assumed causal influence while the numerical values next to each arrow correspond to the factor loadings of each item on the corresponding latent/observed variable. Numerical values next to each arrow correspond to the factor loadings of each (latent/observed) variable on the corresponding latent/observed variable.

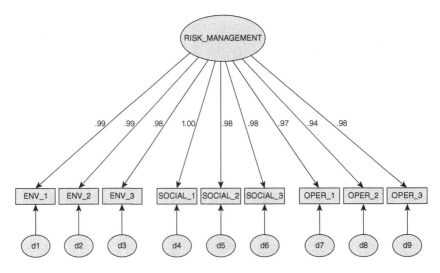

Figure 3.7 Factor loadings resulted from SEM analysis (MODEL A)

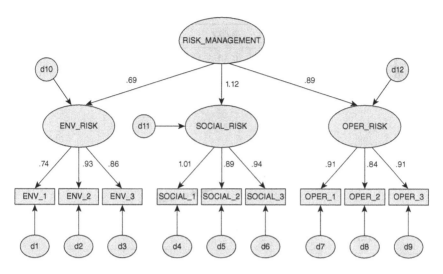

Figure 3.8 Factor loadings resulted from SEM analysis (MODEL B)

The following table (Table 3.3) presents the factor loadings as derived from MODEL B along with the statistical significance (p-value) of each loading.

Finally, we may also test the validity of our model by using several alternative fit statistics, available by the AMOS software (see Table 3.4).

In summary, the following figure (Figure 3.9) presents the relative associations – as derived from the path analysis simulation example – between the three factors hypothesized to affect management risk.

Table 3.3 Factor loadings and corresponding significance (MODEL B)

Directional link	Estimate	p-value
ENV_RISK - → RISK_MANAGEMENT	0.687	<0.001
SOCIAL_RISK → RISK_MANAGEMENT	1.117	<0.001
OPER_RISK → RISK_MANAGEMENT	0.894	<0.001
ENV_1 → ENV_RISK	0.739	0.09
ENV_2 → ENV_RISK	0.929	<0.001
ENV_3 → ENV_RISK	0.862	<0.001
SOCIAL_1 → SOCIAL_RISK	1.007	<0.001
SOCIAL_2 → SOCIAL_RISK	0.891	<0.001
SOCIAL_3 → SOCIAL_RISK	0.938	<0.001
OPER_1 → OPER_RISK	0.908	<0.001
OPER_2 → OPER_RISK	0.839	<0.001
OPER_3 → OPER_RISK	0.912	<0.001

Table 3.4 Results of Goodness-of-fit indices (GOF) for the SEM models

Model	GFI	AGFI	PGFI	RMSEA
Model A	0.724	0.54	0.435	0.208
Model B	0.761	0.552	0.46	0.2

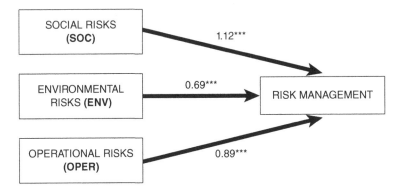

Figure 3.9 Model B path analysis' results

As a next step, in order to obtain an indication of the relative impact of each one of the three risk factors on MR we also calculate scores from SEM analysis as derived by the fit of MODEL B. Within the general framework of factor analysis, it has become common practice to estimate individual factor scores (Bartholomew and Knott, 1999) and utilize them for subsequent analyses. For instance, factor scores for the latent variables can be first predicted and then used as variables in analysis-of-variance (ANOVA) and OLS regression.

In terms of the above, individual factor scores for the latent factor of risk management were derived for the 200 hypothetical respondents and utilized subsequently for deriving a combined measure or score of management risk. This score is essentially the result of the linear combination of the products of weights for each one of the nine observed items obtained by SEM model fit with the original responses, which are given by:

$$MR_{SCORE} = \sum_{i=1}^{9} w_i x_i, \tag{5}$$

where w_i denotes the MR factor's loading on the i-th observed variable, and the x_i is the observed value of the i-th variable (i=1, 2, ..., 9).

In order to illustrate the relative importance of each one of the three latent factors on MR, in addition to the overall MR$_{SCORE}$ based in equation (5), the MR scores based separately on each one of the three latent factors are also calculated. The results are summarized in the following boxplots (Figure 3.10).

The corresponding descriptive statistics on the resulted score values are presented in Table 3.5.

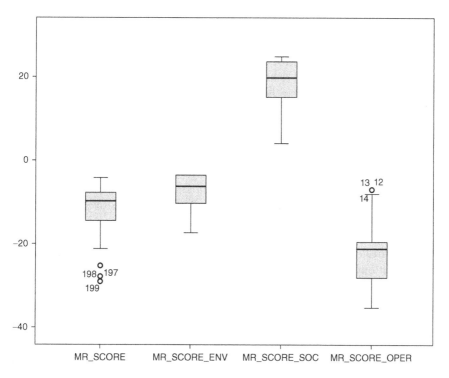

Figure 3.10 Boxplots of MR$_{SCORE}$ values

Table 3.5 Descriptive statistics of MR$_{SCORE}$ values

	average	*min*	*max*	*St.d.*
MR_SCORE	−12.47	−29.16	−4.30	6.64
MR_SCORE_ENV	−7.41	−17.41	−3.48	4.34
MR_SCORE_SOC	17.40	4.01	24.27	6.51
MR_SCORE_OPER	−22.46	−35.32	−7.06	7.72

Discussion

In the current section, the results obtained by the structural equation modeling application for assessing the impact of risks factors on management risk, are discussed in more detail.

Results from MODEL A indicate that all observed items are strongly associated with management risk. However, for a better refinement of the associations between the various single items of social, environmental and operational risks, MODEL B is the most suitable approach. The empirical results of the three-level SEM (MODEL B) were summarized in Figure 3.8.

The factor loadings shown in Figure 3.8 and the corresponding significances indicate that all paths from the three risk factors to MR were found to be significant, with higher associations been found for the risk management-social risk association (beta estimate = 1.12; p-value<0.001). However, there are also noticeable variations. "Social risks" are loading the most on RM, followed by the "operational risks". Less strong is the loading between MR and "environmental risks" (beta estimate = 0.69; p-value<0.001). These results are visually realized and summarized in Figure 3.9.

Regarding the effects of the individual social risks variables on the level of "social risks" factor, it is found that all of the individual items have strong connections with social risks, however the variable of "strikes" is proved to be more significant according to respondents (beta estimate = 1.01) in comparison to religious turmoil (beta estimate = 0.89). The risks associated with the firm's internal practices are all similarly highly positively connected with the operational risks factor.

As concerns the third latent factor of environmental risks, the path diagram indicates that the air pollution (beta estimate = 0.93) is the most important environmental risk for the firm's managers in comparison, e.g., to global warming (beta estimate = 0.74).

Regarding model fit, the fit statistics show that the path analysis structure tested by the two fitted models provides a moderate fit, with MODEL B to fit better the simulated data in comparison to MODEL A, especially according to the comparison made via the GFI and AGFI fit indices (see Table 3.4 results of fit indices for the assessment of model fitting to the data).

Next, it is evident by inspecting Figure 3.10 that the environmental risks are expected to slightly increase the firm's management risk. However, the risk factor that appears to affect the most the management risk is that of social risks, a finding which is in accordance with the results of SEM analysis, whereas the internal risks as conceptualized by the three operational risk items are the ones that contribute less in the increase of management risk, according to these results. Hence, according to these findings, managers should give more emphasis on dealing with the potential social risks in order to address their management risks.

Noticeable however, are also the three outliers of management risk scores based on operational risks in Figure 3.10, which reveal that three managers consider operational risks to be much more important in comparison to the rest of the managers. This also holds for a number of three outlying managers with regard to the total risk management score, with the latter being underestimating the combined measured risk in the company. This could require special attention from the firm's analyst under the hypothetical situation.

Conclusions

To achieve the goal of firm-value maximization, managers should seek to optimize corporate decisions on risk management. In this chapter an attempt was made to illustrate the utilization of standard statistical multivariate methodologies, and specifically structural equation modelling, towards the support of risk analysis planning. Within this framework, we have examined the interactions between key components of risks associated with risk management. The illustrated methodology may assist in identifying the critical risk factors out of many possible risk factors that affect risk management.

Note

1 For SEM implementations based on the Bayesian paradigm, the freely available WinBUGS software (Lunn et al. 2000) is the main option, although Bayesian SEM models can be also fit by Mplus.

References

Akintoye, A.S. and MacLeod, M.J. (1997). Risk analysis and management in construction. International Journal of Project Management, 15(1), 31–38.

Arbuckle, J.L. (2014). Amos 23.0 User's Guide. Chicago, IL: SPSS.

Bartholomew, D.J., and Knott, M. (1999). Latent variable models and factor analysis. (Kendalls Library of Statistics, No.7, 2nd. ed.). New York: Edward Arnold.

Bollen, K.A. (1989). Structural equations with latent variables. New York: Wiley-Interscience.

Dey, P.K. (2010). Managing project risk using combined analytic hierarchy process and risk map. Applied Soft Computing, 10, 990–1000.

Dey, P.K., Clegg, B. and Cheffi, W. (2013). Risk management in enterprise resource planning implementation: a new risk assessment framework. Production Planning and Control, 24(1), 1–14.

Fan, Y. and Stevenson, M. (2018). A review of supply chain risk management: definition, theory, and research agenda. International Journal of Physical Distribution & Logistics Management, 48(3), 205–230.

IBM Corp. Released (2017). IBM SPSS Statistics for Windows, Version 25.0. Armonk: IBM Corp.

Johnson, N.L., Kotz, S. and Balakrishnan, N. (1997). Discrete Multivariate Distributions. New York: Wiley. 68–69.

Jöreskog, K.G. and Sörbom, D. (1979). Advances in factor analysis and structural equation models. New York: University Press of America.

Jöreskog, K.G. and Sörbom, D. (1989). Lisrel 7: A guide to the program and applications. Chicago: SPSS.

Jöreskog, K.G. (1970). A general method for analysis of covariance structures. Biometrika, 57, 239–251.

Jöreskog, K.G. (1994). Structural equation modeling with ordinal variables. Multivariate analysis and its applications. Hayward: Institute of Mathematical Statistics. 297–310.

Karam, E. and Planchet, F. (2012). Operational risks in financial sectors. Advances in Decision Sciences, vol. 2012, Article ID 385387, 57 pages, 2012. doi:10.1155/2012/385387.

Kwak, Y.H. and Anbari, F.T. (2009). Analyzing project management research: perspective from top management. International Journal of Project Management, 27(5), 435–446.

Lloyd-Jones, D.M., Braun, L.T., Ndumele, C.E., Smith, S.C., Sperling, L.S., Virani, S.S. and Blumenthal, R.S. (2019). Use of risk assessment tools to guide decision-making in the primary prevention of atherosclerotic cardiovascular disease. Journal of the American College of Cardiology, 73(24), 3153–3167.

Lunn, D.J., Thomas, A., Best, N. and Spiegelhalter, D. (2000). WinBUGS – A Bayesian modeling framework: concepts, structure and extensibility. Statistical Computing, 10, 325–337.

Maydeu-Olivares, A. (2017). Maximum likelihood estimation of structural equation models for continuous data: standard errors and goodness of fit. Structural Equation Modeling: A Multidisciplinary Journal, 24(3), 383–394.

Muthén, L.K. and Muthén, B.O. (2007). Mplus user's guide (Sixth Edition). Los Angeles: Muthen & Muthen.

Tomarken, A.J. and Waller, N.G. (2005). Structural equation modeling: strengths, limitations and misconceptions. Annual Review of Clinical Psychology, 1, 31–65.

Wu, D.D. and Olson, D.L. (2008). Supply risk, simulation and vendor selection. International Journal of Production Economics, 114 (2), 646–655.

Wu, D.D. and Olson, D.L. (2009). Risk issues in operations: methods and tools. Production Planning Control, 20(4), 293–294.

Zhang, H., Yang, Y. and Jiang, Y. (2012). Study on comprehensive SEM-based enterprise risk management frame. International Journal of Digital Content Technology and its Applications, 6(18), 474–481.

4 Risk management using combined multiple criteria decision-making technique, risk mapping and decision tree

Oil refinery construction project

Prasanta Kumar Dey

Introduction

The success parameters for any project are on-time completion, within specific budget and with requisite performance (technical requirement). The main barriers for their achievement are the changes in the project environment (Chapman 2006). The problem multiplies with the size of the project as uncertainties in project outcome increase with size (Zayed et al. 2008). Oil refinery construction projects are exposed to uncertain environments because of factors such as planning and design complexity, presence of various interest groups (project owner, owner's project group, consultants, contractors, vendors etc.), resources (materials, equipment, funds etc.) unavailability, climatic environment, the economic and political environment and statutory regulations (Dey and Ramcharan 2008). Other risk factors include the complexity of the project, the speed of its construction, the location of the project and its degree of unfamiliarity.

Today's project managers believe that a conventional approach to project management is not sufficient, as it does not enable the project management team to establish an adequate relationship among all phases of the project, to forecast project achievement for building confidence of the project team, to make decisions objectively with the help of an available database, to provide adequate information for effective project management and to establish close co-operation among project team (Dey 2010 and Liang 2009). Although today's organizations appreciate the benefits of managing risks in construction projects, formal risk analysis and management techniques are rarely used due to lack of knowledge and doubts on the suitability of these techniques for construction industry activities (Kwak and Anbari 2009).

The current literature on construction project risk management consists of mainly four types of works – standards for risk management practices, project risk management practices of countries including ranking of risk factors, case studies of organizations and industries using conceptual mathematical models and application of risk responses. The Project Management

Body of Knowledge (PMBoK) of Project Management Institute (PMI 2008) and AS / NZS ISO 31000: 2009 Risk management – Principles and guidelines are the most popular sources for generic project risk management processes. Although they are effective in identifying resources, tools and techniques and outputs, each project needs a customized approach for application. Wang et al. (2004) study risk management practices of developing countries to identify, categorize, evaluate and rank risks, Thuyet et al. (2007) demonstrate risk management practices in oil and gas construction projects in Vietnam, Zayed et al. (2008) show risks that are inherent in Chinese highway projects, and Dey (2010) illustrates risks in Indian construction projects in the oil industry. Although these give the ideas of risk factors, risk management practices and issues and challenges of applications for construction projects in a specific industry and country, do not provide a framework for application. The project risk management literature is very strong in applying quantitative modeling for analyzing risks. Schatteman et al. (2008) use an integrated computerized risk identification and analysis method. However, their model does not integrate mitigating measures. Tuysuz and Kahraman (2006) develop a risk management framework using fuzzy analytic hierarchy process (AHP) and apply it in information technology projects. Tah and Carr (2000) apply fuzzy logic for risk assessment in construction projects. However, they didn't integrate risk assessment with response development. Wang et al. (2004) apply the qualitative risk management model for managing project risk in developing countries, which has very weak integration across risk management processes. There are numerous studies on managing risks as a part of managing overall project (e.g. Shen et al. 2006; Dey 2006; Dey and Ramcharan 2008). Although the conventional risk management methods contribute to furthering risk management practices in the construction industry, a more practical approach is needed. Additionally, according to the author's knowledge there is no other quantitative framework, which integrates risk identification, analysis and response development. Hence, this study bridges the gaps.

The objective of the study is to develop a decision support system which integrates risk identification, analysis and responses development for managing construction project risks. This enables project managers to make the right decisions to accomplish project goals. The organization of the paper is as follows: section II briefs the literature on project risk management practices and frameworks, section III states the methodology, section IV introduces the proposed framework for risk management, section V demonstrates application of the proposed framework, section VI provides the discussion and section VII concludes the study.

Prior studies

As per PMBoK (PMI 2008) risk management is the systematic process of identifying, analyzing and responding to project risk. It includes maximizing the probability and consequences of positive events and minimizing the probability and consequences of adverse events to project objectives. Risk management has six steps. They are risk management planning, risk identification, qualitative risk analysis, quantitative risk analysis, risk response planning, risk monitoring and control (PMI 2008). The AS/NZS ISO 31000: 2009 sets out five steps for risk management – establish the context, identify the risks, analyze the risks, evaluate the risks and treat the risks.

In the past, a number of systematic models have been proposed for use in the risk-evaluation phase of the risk-management process. Kangari and Riggs (1989) classified these methods into two categories: classical models (i.e. probability analysis and Monte Carlo simulation) and conceptual models (i.e. fuzzy-set analysis). They noted that probability models suffer from two major limitations. Some models require detailed quantitative information, which is not normally available at the time of planning, and the applicability of such models to real project risk analysis is limited, because agencies participating in the project have a problem with making precise decisions. The problems are ill defined and vague, and they thus require subjective evaluations, which classical models cannot handle. There is, therefore, a need for a subjective approach to project risk assessment, with there being the necessary objectivity in the methodology. The analytic hierarchy process as shown by Mustafa and Al-Bahar (1991) and Dey et al. (1994) provides both a subjective and objective approach to risk analysis using expert judgement. However, their approaches fail to integrate risk analysis with the project management processes.

Zayed et al. (2008) applied the AHP for assessing risk in Chinese highway projects. The framework prioritizes risk factors and ranks alternative projects. However, their approach doesn't discuss managing projects effectively using risk management methodology as indicated in the project risk management standards (PMI 2008 and AS/NZS ISO 31000: 2009).

Wang et al. (2004) introduce a risk management framework named the 'Alien Eyes' risk model, which shows the hierarchical levels of the risks and the influence relationship among the risks. Build on their findings, a qualitative risk mitigation framework has finally been proposed. This framework also suffers from integrated objective approach to risk management. Schatteman et al.'s (2008) integrated computerized risk management model identifies and quantifies (probability and impact) schedule risk, but fails to integrate mitigating measures. Shen et al. (2006) suggest public-private partnership to manage risks in public sector projects in Hong Kong. While Shen et al.'s study contributes on means for risk management, it helps little to analyze risks. Tuysuz and Kahraman (2006) demonstrate a project risk evaluation method using the fuzzy AHP approach in information technology projects. Like other studies it also doesn't objectively integrate the risk mitigating measures with risk analysis results. Dey (2006) and Dey and Ramcharan (2008) suggest the multiple criteria decision-making method for minimizing risk by selecting the right projects. Although their methods are effective for project selection, they do not tell how to manage risks across the various phases of project. Moreover, they do not determine either probability or impact of risk on project out-comes. Dey (2010) introduces a hierarchical framework for risk analysis. It identifies risks using brainstorming, derives probability using the analytic hierarchy process, and determines impact using risk maps in project, work package and activity level separately. Subsequently it develops risk mitigating measures for each level using collective experience of the project executives. Although this method provides a practical approach to project risk manage-ment, it lacks objective derivation of risk responses.

The proposed approach to project risk management introduces an inno-vative framework which integrates four methods – cause and effect diagram for risk identification, the AHP for determining probability of risks, risk map for deriving risk impact, and decision tree for revealing risk mitigating measures. Therefore, this study bridges the gaps.

Methodology

This study adopts a case study approach. First, a literature review is undertaken to review contemporary approaches to project risk management, the pros and cons of those approaches, and to identify gaps in knowledge. Second, a concep-tual framework for project risk management is proposed. Third, the proposed framework is then applied to a grass-root oil refinery construction project in India. Fourth, the pros and cons of the framework are revealed. Fifth, the prac-tical implication of the proposed framework is validated through a focus group with the representatives of a few executives from the Indian oil industry.

The proposed risk management framework uses a cause and effect diagram to identify risk, the Analytic Hierarchy Process (AHP) to derive probability of occurrence of risk, risk map for determine impact, and deci-sion tree analysis to objectively reveal measures for risk mitigation. The following paragraphs demonstrate the AHP and decision tree briefly.

The analytic hierarchy process (AHP) developed by Saaty (1980) provides a flexible and easily understood way of analyzing complicated problems. It is a multiple criteria decision-making technique that allows subjective as well as objective factors to be considered in the decision mak-ing process. The AHP allows the active participation of decision makers in reaching agreement, and gives managers a rational basis on which to make decisions. AHP is based on the following three principles: decomposition, comparative judgement and synthesis of priorities.

The AHP is a theory of measurement for dealing with quantifiable and intangible criteria that has been applied to numerous areas, such as deci-sion theory and conflict resolution (Vargas 1990). AHP is a problem-solving framework and a systematic procedure for representing the elements of any problem (Saaty 1983).

Formulating the decision problem in the form of a hierarchical structure is the first step of AHP. In a typical hierarchy, the top level reflects the over-all objective (focus) of the decision problem. The elements affecting the decision are represented in intermediate levels. The lowest level comprises the decision options. Once a hierarchy is constructed, the decision maker begins a prioritization procedure to determine the relative importance of the elements in each level of the hierarchy. The elements in each level are compared as pairs with respect to their importance in making the decision under consideration. A verbal scale is used in AHP that enables the decision maker to incorporate subjectivity, experience and knowledge in an intuitive and natural way. After comparison matrices are created, relative weights are derived for the various elements. The relative weights of the elements of each level with respect to an element in the adjacent upper level are computed as the components of the normalized eigenvector associated with the largest eigenvalue of their comparison matrix. Composite weights are then determined by aggregating the weights through the hierarchy. This is done by following a path from the top of the hierarchy to each alternative at the lowest level, and multiplying the weights along each segment of the path. The outcome of this aggregation is a normalized vector of the overall weights of the options. The mathematical basis for determining the weights was established by Saaty (1980). The AHP has numerous applications in project management. Most recently Chen et al. (2010) applied it in R&D strategic alliance partner selection.

Risk management is usually a team effort, and the AHP is one available method for forming a systematic framework for group interaction and group decision making (Saaty 1982). Dyer and Forman (1992) describe the advantages of AHP in a group setting as follows: 1) both tangibles and intangibles, individual values and shared values can be included in an AHP-based group decision process; 2) the discussion in a group can be focused on objectives rather than alternatives; 3) the discussion can be structured so that every factor relevant to the discussion is considered in turn; and 4) in a structured analysis, the discussion continues until all relevant information from each individual member in a group has been considered and a consensus choice of the decision alternative is achieved. A detailed discussion on conducting AHP-based group decision-making sessions including suggestions for assembling the group, constructing the hierarchy, getting the group to agree, inequalities of power, concealed or distorted preferences, and implementing the results can be found in Saaty (1982) and Golden et al. (1989). For problems with using AHP in group decision making, see Islie et al. (1991).

Decision trees use calculations of expected monetary values (EMV) to measure the attractiveness of alternatives. Decision trees, however, use graphical models as well to display several relevant aspects of a decision situation. These graphical models consist of treelike structures (hence the name) with branches to represent the possible action-event combinations. The conditional payoff is written at the end of each branch. A tree gives much the same information as a matrix, but, in addition, it can be used to depict multiple-stage decisions – a series of decisions over

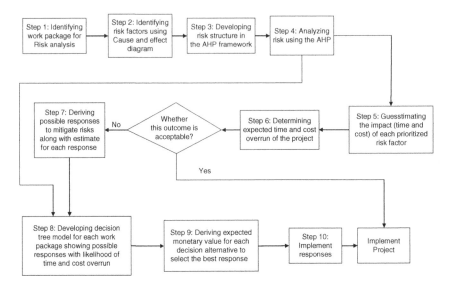

Figure 4.1 Proposed framework for project risk management in a quantitative
framework using a combined AHP and DTA

time (Dilworth 2000). The decision tree approach (DTA) logically structures risk management philosophy by identifying alternative responses in
mitigating risk. It provides a basis for quantitative risk management and
incorporates management perceptions.

The Proposed Risk Management Framework

The proposed risk management framework has the following steps:

1   Identifying the work packages for risk analysis
2   Identifying the factors that affect the time, cost and quality achieve-
    ment of specific work packages using cause and effect diagrams
3   Developing hierarchical risk structure in the AHP framework
4   Analyzing their effect by deriving the likelihood of their occurrences
    using the AHP approach

   4.1 Comparing the risk factors pair wise to determine likelihood of
        their occurrences
   4.2 Comparing the risk subfactors pair wise to determine likelihood of
        their occurrences
   4.3 Comparing the alternative work packages with respect to each sub-
        factor in order to determine likelihood of their failure from the
        risk subfactor
   4.4 Synthesizing the result across the hierarchy in order to derive rela-
        tive overall chance of failure of each work package

5 Determining severity (probability and impact) of failure by guestimating
6 Determining expected time and cost overrun of the project
7 Deriving various alternative responses for mitigating the effect of risk factors and estimating the cost for each alternative
8 Developing a decision tree for each work package showing possible responses with the likelihood of time and cost overrun
9 Deriving expected monetary value (EMV) (cost of risk response in this case) and selecting the best option through statistical analysis
10 Implementing the selected best options.

Figure 4.1 shows the proposed construction risk management framework.

Application

The proposed risk management framework was applied in a case of the construction of a new oil refinery of 7.5 million metric ton capacity in the Central part of India. The project cost was estimated at US $ 600 million.

A risk management group was formed for managing risk for the case study project. The group consisted of one member (with more than 15 years' experience) each from mechanical, electrical, civil, tele-communication and instrumentation, finance, and materials of the project function of the concerned organization. They were entrusted for collecting data, analyzing, interpreting and preparing recommendations with active interactions with the core project implementation team.

The following paragraphs demonstrate each step of the proposed risk management framework.

Step 1: identifying the work packages for risk analysis

The total project scope was hierarchically arranged to form a work break down structure. Figure 4.2 shows the work breakdown structure of the oil refinery construction project under study. The risk management group brainstormed on the complexity in design, implementation and operations of each package using their experience. According to the complexity in achieving time, cost and quality targets, the work packages that were considered for risk management were instrumentation and control room, process equipment and piping, tank farm, pipelines and loading/unloading facilities, and power and utilities.

Step 2: identifying risk factors that affect time, cost and quality achievement of specific work packages using cause and effect diagrams

The risk factors and subfactors were then identified using a cause and effect diagram for each package separately by the project executives, who were actively involved with managing those packages through focus

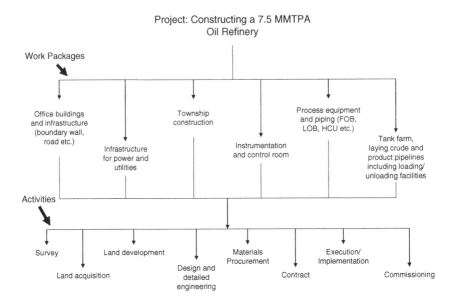

Figure 4.2 Work breakdown structure

group discussions with facilitation of the risk management group. The risk management group then compiled the risk factors and subfactors to develop a consolidated list of risk factors and subfactors for the entire project. Figure 4.3 shows the cause and effect diagram for the risk of the project under study as identified by the risk management group. The following paragraph describes the risk factors briefly.

Oil refinery construction projects deploy complex technologies, which make it vulnerable to failure in terms of time, cost and quality achievement. There is always a chance of changing project scope as an accurate project plan is almost impossible to formulate. The successful accomplishment of projects depends on how effectively these changes are managed. Selecting appropriate technologies, implementing them effectively and operating them efficiently are the key success factors for a business. Other than the above internal factors, projects are always vulnerable from external factors like financial, political and economical risks. Availability of funds throughout project life, and stability of the political and economic environment, impact a project significantly. Additionally, the accuracy of available information for a project cost estimate could also affect project outcomes. There are a number of stakeholders in a megaproject, such as oil refinery construction. Their individual and collective capabilities contribute largely towards successful completion of the projects. Historically, natural hazards are one of the major causes of project failure. These include rain, flood, subsidence, fire and heat for the project under

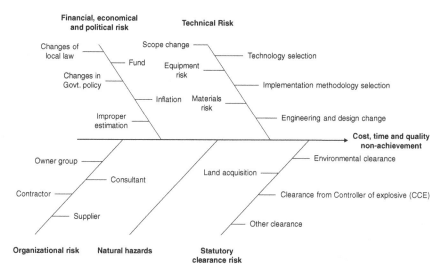

Figure 4.3 Risk identification using cause and effect diagram

study. Large infrastructure projects need regulatory clearances from many competent authorities, which include environment clearance, land acquisition related clearance, explosive fire clearance and construction clearance. On-time approval of those clearances is critical to complete the construction activities on time.

Step 3: developing the risk structure in the AHP framework

The factors as identified in previous steps were arranged hierarchically to form an analytical framework. Figure 4.4 shows the hierarchical structure for risk analysis of oil refinery construction project. The level 1 is the goal i.e. "determining riskiness of project". Levels 2 and 3 are for the factors and subfactors respectively as identified by the project risk management group. Level 4 contains the alternatives i.e. critical work packages as identified by the same group.

Step4 (4.1. to 4.4.): analyzing risk using the AHP

Risk analysis of the project was then carried out using Microsoft Excel. The risk management group in a focus group made pairwise comparisons in both factors (Table 4.1) and subfactor levels (Appendix 1) using the verbal scale (Table 4.2) developed by Saaty (1980) in order to determine the likelihood of their occurrences. For example, for the factor level comparison, the risk management group agreed that "technical risk" was "moderately important / vulnerable" compared to "financial and economical risk". Accordingly, they used "3" in the second row and third column, and "1/3" in the third row and

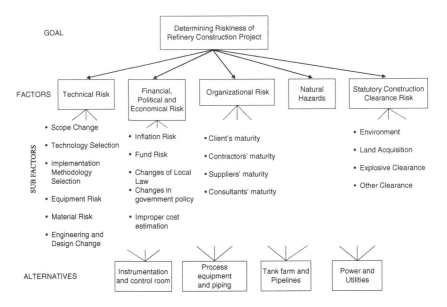

Figure 4.4 The AHP model for determining riskiness of project

second column. Subsequently, they compared each alternative work package with respect to each risk subfactor in order to determine the likelihood of failure of each work package due to occurrences of those risks (Appendix 2). The analysis was done using group consensus. Using Microsoft Excel the results were synthesized across the hierarchy to determine likelihood of failure of the work packages. Table 4.3 shows the outcomes of pairwise comparison in factor, subfactor and alternative levels along with the overall likelihood of failure of each work package. The detailed calculation of the AHP application is available on request.

Table 4.1 Comparison matrices in factor level using information from Table 4.2

Factors	Technical Risk	Financial and Economical Risk	Organizational Risk	Natural Hazards	Statutory Clearance Risk
Technical Risk	1	3	4	5	5
Financial and Economical Risk	1/3	1	2	3	3
Organizational Risk	1/4	1/2	1	2	2
Natural Hazards	1/5	1/3	1/2	1	1
Statutory Clearance Risk	1/5	1/3	1/2	1	1
Column sum	1.9833	5.1667	8	12	12

Normalized matrix to determine likelihood of failure from each factor (each cell has been divided by the column sum to form normalized matrix; likelihood of risk factors is then determined by averaging across each row)

Factors	Technical Risk	Financial and Economical Risk	Organizational Risk	Natural Hazards	Statutory Clearance Risk	Likelihood
Technical Risk	0.504*	0.581	0.500	0.417	0.417	0.484
Financial and Economical Risk	0.168	0.194	0.250	0.250	0.250	0.222
Organizational Risk	0.126	0.097	0.125	0.167	0.167	0.136
Natural Hazards	0.101	0.065	0.063	0.083	0.083	0.079
Statutory Clearance Risk	0.101	0.065	0.063	0.083	0.083	0.079

*1/1.9833 = 0.504

λmax = 1.9833 × 0.484 + 5.1667 × 0.222 + 8 × 0.136 + 12 × 0.079 + 12 × 0.079 = 5.0914;

Consistency Index, CI = (λmax – 1)/ (n – 1); CI = (5.0914 – 5)/(5 – 1) = 0.02286

Consistence Ratio = (Consistency Index / Randon Index) X 100; Random Index = 1.12

Consistency Ratio: 2.04, which is less than 10%. Hence it is acceptable.

Table 4.2 Scale of relative importance for pair-wise comparison (Saaty 1980)

Intensity	Definition	Explanation
1	Equal importance	Two activities contribute equally to the object
3	Moderate importance	Slightly favors one over another
5	Essential or strong importance	Strongly favors one over another
7	Demonstrated importance	Dominance of the demonstrated importance in practice
9	Extreme importance	Evidence favoring one over another of highest possible order of affirmation
2, 4, 6, 8	Intermediate values	When compromise is needed

Table 4.3 Likelihood of risk in project

Factors	Likelihood	Subfactors	Likelihood		Instrumentation and control room		Process Equipment and piping		Tank farm and pipelines		Power and utilities	
			LP	GP	LP	GP	LP	GP	LP	GP	LP	GP
Technical Risk	0.484*	Scope change	0.354**	0.171	0.162	0.028	0.438	0.075	0.251	0.043	0.149	0.025
		Technology selection	0.129	0.062	0.277	0.017	0.161	0.010	0.096	0.006	0.466	0.029
		Implementation methodology	0.150	0.073	0.469***	0.034	0.280***	0.020	0.136***	0.010	0.115***	0.008
		Equipment	0.057	0.027	0.373	0.010	0.209	0.006	0.209	0.006	0.209	0.006
		Materials	0.067	0.032	0.097	0.003	0.432	0.014	0.287	0.009	0.184	0.006
		Engineering and design change	0.243	0.118	0.362	0.043	0.326	0.038	0.148	0.017	0.163	0.019
Financial & economical risk	0.222	Inflation	0.112	0.025	0.250	0.006	0.250	0.006	0.250	0.006	0.250	0.006
		Fund	0.504	0.112	0.250	0.028	0.250	0.028	0.250	0.028	0.250	0.028
		Local law	0.068	0.015	0.167	0.003	0.167	0.003	0.333	0.005	0.333	0.005
		Policy	0.078	0.017	0.250	0.004	0.250	0.004	0.250	0.004	0.250	0.004
		Cost estimate	0.239	0.053	0.351	0.019	0.351	0.019	0.189	0.010	0.109	0.006
Organizational risk	0.136	Client	0.111	0.015	0.379	0.006	0.358	0.005	0.179	0.003	0.085	0.001
		Contractor	0.280	0.038	0.385	0.015	0.415	0.016	0.126	0.005	0.074	0.003
		Supplier	0.452	0.062	0.227	0.014	0.227	0.014	0.423	0.026	0.123	0.008
		Consultant	0.156	0.021	0.305	0.006	0.490	0.010	0.126	0.003	0.079	0.002
Natural Hazards	0.079				0.333	0.026	0.333	0.026	0.167	0.013	0.167	0.013
Clearance risk	0.079	Environmental	0.079	0.006	0.100	0.001	0.300	0.002	0.300	0.002	0.300	0.002
		Land acquisition	0.519	0.041	0.250	0.010	0.250	0.010	0.250	0.010	0.250	0.010
		Explosive	0.201	0.016	0.109	0.002	0.189	0.003	0.351	0.006	0.351	0.006
		Other	0.201	0.016	0.110	0.002	0.230	0.004	0.302	0.005	0.358	0.006
		Overall Risk Ranking				0.276		0.314		0.217		0.193

LP: Local percentage; GP: Global Percentage

*from table 1; **from appendix 1; ***from appendix 2; GP for subfactor is derived by multiplying factor likelihood and subfactor LP; GPs of alternatives are determined by multiplying GP of subfactor and LP of each alternative.

The following paragraphs describe the observations from risk analysis:

i) Technical risk was the major risk category for time and cost overrun of the project. Among the technical risks, scope change, engineering and design change, technology and implementation methodology selections were the major causes of project failure. The "process equipment and piping" and "tank farm and pipelines" work packages were vulnerable from scope change. Technology selection was vital for "instrumentation and control room" and "power and utility" work packages. Engineering and design change was quite likely to occur for the "instrumentation and control room" and "process equipment and piping" work packages. Prior selection of implementation methodology was crucial for the "instrumentation and control room" packages, as improper selection could cause major time and cost overrun. Unavailability of pipe materials and delayed delivery of pumping unit could result in considerable time overrun.

ii) Other major risks in project achievement were financial, economic and political risk and organizational risk. Within this category fund flow problems and improper estimation were the major causes of concern. All the packages were equally vulnerable from fund flow problems. However, the "instrumentation and control room" and "process equipment and piping" packages were prone to improper estimates due to more uncertainties in design and implementation methodology selection. Although the organizational risk was less vulnerable for the project under study, the consultant and contractor's capabilities were a bit of a concern to the management of the project. The "instrumentation and control room" work package was the most susceptible to the consultant and contractor's performance. The capability of the owner's project group was required for achievement of all the work packages.

iii) Although the project under study was not very vulnerable from statutory clearance risk, care should be taken for getting environmental clearance and explosive clearance on time for trouble free implementation.

iv) Natural hazards were the part and parcel of the oil refinery project as it was exposed for many seasons. Although it had less priority compared to other risk factors, almost all the work packages were vulnerable. Accordingly, appropriate contingency plans were developed for each package.

v) The "process equipment and piping" work package was the most risky package with a probability of failure of 0.314. The major factors for possible failure were changes in scope, change in engineering and design, fund availability, vendor capability, natural hazard and clearance for land acquisition. The "instrumentation and control room" work package with probability of failure 0.276 came next. The main contributing factors were scope change, implementation methodology selection, engineering and design change, and improper estimate thereon. The "tank farm and pipelines" and "power Utility" work packages were relatively less vulnerable and they had relative failure chances of 0.217 and 0.176 respectively.

Step 5: guesstimating the impact (time and cost) of each prioritized risk factor

The risk management group then guestimate the impact of the risk factors in terms of time and cost overrun. Table 4.4 shows the probability (from Table 4.3) and impact of all risk factors.

Step 6: determining expected time and cost overrun of the project

The above results were then used to derive the expected time and cost overrun (Canavos 1984) using

$$\mu'_r = E(X^r) = \Sigma x^r p(x) \qquad \text{if X is discrete, or } \ldots .1$$

Table 4.4 Probability and severity of risk factors

Risk Factors	Probability *	Impact	
		Time over-run (in months) **	Cost Over-run (in Million US $) **
Scope change	0.171	8	90
Technology selection	0.062	6	20
Implementation methodology	0.073	3	0
Equipment	0.027		
Materials	0.032	3	0
Engineering and design change	0.118	5	30
Inflation	0.025		
Fund	0.112	2	0
Local law	0.015		
Policy	0.017		
Cost estimate	0.053	2	0
Client	0.015		
Contractor	0.038	6	30
Supplier	0.062	8	30
Consultant	0.021		
Natural hazards	0.079	12	90
Environmental	0.006		
Land acquisition	0.041	4	0
Explosive	0.016		
Other	0.016		

*the figures are from Table 4.3 column 5

**the figures are guessestimation by the risk management from previous project experience

Table 4.5 Expected time and cost overrun

Risk Factors	Probability	Impact			
		Time overrun	Cost overrun	Expected time overrun	Expected cost overrun
Scope change	0.171	8	90	1.37	15.41
Technology selection	0.062	6	20	0.37	1.25
Implementation methodology	0.073	3	0	0.22	0.00
Equipment	0.027			0.00	0.00
Materials	0.032	3	0	0.10	0.00
Engineering and design change	0.118	5	30	0.59	3.53
Inflation	0.025			0.00	0.00
Fund	0.112	2	0	0.22	0.00
Local law	0.015			0.00	0.00
Policy	0.017			0.00	0.00
Cost estimate	0.053	2	0	0.11	0.00
Client	0.015			0.00	0.00
Contractor	0.038	6	30	0.23	1.15
Supplier	0.062	8	30	0.49	1.85
Consultant	0.021			0.00	0.00
Natural hazards	0.079	12	90	0.95	7.10
Environmental	0.006			0.00	0.00
Land acquisition	0.041	4	0	0.16	0.00
Explosive	0.016			0.00	0.00
Other	0.016			0.00	0.00
Overall expected time and cost overrun				4.81	30.28

Table 4.5 shows the overall time and cost overrun of the project.

The analysis revealed that the project was expected to have experienced 4.81 months delay and US$ 30.28 million cost overrun.

Step 7: deriving possible risk responses

Risk analysis results derived a few risk responses in line with the principles to avoid, to reduce, to transfer and to absorb. The risk management group through a brainstorming session derived the following responses for the project under study:

1 Carrying out a detailed survey with the objective of minimum scope and design change
2 Selecting technology and implementation methodology on the basis of owner's / consultant's expertise, availability of contractors and vendors and lifecycle costing

3   Executing design and detailed engineering on the basis of selected technology and implementation methodology and detailed survey
4   Selecting superior contractors, consultants and vendors on the basis past performance
5   Scheduling the project by accommodating seasonal calamities
6   Planning contingencies and acquiring insurance
7   Ensuring the availability of all statutory clearance before design and detailed engineering.

Table 4.6 shows the estimated cost of the above risk responses for each work package. Sources for cost data are the detailed feasibility report and cost estimate for the project concerned, based on other recently completed projects and quotations from vendors and contractors.

Step 8: developing the decision tree model

The next step was to form a decision tree for each work package with the consideration of probability and severity of failure and various possible responses.
    The group decided the following decision alternatives:

• Do nothing
• Carry out detailed survey (additional)
• Use superior technology
• Engage expert project team
• Implement all responses as indicated in the Table 4.6.

The decision trees were formed for the work packages (tank farm and pipelines laying, process equipment and piping, instrumentation and control room, and power utilities) of the oil refinery construction project under study. The probability and impact (time and cost) for each decision alternative were derived from the risk analysis study of each package and expert opinion through brainstorming.

Step 9: deriving expected monetary value for each decision alternative to select the best response

The expected monetary values (EMV) were then calculated for each alternative decision for all the packages. Figure 4.5 shows the decision tree for the work package 'tank farm and pipelines'. The tables 4.7, 4.8, 4.9 and 4.10 show the expected monetary values for various decision options of the work packages "tank farm and pipelines", "process equipment and piping", "instrumentation and control room" , and "power and utilities" respectively. The cost figures have been taken from the Table 4.6, the

Table 4.6 The cost data (million US $) for each package against various responses

Responses	Tank farm and pipelines	Process equipment and piping	Instrumentation and control room	Power and utilities	Office buildings and plant infrastructure
Carrying out detailed survey with the objective of minimum scope and design change	12*	6	6	3	3
Selecting technology and implementation methodology on the basis of owner's / consultant's expertise, availability of contractors and vendors and lifecycle costing	3	6	3	1.5	1.5
Executing design and detailed engineering on the basis of selected technology and implementation methodology and detailed survey	1	1	1	1	1
Selecting superior contractors, consultants and vendors on the basis past performance	22	16	10	2	2
Scheduling project by accommodating seasonal calamities	6	–	5	–	–
Planning contingencies and acquiring insurance	11	2	6	1	1
Ensuring the availability of all statutory clearance before design and detailed engineering	1	1	1	1	1
Total	56	32	32	10	10
Grand total	**140**				

*14 km pipeline survey, soil testing for entire tank farm, control valves and control room design and implementation planning cost US$4 million, US$3 million, US$ 5 million respectively.

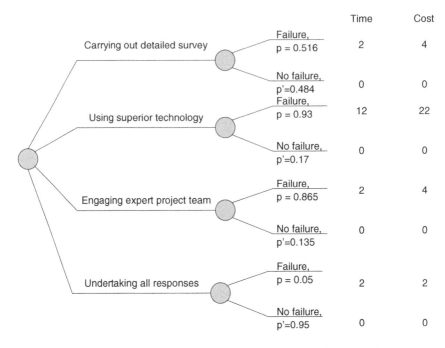

Figure 4.5 Decision tree for the work package 'tank Farm and Pipelines'

probability figures are from the Table 4.3 (it was assumed that if all the responses were undertaken the probability of residual risk would be 5%), and the effects of risk factors on time and cost after risk response have been estimated from cumulative experience of the risk management group through focus group.

Table 4.11 shows the decisions emerge from the decision tree approach of risk management.

Total cost for risk responses was US$56 million which was much lower than US$140 million (if every response as indicated in the Table 4.6 is implemented).

Step 10: implementing the responses

The responses as indicated in Table 4.11 were implemented.

The project was completed in early 2004 with no time and cost overrun. There were a few issues related to engineering and design changes, procurement and commissioning of the hydrocracker units. However, those issues were managed effectively by the project team in collaboration with both the contractors and suppliers. The project team realized the benefit of using

Table 4.7 The EMV for "tank farm and pipelines"

Decision Alternatives	Cost (million US$)	Probability of failure	Effect		Expected Value		EMV* (million US$)
			Time (months)	Cost (million US$)	Time (months)	Cost (million US$)	
Do nothing	0				4.81	30.28	66.36
Carrying out detailed survey	**12**	**0.516****	**2**	**4**	**1.03**	**2.06**	**21.80**
Using superior technology	3	0.93***	12	22	11.16	20.46	107.16
Engaging expert project team	22	0.865****	2	4	1.73	3.46	38.44
Taking all responses as indicated in table	56	0.05	2	2	0.10	0.10	56.85

*Sample EMV calculation for "do nothing") = 0 + 4.81 X 7.5 + 30.28 = 66.36
(Return on investment is 7.5 million US $ per month i.e. 15% of 600 million US $ per annum)
**It is assumed that the "technical risk" will reduce to zero if "carrying out detailed survey" responses were undertaken. Therefore, the probability of project failure would be (1 – 0.484) i.e. 0.516.
***Similarly, "superior technology" response will reduce probability of failure by 0.07. Therefore, the probability of project failure would be (1 – 0.07), e.i. 0.93.
****Similarly, "engaging expert project team" response will reduce probability of failure by 0.136. Therefore, the probability of project failure would be (1 – 0.136) i.e. 0.864.

Table 4.8 The EMV for "process equipment and piping"

Decision Alternatives	Cost (million US$)	Probability of failure	Effect		Expected Value		EMV* (million US$)
			Time (months)	Cost (million US$)	Time (months)	Cost (million US$)	
Do nothing	0				4.81	30.28	66.36
Carrying out detailed survey	6	0.516	15	40	7.74	20.64	84.69
Using superior technology	6	0.93	15	40	13.95	37.20	147.83
Engaging expert project team	16	0.865	8	20	6.92	17.30	85.20
Taking all responses as indicated in table	**32**	**0.05**	**2**	**8**	**0.10**	**0.40**	**33.15**

Table 4.9 The EMV for "Instrumentation and control room"

Decision Alternatives	Cost (million US$)	Probability of failure	Effect		Expected Value		EMV* (million US$)
			Time (months)	Cost (million US$)	Time (months)	Cost (million US$)	
Do nothing	0				4.81	30.28	66.36
Carrying out detailed survey	6	0.516	6	12	3.10	6.19	35.41
Using superior technology	3	0.93	6	12	5.58	11.16	56.01
Engaging expert project team	**10**	**0.865**	**2**	**4**	**1.73**	**3.46**	**26.44**
Taking all responses as indicated in table	32	0.05	2	4	0.10	0.20	32.95

Table 4.10 The EMV for "power and utilities"

Decision Alternatives	Cost (million US$)	Probability of failure	Effect		Expected Value		EMV (million US$)
			Time (months)	Cost (million US$)	Time (months)	Cost (million US$)	
Do nothing	0				4.81	30.28	66.36
Carrying out detailed survey	3	0.516	2	2	1.03	1.03	11.77
Using superior technology	1.5	0.93	2	2	1.86	1.86	17.31
Engaging expert project team	**2**	**0.865**	**0.5**	**1**	**0.43**	**0.87**	**6.11**
Taking all responses as indicated in table	10	0.05	0.5	1	0.03	0.05	10.24

the proposed risk management framework. However, they pointed out that identifying the relevant risk factors was very critical. The risk management group had faced the highest complexity while comparing each alternative work package with respect to the risk subfactors under a particular primary

Table 4.11 The decisions emerge from the decision tree approach of risk management for each work package

Work package	Risk response
Tank farm and pipelines	Carrying out detailed survey
Process equipment and piping	Taking all responses as indicated in Table 4.5
Instrumentation and control room	Engaging expert project team
Power and utilities	Engaging expert project team

risk criterion. They suggested that a more detailed training workshop about the AHP theory and application prior to the exercise would greatly reduce the duration of the exercise.

Validation of the framework

The proposed risk management framework was validated for universal application through a questionnaire survey among a few key executives across the Indian oil industry. This was carried out in order to reveal the following:

- the overall significance and importance of risk management for managing projects
- the acceptability of the method
- the usability of the proposed framework
- the comprehensibility of the framework
- the implementability of the outcomes (the responses)
- the acceptability of the research findings
- the adequacy of the stakeholders' involvement in the process of analysing risk
- the applicability of the methodology and the risk management framework in other projects
- future improvement of the model and method of application.

Twenty executives were contacted from ten companies for the validation survey, out of which 14 executives agreed to be interviewed. They were briefed about the proposed risk management framework and its application before asking the questions as stated above. In overall response the participants had been fairly positive about the framework. They had also been in favour of adopting the framework for their project management practices. They have indicated that the basic principle and application of the framework is quite user friendly, the steps are easy to implement and helpful as they consider the decisions of individual stakeholders before reaching a consensus. However, they have agreed that the success of its use would largely depend on the number of stakeholders involved and collective utilisation of their experiences.

Discussion

The proposed risk management framework using the cause and effect diagram, the AHP risk map and DTA helps project executives to make decisions dynamically during the project planning phase. This provides an effective monitoring and control mechanism of projects across various levels of management of the organization. The proposed decision support system is a computerized model that uses Microsoft Excel Spreadsheet to analyze the decision situation. Additionally, the sensitivity utility of the AHP provides an opportunity to the risk management group to observe the nature of the model outcome in different alternative decision situations. DTA helps in selecting one among various decision alternatives in a quantitative framework. The following additional benefits are derived from the designed DSS using the AHP:

1   The AHP provides a flexible and easily understood way to analyze each risk factor with respect to project achievement.
2   The AHP calls for active involvement of project stakeholders in risk analysis and provides a rational basis for probability of project failure.
3   Risk management using the AHP integrates all project stakeholders. Hence, this not only involves them in making group decisions, but also improves team spirit and motivation.
4   The AHP is a suitable approach for reaching a consensus in controversial decisions. Despite the existence of diverging interests, AHP evoked collectively judgements based on a reasonable compromise or consensus.
5   In general, the AHP is used either by application of a questionnaire or by group decision processes. The combination of both uses in this study proved to provide an additional value. The decision makers can systematically deliberate the specific comparisons before group decision processes take place. This creates an efficient dialogue in order to reach compromises.
6   The collective judgements after group decision processes often deviated from the computed group means of the initial individual judgements. This deviation is an indicator for a high-quality collective decision (Sniezek and Henry 1989)
7   Although the DTA in deciding a specific course of action is not a new method, it logically structures the risk management philosophy by identifying alternative responses in mitigating risk and incorporates management perceptions.

Therefore, risk management using a combined cause and effect diagram, the AHP, risk map and DTA provides an effective means for managing a complex project against time, cost and quality non-achievement.

The proposed risk management using the AHP suffers the following shortcomings:

- Though this study makes an effort to quantify risk by modeling the probability and severity of risk in line with the perception of the experienced project executives, subjectivity could not be reduced to zero.

- A limitation of AHP is its inability to indicate those judgements that need to be revised. Expert choice gives a recommended revision regardless of whether the recommended value fits within the 9-point scale of AHP. An additional approach is recommended. The study of Genset and Zhang (1996) can be a first instigator for a surveyable approach.

- The choice of the scale and whether or not to use normalizations, are important issues which should be seen as practical procedural choices whose consequences need to be understood. Although discrete ratio scales such as the 1–9 scale of the the AHP can be very helpful in preference elicitation, they are nevertheless problematic as they severely restrict the range and distribution of possible priority vectors. The balanced scale proposed by Salo and Hamalainen (1997) provides an essential improvement in this matter. Even so, the assumption that verbal expressions can be mapped onto numbers in the same way, no matter who is responding and in what context, must be regarded with due caution. The implication of scale selections must be considered explicitly, especially if the results are to be used in a normative sense. Risks associated with scale selection can be mitigated through software tools, which allow the practitioner to compare results based on different scales.

- The real problem with the AHP is the way it aggregates over levels of the hierarchy. This has been well documented in the work of Barzilai (1998), Finan and Hurley (2002), and Belton and Gear (1983). Belton and Gear (1983) introduced the rank reverse phenomenon and most researchers agree that it poses a serious challenge to the AHP.

Nevertheless, on the whole, the AHP has been a useful tool in dealing with multiple factors on different qualitative domains.

The findings and recommendations would be varying across projects, risk perceptions of the managers, organization's objectives and policies and business environment.

Managing risk across various phases of a project ensures effective management of entire oil refinery construction projects. Although various tools and techniques are available and being practised for risk identification, analysis and developing responses, an integrated framework helps managing risk effectively, as it provides an analytical framework in a group decision-making framework. Risk identification using experts' opinions, analyzing risk using the analytic hierarchy process and statistical analysis, and selecting the best responses using the decision tree approach establishes an integrated cost effective project risk management framework. Large scale construction projects like oil refinery construction, where stakeholders (owner, contractors, suppliers, etc.) experience lots of uncertainties during the planning and implementation phases, get considerable benefits using the proposed risk management framework.

Conclusion

Oil refinery construction projects are large, use complex technology, involve many stakeholders and have both considerable environmental and social impacts. Therefore, time, cost and quality achievement of these projects are not always certain. In this circumstance, in order to ensure successful implementation of projects along with conventional project scope, time and cost management, organizations need to formally adopt project risk management practice. The proposed combined qualitative and quantitative approach using cause and effect diagram, the AHP, risk map and decision tree analysis helps integrate every process of risk management namely identification, analysis and response development. The risk management framework involves project personnel for analysis and decision making in a group decision-making framework. It additionally desires management commitment for implementing the responses on risk mitigating measures. This study shows evidence of its successful use for effective project management in the Indian oil industry.

References

AS/NZS ISO 31000 (2009), Risk management – Principles and guidelines, Sydney, NSW.

Barzilai, J. (1998), On the decomposition of value functions, *Operations Research Letters*, 22, 159–170.

Belton, V. and Gear, A.E. (1983), On shortcoming of Saaty's method of analytical hierarchies, *Omega*, 11, 227–230.

Canavos, G.C. (1984), *Applied Probability and Statistical Methods*. Portland: Little, Brown and Company.

Chapman, C. (2006), Key points of contention in framing assumptions for risk and uncertainty management, *International Journal of Project Management*, Vol. 24, 303–313.

Chen, S.H., Wang, P.W., Chen, C.M. and Lee, H.T. (2010), An analytic hierarchy process approach with linguistic variables for selection of a R&D strategic alliance partner, *Computer and Industrial Engineering*, 58 (2), 278–287.

Dey, P.K., Tabucanon, M.T. and Ogunlana, S.O. (1994), Planning for project control through risk analysis; a case of petroleum pipeline laying project, *International Journal of Project Management*, Vol. 12 no. 1, 23–33.

Dey, P.K. (2006), Integrated approach to project selection using multiple attribute decision-making technique, *International Journal of Production Economics*, Vol. 103, 90–103.

Dey, P.K. and Ramcharan, E. (2008), Analytic hierarchy process helps select site for limestone quarry expansion in Barbados, *Journal of Environmental Management*, 88 (4), 1384–1395.

Dey, P.K. (2010), Project risk management using combined analytic hierarchy process and risk map, *Applied Soft Computing*, proof read, available on-line.

Dilworth, J.B. (2000), *Operations Management: Providing Value in Goods and Services*. New York: The Dryden press.

Dyer, R.F. and Forman, E.H. (1992), Group decision support with the analytic hierarchy process, *Decision Support Syst.*, no. 8, 99–124.

Finan, J.S. and Hurley, W.J. (2002), The analytic hierarchy process: can wash criteria be ignored?, *Computers and Operations Research*, 29(8), 1025–1030.

Forman, E.H. and Saaty, T.L. (1983), *Expert Choice Expert Choice*, Pittsburgh, PA USA.

Golden, B.L., Wasli, E.A. and Harker, P.T. (1989), *The Analytic Hierarchy Process: Applications and Studies*. New York: Springer Verlag.

Genset, C and Zhang, S.S. (1996), A graphical analysis of ratio – scaled paired conversion data, *Management Science*, 42 (3), 335–349.

Islie, G., Lockett, G., Cox, B. and Stratford, M. (1991), A decision support system using judgmental modeling: A case of R&D in the pharmaceutical industry, *IEEE Trans. Engineering Manage.*, vol. 38, August, 202–209.

Kangari, R. and Riggs, L.S. (1989), Construction risk assessment by linguistics, *IEEE Trans. Eng. Manag.*, Vol. 36 No. 2, 126–131.

Kwak, Y.H., Anbari, F.T. (2009), Analyzing project management research: Perspectives from top management journal, *International Journal of Project Management*, Vol. 27, 435–446.

Liang, T. (2009), Fuzzy multi-objective project management decisions using two phase fuzzy goal programming approach, *Computer and Industrial Engineering*, 57 (4), 1407–1416.

Mustafa, M.A. and Al-Bahar, J.F. (1991), Project risk assessment using the Analytic Hierarchy Process, *IEEE Trans. Eng. Manag.*, Vol. 38 No. 1, 46–52.

PMI (Project Management Institute). (2008). *A guide to project management body of knowledge, Guide*. 4th ed. Upper Darby, PA: PMI.

Saaty, T.L. (1980), *The Analytic Hierarchy Process*. New York: McGraw-Hill.

Saaty, T.L. (1982), *Decision Making for Leaders*. New York: Lifetime Learning.

Saaty, T.L. (1983), Priority setting in complex problems, *IEEE Trans. Engineering Manage.*, vol. EM-30, August, 140–155.

Salo, A.A. and Hamalainen, R.P. (1997), On the measurement of performances in the analytic hierarchy process, *Journal of Multi-Criteria Decision Analysis*, 6, 309–319.

Schatteman, D., Herroelen, W., Vonder, S. and Boone, A. (2008), Methodology for integrated risk management and proactive scheduling of construction projects, *Journal of Construction Engineering and Management*, Vol. 134 (11), 885–893.

Shen, L.Y., Platten, A. and Deng, X.P. (2006), Role of public private partnership to manage risks in public sector projects in Hong Kong, *International Journal of Project Management*, 24 (1), 53–65.

Sniezek, J.A. and Henry, R.A. (1989), Accuracy and confidence in group judgement, *Organizational Behaviour and Human Decision Processes*, 43, 1–28.

Tah, J.H.M and Carr, V. (2000), A proposal for construction project risk assessment using fuzzy logic, *Construction Management and Economics*, Vol. 18, 491–500.

Thuyet, N.V., Ogunlana, S.O. and Dey, P.K. (2007), Risk management in oil and gas construction projects in Vietnam, *International Journal of energy Sector Management*, Vol. 1 No. 2, 175–194.

Tuysuz, F. and Kahraman, C. (2006), Project risk evaluation using fuzzy AHP: An application to information technology projects, *International Journal of Intelligent Systems*, 21 (6), 559–584.

Vargas, L.G. (1990), An overview of the analytic hierarchy process and its applications, *Europe J. Operation Research.*, Vol. 48 No. 1, 2–8.

Wang, S.Q., Dulaimi, M.F. and Aguria, M.Y. (2004), Risk management framework for construction projects in developing countries, *Construction Management and Economics*, Vol. 22 (3), 237–252.

Zayed, T., Amer, M. and Pan, J. (2008), Assessing risk and uncertainty inherent in Chinese highway projects using AHP, *International Journal of Project Management*, Vol. 26, 406–419.

Appendix 1 (Sample pairwise comparison in subfactor level)

Subfactor Level: Technical Risk

	Scope change	Technology selection	Implementation methodology	Equipment risk	Material risk	Engineering and design change
cope change	1.00	3.00	3.00	5.00	5.00	2.00
Technology selection	0.33	1.00	0.50	3.00	3.00	0.50
Implementation methodology	0.33	2.00	1.00	2.00	4.00	0.33
Equipment risk	0.20	0.33	0.50	1.00	0.50	0.33
Material risk	0.20	0.33	0.25	2.00	1.00	0.33
Engineering and design change	0.50	3.00	3.00	3.00	3.00	1.00
Column sum	2.57	9.67	8.25	16.00	16.50	4.50

Normalized Matrix

	Scope change	Technology selection	Implementation methodology	Equipment risk	Material risk	Engineering and design change	Risk priority*	λ max
Scope change	0.39	0.31	0.36	0.31	0.30	0.44	**0.354**	6.50
Technology selection	0.13	0.10	0.06	0.19	0.18	0.11	**0.129**	CI
Implementation methodology	0.13	0.21	0.12	0.13	0.24	0.07	**0.150**	0.10
Equipment risk	0.08	0.03	0.06	0.06	0.03	0.07	**0.057**	
Material risk	0.08	0.03	0.03	0.13	0.06	0.07	**0.067**	CR=(CI/RI)
Engineering and design change	0.19	0.31	0.36	0.19	0.18	0.22	**0.243**	8.08

* The risk priority figures go to Table 4.3 in the 'Likelihood LP' column. Similarly, other subfactors were pair wise compared and their relative risk priorities were determined. The results have been shown in Table 4.3.

Appendix 2 (Sample pairwise comparison in alternative level)

Alternative level: Implementation Methodology

	Instrumentation & control room	Process equipment	Tank farm	Power & utilities
Instrumentation & control room	1.000	2.000	3.000	4.000
Process equipment	0.500	1.000	2.000	3.000
Tank farm	0.333	0.500	1.000	1.000
Power & utilities	0.250	0.333	1.000	1.000
Column sum	2.083	3.833	7.000	9.000

Normalized Matrix

	Instrumentation & control room	Process equipment	Tank farm	Power & utilities	Risk Priority*	λmax	CI	CR=(CI/RI)
Instrumentation & control room	0.480	0.522	0.429	0.444	0.469	4.039	0.013	1.462
Process equipment	0.240	0.261	0.286	0.333	0.280			
Tank farm	0.160	0.130	0.143	0.111	0.136			
Power & utilities	0.120	0.087	0.143	0.111	0.115			

* The risk priority figures go to Table 4.3 against 'Implementation Methodology' corresponds to four alternative work packages. Similarly, the alternatives were pair wise compared with respect to other subfactors and their relative risk priorities were determined. The results have been shown in Table 4.3.

5 Risk management practices in the construction industry using Bayesian belief networks approach

Van Truong Luu, Soo-Yong Kim,
Nguyen Van Tuan and Stephen O. Ogunlana

Highlights

- Sixteen significant factors causing delays in construction projects in Vietnam are identified.
- Eighteen cause-and-effect relationships, out of which, 16 significant factors were established based on the expert survey. The relationship between "owners' site clearance difficulties (x_{13})" and "slow site hand over (x_5)" ranks first.
- The top main causes of delay in building and industrial construction projects in Vietnam include owners' financial difficulties, inadequate experience and financial difficulties of contractors, shortage of materials, slow site handover, inappropriate construction methods, defective works and reworks, and lack of management capacity by owners/project managers.
- Based on the set of 16 significant delay factors and 18 cause-and-effect relationships identified, the BBN-based model was developed and applied to two cases of building projects in Ho Chi Minh city in order to validate the model. The model performed well in predicting the probability of construction delay in both case studies.
- The result of sensitivity analysis using the BBN-based model indicated that construction delay is extremely sensitive to the factors "shortage of materials", "defective construction work" and "slow site handover".

Introduction: significance, problem statement and research questions/objectives

Along with other Artificial Intelligence (AI) technologies such as expert systems, Artificial Neural Networks (ANNs), Genetic Algorithms

(GAs), and Fuzzy Logic, Bayesian Belief Networks (BBNs) approach is an approach mainly based on the conditional probability to predict or diagnose a phenomenon, a problem which has been and is going to happen again. For example, in order to forecast floods or storms for a certain area, we can use the data of the occurred storms, previous floods and relevant current evidence so as to build BBNs models and from which we can predict the probability of floods or storms.

Schedule delays are common in various construction projects and cause considerable losses to project parties. It is widely accepted that a construction project schedule plays a key role in project management due to its influence on project success. Therefore, it is important to quantify probabilities of schedule delays when managing a construction project. To serve the need for proactive project management, the need has emerged for the development of straightforward methods to evaluate the probability of construction time overruns.

In this chapter, the process of building a BBNs model and practical case studies will be given to demonstrate its usefulness for risk management practices in the construction industry. Case studies of the chapter apply BBNs method to predict schedule delay probability on construction projects in Vietnam. Predicting the possible construction schedule delays is an effective step towards improving the chances for success on construction projects.

Literature review

Application of BBNs

BNNs have been widely and effectively applied in many fields that require highly logical descriptions and probabilistic quantification of complicated relationships among diverse variables[1], including clinical diagnoses[2], biomedicine and health care[3], geographic information processing[4], reliability analysis of complex systems[5], ecosystem service modeling[6], marine planning[7], validation framework for expert elicitation[8], fault analysis in a centrifugal compressor[9], relationships between environmental stressors and stream condition in urban and agricultural land uses[10], public policy and government settings[11], real-time crash prediction on the basic freeway segments of urban expressways[12], tourism loyalty[13], customer churn analysis of the telecom industry of Turkey[14], potential compatibilities and conflicts between development and landscape conservation[15], reliability of military vehicles[16], operational risk in financial institutions[17], impacts of commercializing non-timber forest products on livelihoods[18], analysis of information system network risk[19], failure prediction for water mains[20], risk analysis for large projects[21], project cost estimation[22], information synthesis in complex systems[23], selecting new technology investment projects[24], accident rates of electrical and mechanical (E&M) work[25], engineering reliability assessment[26]. Furthermore, they have been proven accurate through use in reliability analyses ([[27],[28]). In construction, the use of BBNs has focused mainly on the improvement of construction operations[29], diagnosing upsets in an anaerobic wastewater treatment system[30], the false-work erection productivity[31], inferences in highway construction costs[32], pessimistic and optimistic values of activity durations based on project characteristics[33], risk in construction contracts[34], risk of erosion in peat soils[35], workplace accidents caused by falls from a height[36], water resources management in Spain[37], leak prediction for water distribution networks[38], predicting safety risk of working at heights using Bayesian Networks[39], cash flow forecasting[40], lifecycle cost risk analysis for infrastructure projects[41], construction performance monitoring[42], schedule risk analysis in construction project[43], prediction of time performance in building construction project[44], predicting construction worker accident[45], schedule risk paths in submarine pipeline projects[46].

Introduction about BBNs

BBNs are also known by many names such as Belief Networks, Causal Probabilistic Networks, Causal Nets, Graphical Probability Networks, and Probabilistic Cause Effect Models, and were first developed at Stanford University in the 1970s[29]. BBNs describe cause-effect relationships among variables through graphical models. They are directed acyclic graphs composed of nodes and arrows. The nodes represent variables (ideally, mutually exclusive and collectively exhaustive); the arrows represent probabilistic dependence relationships between the variables[47]. BBNs have the following principal advantages as follows[1]:

- BBNs can achieve precise results when assessing problems with numerous subjective factors and uncertain information.
- BBNs models can integrate new evidence and historical information as well as experiential knowledge.
- BBNs can reflect the important characteristics of continuity and accumulation of risk assessment. According to the given situation, BN analyses can be applied repeatedly during an assessment, making decision making more faultless and scientific.
- BBNs can be used in dynamic risk assessments.

Some BBNs are illustrated as Figure 5.1 below[48]:

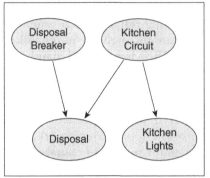 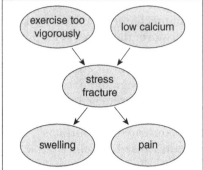

Figure 5.1 Some Bayesian belief networks[48]

Conditional probabilities and Bayes' theorem

A conditional probability is a probability that a variable is dependent on the state of another variable. BBNs are based on conditional probability theory which was developed in the late 1700s by Thomas Bayes. He discovered a basic law of probability which was then called Bayes' theorem[48]. Bayes' theorem is stated mathematically as the following equation[49]:

$$P(A \ / \ B) = P(B \ / \ A) \times \frac{P(A)}{P(B)}$$

Where A and B are events.

P(A) and P(B) are the probabilities of A and B independent of each other.

P(A/B), a conditional probability, is the probability of A given that B is true.

P(B/A), is the probability of B given that A is true.

For example:

The entire output of a factory is produced on three construction machines. The three construction machines account for 20%, 30% and 50% of the output, respectively. The fraction of defective items produced is this: for the first machine, 5%; for the second machine, 3%; for the third machine, 1%. If an item is chosen at random from the total output and is found to be defective, what is the probability that it was produced by the third construction machine?

A solution is as follows. Let A_i denote the event that a randomly chosen item was made by the ith machine (for $i = 1,2,3$). Let B denote the event that a randomly chosen item is defective. Then, we are given the following information:

$$P(A_1) = 0.2, P(A_2) = 0.3, P(A_3) = 0.5.$$

If the item was made by machine A_1, then the probability that it is defective is 0.05; that is, $P(B/A_1) = 0.05$. Overall, we have

$$P(B/A_1) = 0.05, P(B/A_2) = 0.03, P(B/A_3) = 0.01.$$

To answer the original question, we first find P(B). That may be done in the following way:

$$P(B) = \Sigma_i P(B/A_i).P(A_i) = (0.05)(0.2) + (0.03)(0.3) + (0.01)(0.5) = 0.024.$$

Therefore, 2.4% of the total output of the factory is defective.

We are given that B has occurred, and we want to calculate the conditional probability of A_3. By Bayes' theorem,

$$P(A_3 \,/\, B) = P(B \,/\, A_3) \times \frac{P(A_3)}{P(B)} = \frac{0.01 \times 0.50}{0.024} = \frac{5}{24}$$

Given that the item is defective, the probability that it was made by the third machine is only 5/24. Although machine 3 produces half of the total output, it produces a much smaller fraction of the defective items. Hence the knowledge that the item selected was defective enables us to replace the prior probability $P(A_3) = 1/2$ by the smaller posterior probability $P(A_3/B) = 5/24$.

The extended form of Bayer's theorem is described as follows:

$$P(A_k \,/\, F) = \frac{P(A_k).P(F \,/\, A_k)}{P(F)} = \frac{P(A_k).P(F \,/\, A_k)}{\sum_{i=1}^{n} P(A_i).P(F \,/\, A_i)}$$

Where $P(F) = \sum_{i=1}^{n} P(A_i).P(F \,/\, A_i)$

Structure of a Bayesian belief network

BBNs are graphical models where each variable is represented by a node, and causal relationships are denoted by an arrow, called an edge as a labeled oval. Direction of the arrow is from cause-nodes "*parent nodes*" to effect-nodes "*child nodes*". Child nodes are conditionally dependent upon their parent nodes. Each node has the states depending on its characteristics. Specifically, according to Figure 5.2, the node of *Precipitation* is cause-node affecting effect-node "*Road Conditions*" and they have the corresponding states.

A structure of BNN in the construction is presented as Figure 5.3 below.

Figure 5.4 shows the overall structure of BNN with many more nodes and arrows of link.

Conditional probability tables (CPT)

Every node also has a conditional probability table (CPT), associated with it. Conditional probabilities represent likelihoods based on prior information

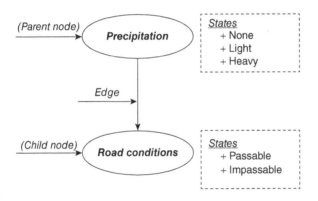

Figure 5.2 A simple structure of BNN in the nature[48]

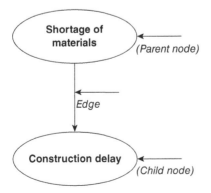

Figure 5.3 A simple structure of BNN in the construction delay

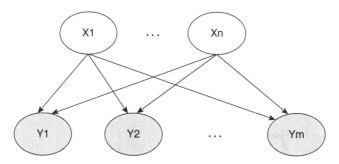

Figure 5.4 The overall structure of BNN

or past experience. As example of BBN in Figure 5.2, CPT of variables is shown as follows (please see Table 5.1)[48]:

The first cell of the CPT shown in Table 5.1 for the *Road Conditions* node can be read as: "If parent *Precipitation* is in state *None,* then the probability that *Road Conditions* will be in the state *Impassable* is 5%." Each cell of the CPT can be read in this way[48]. In the BBNs, Nodes with no parents are called as root nodes. The CPT of these nodes is called as prior probability. According to Figure 5.2, the CPT of *Precipitation* node is shown as Table 5.2.

Table 5.1 Conditional probability table for a child node

Parent node	Child node	
Precipitation	Road Conditions	
	Impassable	Passable
None	0.05	0.95
Light	0.10	0.90
Heavy	0.70	0.30

Table 5.2 Conditional probability table for a parent node

Precipitation		
None	Light	Heavy
0.800	0.150	0.005

Calculation software for BBNs

There is much software supporting in the calculation of BNNs such as BNet. Builder, Hugin Explorer, MSBNx, and etc. They can be downloaded from the following addresses:

- www.research.microsoft.com/adapt/MSBNx/;
- www.kdnuggets.com/software/bayesian.html/;
- www.hugin.dk/;
- www.cs.cmu.edu/~javebayes/;

This chapter uses software of Microsoft Bayesian Belief Networks Tools (MSBNX) downloaded at: www.research.microsoft.com/adapt/MSBNx/.

Illustration of BBNs

Consider Figure 5.5 as below:

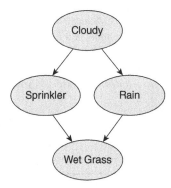

Figure 5.5 The structure of BBN in "Wet Grass"

Suppose that after the survey and analysis we obtain the conditional probability tables (CPT) as follows (Table 5.3, Table 5.4, Table 5.5, Table 5.6):

Table 5.3 CPT of "**Cloudy**" node

Cloudy	
True	*False*
0.50	0.50

These conditional probability tables (Figure 5.6, Figure 5.7, Figure 5.8, Figure 5.9) are entered into the software of MSBNX, respectively, as follows:

Table 5.4 CPT of "**Spinkler**" node

Parent nodes	Child node	
Cloudy	Sprinkler	
	True	False
True	0.10	0.90
False	0.50	0.50

Table 5.5 CPT of "**Rain**" node

Parent nodes	Child node	
Cloudy	Rain	
	True	False
True	0.80	0.20
False	0.20	0.80

Table 5.6 CPT of "**Wet Grass**" node

Parent nodes		Child node	
Sprinkler	Rain	Wet Grass	
		True	False
True	True	0.99	0.01
	False	0.90	0.10
False	True	0.90	0.10
	False	0.00	1.00

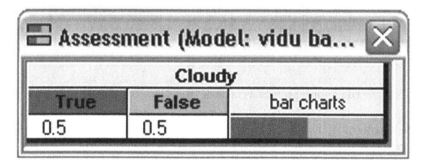

Figure 5.6 The conditional probability table of "Cloudy" in MSBNX

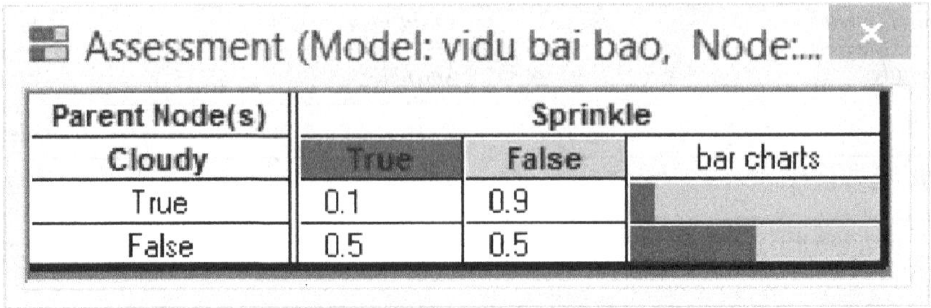
Assessment (Model: vidu bai bao, Node:...

Parent Node(s)	Sprinkle		
Cloudy	True	False	bar charts
True	0.1	0.9	
False	0.5	0.5	

Figure 5.7 The conditional probability table of "Spinkler" in MSBNX

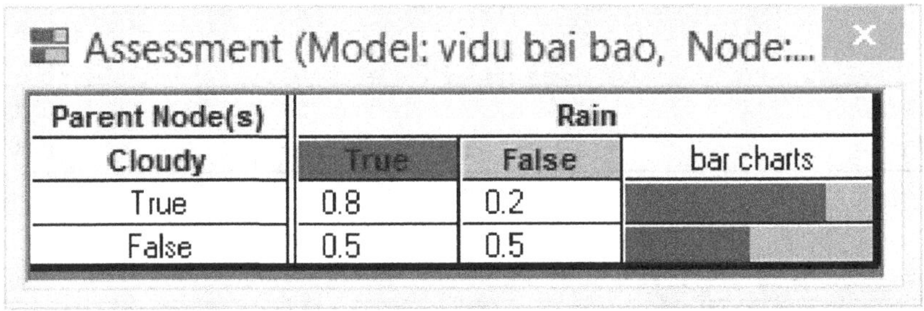
Assessment (Model: vidu bai bao, Node:...

Parent Node(s)	Rain		
Cloudy	True	False	bar charts
True	0.8	0.2	
False	0.5	0.5	

Figure 5.8 The conditional probability table of "Rain" in MSBNX

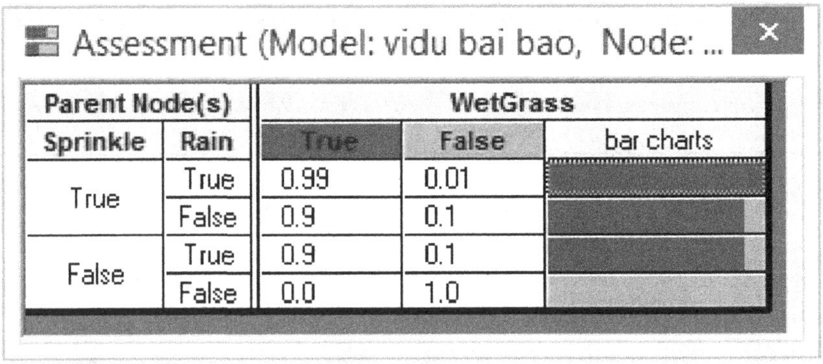
Assessment (Model: vidu bai bao, Node: ...

Parent Node(s)		WetGrass		
Sprinkle	Rain	True	False	bar charts
True	True	0.99	0.01	
	False	0.9	0.1	
False	True	0.9	0.1	
	False	0.0	1.0	

Figure 5.9 The conditional probability table of "Wet Grass" in MSBNX

The results of analysis are presented as Figure 5.10. Based on Figure 5.10, we can see that probability of "Wet Grass" at the state of True is 0.6471, whereas at the state of False is 0.3529.

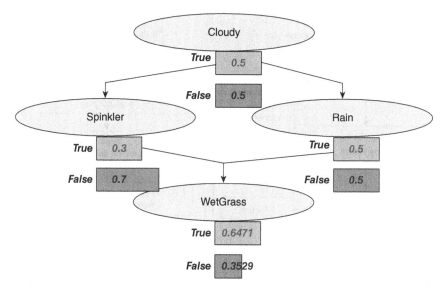

Figure 5.10 Calculation results of BBN

The steps of building BBNs model

- Identify variables and their states in order to enter them into model;
- Identify cause-effect relationships among variables based on logical inference, prior information or past experience, etc.;
- Based on those cause-effect relationships in order to build linked model or influence diagram for application in BBNs;
- Set up the conditional probability tables for each variable, CPTs could be identified from experience of experts, or from results of other models, etc.;
- Enter data into software of MSBNX.

Methodology

In order to develop a well-designed BBN research framework, the study adopts, adjusts and breakdowns steps of a belief network model from McCabe et al.[29] because of the conditions of case studies in Vietnam.

The procedure of BBN-based model building for predicting schedule delay probability on construction projects was drawn in step by step as shown in Figure 5.11. This procedure consists of two phases: a qualitative phase and a quantitative phase.

The research methodology was drawn in step by step as shown in Figure 5.11.

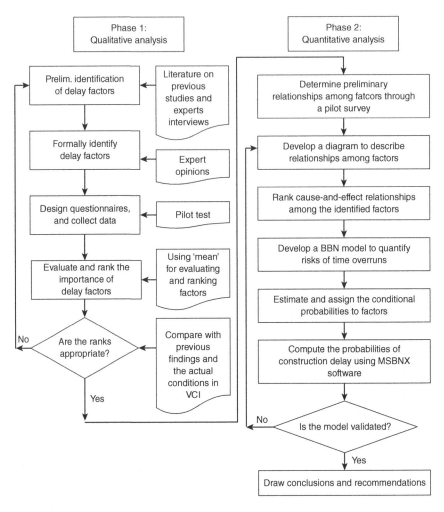

Figure 5.11 The procedure of BBN-based model building

The purpose of phase 1 was to identify significant delay factors being applicable to the context of construction projects in Vietnam. A set of factors was initially identified from the published literature and brainstorming sessions with experts. Unstructured interviews were conducted with an expert group comprising ten construction practitioners and four academicians to identify the delay factors experienced in the Vietnam Construction Industry. The purposes of phase 2 are to determine the cause-and-effect relationships among factors identified in the qualitative phase, to develop the BBN-based model, and to estimate the probability of delays in construction.

Data analysis and results

Building BBN-based model for predicting schedule delay probability on construction projects

As the outcome of phase 1, 42 preliminary delay factors impacting construction projects in Vietnam were extracted and 16 significant factors having the strongest effect to schedule delay were identified as shown in Table 5.7 and Table 5.8.

Table 5.7 42 preliminary delay factors impacting construction projects (refer to the article "Quantifying schedule risk in construction projects using Bayesian belief networks" of the author[50])

No.	Delay factors	No.	Delay factors
1	Types of Owners (state own, private enterprise . . .)	22	Lack of capable and responsible site supervisors
2	Owners' financial difficulties	23	Ungenerous payment for workers
3	Delay in making decision by owners	24	Man-power not available
4	Delays in progress payments by owners	25	Lack of co-operation between workers
5	Lack of capable owners/project managers	26	Un-smooth relationship between site supervisors and workers
6	Unrealistic term of contract	27	Unskilled workers
7	Owners' changes during construction	28	Accident at work
8	Slow site handover	29	Shortage of materials
9	Contractors' financial difficulties	30	Contract changes
10	Inadequate contractors' experience	31	Unclear responsibilities in contract
11	Inappropriate construction methods	32	Lack of information between related parties
12	Defective works and reworks	33	Transportation's difficulties
13	Shortage of equipment	34	Restriction of allowable construction time
14	Equipment in bad condition	35	Narrow construction layout
15	Unrealistic schedule estimation	36	Inclement weather
16	Low awarded bid prices	37	Owners' site clearance difficulties
17	Carrying out too many projects at the same time (contractors)	38	Unavailable material on market
18	Insufficient information in construction drawings	39	Regulation changes
19	Delay in responses of related parties	40	Material price fluctuations
20	Designers' inadequate experience and capability	41	Complicated geological condition
21	Design and execution at the same time	42	Surrounding problems (noise, sanitary, . . .)

Table 5.8 16 significant factors having the strongest effect on schedule delay[50]

Causes of delay	Overall		Owners		Site supervisors		Designers		Contractors	
	Mean	Rank	Mean	Rank	Mean	Rank	Mean	Rank	Mean	Rank
Owners' financial difficulties (x_1)	4.23	1	4.26	4	4.19	2	4.25	1	4.23	1
Inadequate contractors' experience (x_2)	4.13	2	4.29	3	4.27	1	3.82	4	4.14	2
Shortage of materials (x_3)	4.07	3	4.35	1	3.92	3	3.89	3	4.07	3
Contractors' financial difficulties (x_4)	3.98	4	4.35	1	3.85	4	3.79	5	3.94	7
Slow site handover (x_5)	3.94	5	3.87	7	3.81	5	3.96	2	4.00	5
Delays in progress payments by owners (x_6)	3.80	6	3.48	15	3.69	9	3.46	13	4.07	3
Low awarded bid prices (x_7)	3.73	7	3.52	14	3.27	14	3.75	6	3.96	6
Inappropriate construction methods (x_8)	3.70	8	3.81	10	3.54	12	3.61	7	3.74	9
Defective works and reworks (x_9)	3.67	9	3.87	7	3.58	10	3.54	8	3.68	11
Material price fluctuations (x_{10})	3.66	10	3.87	7	3.15	16	3.50	11	3.79	8
Lack of capable and responsible site supervisors (x_{11})	3.64	11	4.00	5	3.42	13	3.29	14	3.70	10
Inclement weather (x_{12})	3.63	12	3.61	13	3.81	5	3.54	8	3.62	13
Owners' site clearance difficulties (x_{13})	3.57	13	3.39	16	3.73	7	3.54	8	3.59	14
Lack of capable owners/project managers (x_{14})	3.56	14	3.71	11	3.58	10	3.50	11	3.52	16
Designers' inadequate experience and capability (x_{15})	3.55	15	3.97	6	3.23	15	3.11	15	3.64	12
Shortage of equipment (x_{16})	3.52	16	3.68	12	3.73	7	3.11	15	3.53	15

Note: x_i=encoding of the factor i

As a first step of phase 2, the questionnaire was developed to determine the cause-effect relationships among the 16 identified factors. The questionnaire consists of two subtypes. The first-subtype questionnaire was designed in matrix form, in which the left column presented 16 identified factors as causes and the top row presented those factors as effects. Nine experts were involved in this procedure. Each expert was asked to rate the factors using the following number values: '="very strong relationship" – 4, "strong relationship" – 3, "somewhat relationship" – 2, "weak relationship" – 1, "no relationship" – 0. In order to reduce every combination pair of variables in the matrix, pairs of factors having either no relationship or weak relationship should be eliminated. For example, "inclement weather" has no relationship with "owners' financial difficulties" or "contractors' financial difficulties".

To obtain conditional dependence relationships among those factors, the process of analysis adopted a procedure consisting of nine logical tests using two statistical values, namely the average and the skewness. Nine experts were involved in this procedure. Details of this procedure are available in Nasir et al.[51]. The procedure yielded four parent nodes that directly affect schedule delay, namely "inclement weather", "defective works and reworks", "slow site handover" and "shortage of materials"; and the child node, namely "construction delay". Expert's judgements have reduced the number of factor pairs significantly to 20. This resulted in the second-subtype questionnaire, which was a graphical representation of conditional dependence among the identified factors. The second-subtype questionnaire is described as Figure 5.12.

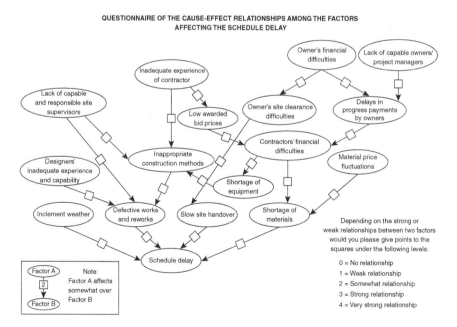

Figure 5.12 The second-subtype questionnaire

The collected data of the 20 cause-effect relationships among the factors were then averaged. There were two cause-effect relationships that their means were less than two continuously were discarded as relationship between x_{11} and x_9, and relationship between x_2 and x_7. The 18 remaining cause-effect relationships and the BBN-based conceptual model are presented as Table 5.9 and Figure 5.13. Finally, the developed model used two construction projects as case studies to validate the model. Details are available in the section on case studies.

Application of case studies

Description of the case studies

Two case studies were used to validate the BBN-based model. Table 5.10 presents brief information on both projects. The two projects were funded by private owners and were high-rise buildings in Ho Chi Minh City – Vietnam. Project A was under construction while project B was completed at the time of the study.

Verifying variables and cause-and-effect relationships

Before applying the BBN-based model, the verification of variables and cause-and-effect relationships are indispensable for the model's validation.

Table 5.9 18 cause-effect relationships among variables

Rank	Cause-effect relationships	Model variables	Mean	SD
1	$x_{13} - x_5$	X_{13-5}	3.27	0.854
2	$x_5 - y$	X_{5-y}	3.23	1.003
3	$x_3 - y$	X_{3-y}	3.20	0.860
4	$x_4 - x_3$	X_{4-3}	3.14	0.899
5	$x_1 - x_6$	X_{1-6}	3.08	1.127
6	$x_2 - x_8$	X_{2-8}	3.03	0.877
7	$x_6 - x_4$	X_{6-4}	2.85	0.824
8	$x_9 - y$	X_{9-y}	2.83	1.031
9	$x_8 - x_9$	X_{8-9}	2.67	0.956
10	$x_{14} - x_6$	X_{14-6}	2.63	0.835
11	$x_1 - x_{13}$	X_{1-13}	2.51	1.213
12	$x_{10} - x_3$	X_{10-3}	2.38	1.138
13	$x_{15} - x_9$	X_{15-9}	2.28	0.909
14	$x_{11} - x_8$	X_{11-8}	2.23	0.867
15	$x_4 - x_{16}$	X_{4-16}	2.17	0.746
16	$x_{16} - x_8$	X_{16-8}	2.07	1.070
17	$x_7 - x_4$	X_{7-4}	2.05	1.209
18	$x_{12} - y$	X_{12-y}	2.02	0.994

Note: SD=standard deviation

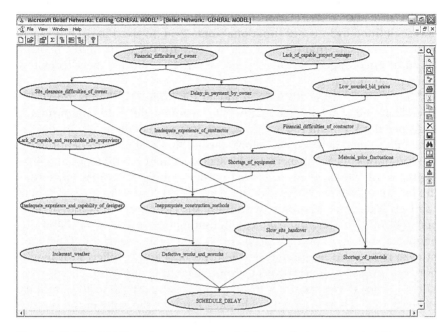

Figure 5.13 The BBN-based conceptual model for predicting the probability of construction delay

Table 5.10 Brief information about two projects

Description	Project	
	Project A	*Project B*
Project type	Building project	Building project
Owner	Private sector	Private sector
Project scope	22 floors including one basement	10 floors including one basement
Commencement date	April 2004	March 2005
Completion date based on construction contract	June 2006	November 2005
Original duration based on contract	26 months	8 months
Actual completion date	August 2006	December 2005
Time overrun duration	02 months	01 month
Percentage of time overrun	7.7%	12.5%

Therefore, the model was insightfully verified and applied into the construction phase of the two case studies. The following work was conducted to present the model validation and to demonstrate its application.

The conceptual model was then refined by professionals in the two projects. Cause-and-effect relationships among variables were also reviewed and re-defined by line and functional managers involved in each project. In order to ensure that the model is suitable for each case study, managers were encouraged to add new factors or to remove others from the conceptual model.

After reviewing, the BBN-based model for project A (Figure 5.14) differed from the conceptual model in "site clearance difficulties of owners".

As per the Vietnamese Land Law, owners must completely compensate for land use on a new project before starting site clearance. Depending on the conditions of each project, the local government will then permit clearing the whole or partial project area in order to mitigate adverse impacts to the local community. Thus, slow site handover of the remaining area may cause critical delays. In these case studies, "site clearance difficulties of owners" are problems deriving from compensating for land use while "slow site handover" not only concerns problems caused by site clearance difficulties of owners but also by slow government permits and conflicts with neighbors. For project A, a new project, although compensating for land use was fully completed, the local government only allowed 60% of the site area to be cleared. Therefore, although "site clearance difficulty of the owner" was absent from model A, "slow site handover" was still cited perhaps due

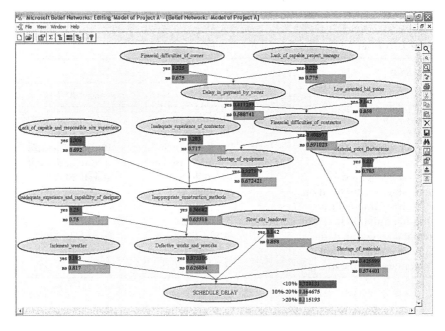

Figure 5.14 The BBN-based model and base-run probabilities for variable's states (project A)

to slow granting of permits by the municipality for clearance and possible problems with neighbors.

The BBN-based model for project B differed from the conceptual model in "site clearance difficulties of owners", "lack of capable project managers" and "slow site handover". It should be noted that project B is concerned with the re-construction of a very old building. Thus, there were very few problems of site clearance. In addition, the owner of project B employed a professional firm of project management consultants. As a result, "lack of capable project managers" did not feature in the model for project B.

Computerized model building

Based on the variables and cause-and-effect relationships identified by line and functional managers, the BBN-based model was built using the free download software MSNBX from Microsoft Inc. In a BBN, each variable has four characteristics as follows: (1) the variable name; (2) the variable status; (3) its relationships (parent node) with other variables (child nodes); (4) its data table (table of condition probabilities). Since the determination of a variable's state is important for computerized applications, each variable is assigned its appropriate status to reflect its conditions in regard to the risk during the construction phase. For example, the variable 'schedule delay' is assigned three states, namely "<10%", "10%–20%", and ">20%" whereas the remaining variables have two opposite states: "YES" and "NO". The assignments "<10%", "10%–20%", and ">20%" are defined as time overrun durations which are less than 10%, between 10 and 20%, and greater than 20% compared to the original duration is stipulated in the construction contract, respectively. It should be noted that the assigned states for each variable was advised by project managers who were directly in charge of the construction tasks in both projects. Data tables for the variable "schedule delay" (Figure 5.15) and other variables (Figure 5.16, for example) are inputs from the judgements of line and functional managers

Assessment (Model: Model of Project A, Node: SCHEDULE_DELAY)

Inclement_weather	Defective_works_and_reworks	Slow_site_handover	Shortage_of_materials	<10%	10%-20%	>20%	bar charts
yes	yes	yes	yes	0.417	0.2	0.383	
			no	0.533	0.25	0.217	
		no	yes	0.433	0.233	0.334	
			no	0.55	0.283	0.167	
	no	yes	yes	0.567	0.2	0.233	
			no	0.625	0.167	0.208	
		no	yes	0.667	0.225	0.108	
			no	0.825	0.158	0.017	
no	yes	yes	yes	0.758	0.142	0.1	
			no	0.658	0.183	0.159	
		no	yes	0.642	0.183	0.175	
			no	0.725	0.15	0.125	
	no	yes	yes	0.575	0.225	0.2	
			no	0.683	0.175	0.142	
		no	yes	0.65	0.183	0.167	
			no	0.884	0.108	0.008	

Figure 5.15 Full data table for the output variable, namely "schedule delay"

⊞ Assessment (Model: Model of Project A, Node: Shortage_of_materials)				⊠
Parent Node(s)		**Shortage_of_materials**		
Material_price_fluctuations	**Financial_difficulties_of_contractor**	**yes**	**no**	bar charts
yes	yes	0.788	0.212	
	no	0.5	0.5	
no	yes	0.513	0.487	
	no	0.275	0.725	

Figure 5.16 Full data table for the variable "shortage of materials"

in the cases. This technique is referred to as Bayesian belief networks or belief networks because of the aforementioned manner of data input.

Computerized model input

After the computerized model was built in the MSBNX environment and verified by managers, a questionnaire, shown in part in Table 5.11, was developed to obtain the conditional probabilities for each variable from line and functional managers.

Eight experts participated in the survey. Generally, experts were asked to judge frequencies of a variable's states based on certain states of other variables that have cause-and-effect relationships with that variable. For example, given that: (1) "inclement weather" and "defective works and reworks" are NO; (2) "slow site handover" and "shortage of materials" are YES, the frequencies of "<10%", "10%–20%" and ">20%" states of "schedule delay" were judged as 57.5%, 22.5% and 20%, respectively (Figure 5.15). These frequencies were facilitated in practice by a built-in function of MSBNX. Based on expert judgement, a similar procedure is applied to each state of this variable – "schedule delay" – in other conditions and to the other variables.

Table 5.11 A part of the questionnaire to quantify the conditional probabilities for each variable

Parent nodes		Child node	
If		*Defective works and reworks*	
Inappropriate construction methods	Designers' inadequate experience and capability	Yes	No
Yes	Yes		
	No		
No	Yes		
	No		

Table 5.12 The BBN-based model outputs of projects A and B

The states of construction time overrun (compared with original duration based on contract)	The probabilities of time overrun (%)	
	Project A	Project B
"<10%"	72	20
"10%–20%"	16	67
">20%"	12	13

Model output and evaluation

After the data input facilitated by MSBNX, the model could be used to identify and evaluate the probabilities of each variable's states. Table 5.12 illustrates BBNs model outputs of projects A and B.

It is estimated that the likelihood of construction delay in project A is approximately 72% for the state of "<10%", whereas it is 16% for the state of "10%–20%" and 12% for the state of ">20%". Consequently, the construction duration of project A tends to extend with time overrun duration less than 10% of the original duration. Regarding project B, the likelihood of construction delay is approximately 67% for the state of "10%–20%" whereas it is 20% for the state of "< 10%" and 13% for the state of ">20%". Consequently, the construction duration of project B tends to extend with time overrun duration between 10% and 20% of the original duration.

In reality, there was a two-month delay by the contractor on project A completed in August 2006. The delay on project A is 7.7% of the original duration (26 months). Interestingly, the model predicted that the probability of construction delay in project A is nearly 72% with time overrun less than 10% of project A's original duration. This implied that the BBN-based model gave very good prediction for the construction delay in project A. For project B, there is a one-month delay, which is approximately 12% of the project original duration. The BBN-based model for project B forecasted that the probability of construction delay is approximately 67% with the time overrun between 10% and 20% of the original duration. In summary, the BBN-based model provided a good prediction for quantifying construction delay in both case-study projects.

Discussion

Each project stakeholder can identify its capability to eliminate the possibility of time overrun. For example, on project A, contractors should understand their responsibility to provide materials on time and should be well prepared for their finance to avoid time overrun since the probability of "shortage of materials" and "financial difficulties by contractors" are 43% and 41%, respectively (Figure 5.14). Owners should recognize their duty in

Table 5.13 The probability distributions of construction time overrun under different scenarios (project A)

Scenarios of "non-available materials on time"	Probability distribution (%) of construction time overrun		
	"<10%"	"10%–20%"	">20%"
YES	63	19	18
Base-run	72	16	12
NO	79	14	7

monthly payments on a timely basis to contractors as an effective solution to eliminate time overrun.

The model can be used to evaluate the effects of each variable's state on the distribution of the other variables. This technique is called sensitivity analysis which can help project managers to make right decisions in order to avoid construction delay. For example, regarding project A, when "shortage of materials" is NO (probability of NO state is 1), the probability of time overrun being less than 10%, between 10% and 20% and greater than 20% is 79%, 14% and 7%, respectively (Table 5.13).

Conversely, the probabilities of "10%–20%" and ">20%" increases to 19% and 18%, respectively when "shortage of materials" is YES (probability of YES state is 1). It can be said that the probable time overrun of project A is very sensitive to the "shortage of materials". Most other factors are insensitive in the case study A. Similarly, the results of sensitive analysis indicated that the probability of project B's time overrun is very sensitive to the "shortage of materials". Therefore, the probable time overrun is very sensitive to the "shortage of materials". It is recommended that all project parties should focus on eliminating shortage of materials in order to avoid time overrun.

The model developed may be used to evaluate the effects of each variable's state on the distribution of other variables. The result of sensitivity analysis using the BBN-based model indicated that construction delay is extremely sensitive to the factors 'shortage of materials', 'defective construction work' and 'slow site handover'.

It is recommended that contractors should clearly understand their responsibility to provide materials on time and be well-prepared for this financial responsibility in order to avoid time overrun. Owners need to focus on their responsibility for monthly timely payment to contractors as an effective solution to eliminate delay in construction projects. Moreover, it should be noted that all project parties should focus on the shortage of materials as a way of preventing time overrun.

Conclusion

In this chapter we have looked at the Bayesian belief networks approach for risk management practices in the construction industry. An overview of BBNs, BBN-based model building and application of case studies are presented in detail in the chapter. The risk assessment model for construction delay in building projects using BBNs has been developed and the potential demonstrated in this study. The results proved that belief networks are very useful when historical data are insufficient. In such cases, the experts' judgement plays a vital role. Belief networks are an expressive graphical language for representing uncertain knowledge about causal and associational relations among construction schedule variables. The results encourage practitioners to benefit from the power of BBNs.

This study only used two case studies to validate the proposed framework and model. Thus, generalization for the other similar projects may be too bold. Further studies could use the BBN-based model in other construction projects in order to explore its utility. However, the approach in this research is still general, and as such, it may be applied to other construction projects with minor modifications.

References

1 Hong-bo Zhou, Hui Zhang (2011). Risk assessment methodology for a deep foundation pit construction project in Shanghai, China. *Journal of Construction Engineering and Management*, ASCE 2011, 137(12), 1185–1194.
2 Spiegelhalter, D. J., Dawid, A. P., Lauritzen, S. L. and Cowell, R. G. (1993). Bayesian analysis in expert system. *Stat. Sci.*, 8(3), 219–283.
3 Lucas, P. J. F., van der Gaag, L. C. and Abu-Hanna, A. (2004). Bayesian networks in biomedicine and health-care. *Artificial Intelligence in Medicine*, 30(3), 201–214.
4 Stassopoulou, A., Petrou, M. and Kittler, J. (1996). Bayesian and neural networks for geographic information processing. *Pattern Recognition Letters*, 17(13), 1325–1330.
5 Torres-Toledano, J. G. and Sucar, L. E. (1998). Bayesian networks for reliability analysis of complex systems. The 6th Ibero-American Conf. on AI, Springer Verlag, Berlin, 195–206.
6 Landuyt, D., Broekx, S., D'hondt, R., Engelen, G., Aertsens, J. and Goethals, P.L.M. (2013). A review of Bayesian belief networks in ecosystem services modeling. *Environmental Modeling & Software*, 46(August 2013), 1–11.
7 Stelzenmuller, V., Lee, J, Ganacho, E. and Rogers, S.I. (2010). Assessment of Bayesian Belief Network-GIS framework as a practical tool to support marine planning. *Marine Pollution Bulletin*, 60(10), 1743–1754.
8 Pitchforth, J. and Mengersen, K. (2013). A proposed validation framework for expert elicited Bayesian Networks. *Expert System with Applications*, 40(1), 162–167.
9 Jun, Hong-Bea and David, Kim (2017). A Bayesian network-based approach for fault analysis. *Expert System with Applications*, 40(15 September 2017), 332–348.
10 Allan, J. D., Yuan, L. L., Black, P., Stockton, T., Davies, P. E., Magierowski, R. H., et al. (2012). Investigating the relationships between environmental stressors and stream condition using Bayesian belief networks. *Freshwater Biology*, 57, 58–73.
11 Fienberg, S. E. (2011). *Bayesian models and methods in public policy and government settings*. https://arxiv.org/pdf/1108.2177.pdf.
12 Hossain, M. and Muromachi, Y. (2012). A Bayesian network based framework for real-time crash prediction on the basic freeway segments of urban expressways. *Accident Analysis and Prevention*, 45, 373–381.

13 Hsu, C. I., Shih, M. L., Huang, B. W., Lin, B. Y. and Lin, C. N. (2009). Predicting tourism loyalty using an integrated Bayesian network mechanism. *Expert Systems with Applications*, 36(9), 11760–11763.

14 Kisioglu, P. and Topcu, Y. I. (2011). Applying Bayesian belief network approach to customer churn analysis: A case study on the telecom industry of Turkey. *Expert Systems with Applications*, 38(6), 7151–7157.

15 McCloskey, J. T., Lilieholm, R. J. and Cronan, C. (2011). Using Bayesian belief networks to identify potential compatibilities and conflicts between development and landscape conservation. *Landscape and Urban Planning*, 101(2), 190–203.

16 Neil, M., Fenton, N., Forey, S. and Harris, R. (2001). Using Bayesian belief networks to predict the reliability of military vehicles. *Computing & Control Engineering Journal*, 12(1), 11–20.

17 Neil, M., Hager, D. and Andersen, L. B. (2009). Modeling operational risk in financial institutions using hybrid dynamic Bayesian networks. *Journal of Operational Risk*, 4(1), 3–33.

18 Newton, A. C., Marshall, E., Schreckenberg, K., Golicher, D., te Velde, D. W., Edouard, F. and Arancibia, E. (2006). Use of a Bayesian belief network to predict the impacts of commercializing non-timber forest products on livelihoods. *Ecology and Society*, 11(2): 24. [online] URL:www.ecologyandsociety.org/vol11/iss2/art24/.

19 Staker, R. J. (1999). Use of Bayesian belief networks in the analysis of information system network risk. *Information, Decision and Control, 1999. IDC 99. Proceedings*, 145–150.

20 Kabir, G., Demissie, G., Sadiq, R. and Tesfamariam, S., (2015). Integrating failure prediction models for water mains: Bayesian belief network based data fusion. *Knowledge-Based Systems*, 85(September 2015), 159–169.

21 Fineman, M. (2010). *Improved Risk Analysis for Large Projects: Bayesian Networks Approach*. Dissertation (Ph.D). Queen Mary, University of London.

22 Raoofpanah, H. and Hassanlou K. (2013). A probabilistic approach for project cost estimation using Bayesian networks. *Life Science Journal*, 10(6s), 342–349.

23 Pitchforth, J. O. (2015) *Bayesian networks for information synthesis in complex systems*. PhD by Publication, Queensland University of Technology.

24 James, S. A. (2018). *A Bayesian Network Tool for Selecting New Technology Investment Projects*. Dissertation (Doctor of Engineering). George Washington University

25 Chan, A. P. C., Wong, F. K. W., Hon, C. K. H., and Choi, T. N. Y. (2018) A Bayesian network model for reducing accident rates of electrical and mechanical (E&M) work. *International Journal Environmental Research Public Health*, 15(11), 2496.

26 Sou-Sen Leu and Quang-Nha Bui (2011). A novel Bayesian Network approach for engineering reliability assessment [online]. Proceedings of the 28th International Symposium on Automation and Robotics in Construction, ISARC 2011, Seoul, Republic of Korea, 1190–1195. Available from: https://pdfs.semanticscholar.org/9c10/16bc3095c5fe2a6b4d1babf080b437dea8b9.pdf.

27 Pan, M. Q., Hui, K. X., Zhang, Y. L. and Wang, Q. (2003). The application of Bayesian to evaluate relay reliability. *RELAY*, 31(5), 27–29.

28 Fu, Y., Wu, X. P., and Yan, C. H. (2006). The method of information security risk assessment using Bayesian networks. *J. Wuhan Univ. (Nat.Sci. Ed.)*, 52(5), 631–634.

29 McCabe, B., AbouRizk, S. M. and Goebel, R. (1998). Belief networks for construction performance diagnostics. *J Comput Civil Eng ASCE*, 12(2), 93–100.

30 Sahely, B. S. G. E. and Bagley, D. M. (2001). Diagnosing upsets in anaerobic wastewater treatment using Bayesian belief networks. *J Environ Eng ASCE*, 127(4), 302–310.

31 Tischer, T. E. and Kuprenas, J. (2003). Bridge falsework productivity-measurement and influences. *J Construct Eng Manage ASCE*, 129(3), 243–250.
32 Attoh-Okine, N. O. (2002). Probabilistic analysis of factors affecting highway construction costs: a belief network approach. *Can J Civil Eng*, 29(3), 369–374.
33 Nasir, D., McCabe, B. and Hartono, L. (2003). Evaluating risk in construction–schedule model (ERIC-S): construction schedule risk model. *J Construct Eng Manage ASCE*, 129(5), 518–527.
34 Adams, F. (2006). Expert elicitation and Bayesian analysis of construction contract risks: an investigation. *Construction Management Economics*, 24(1), 81–96.
35 Aalders, I., Hough, R. L. and Towers, W. (2011). Risk of erosion in peat soils – an investigation using Bayesian belief networks. *Soil Use and Management*, 27(4), 538–549.
36 Martin, J. E., Rivas, T., Matias, J. M., Taboada, J. and Arguelles, A. (2009). A Bayesian network analysis of workplace accidents caused by falls from a height. *Safety Science*, 47(2), 206–214.
37 Zorrilla, P., Carmona, G., De la Hera, A., Varela-Ortega, C., Martinez-Santos, P., Bromley, J., et al. (2010). Evaluation of Bayesian networks in participatory water resources management, Upper Guadiana Basin, Spain. *Ecology and Society*, 15(3), 12.
38 Leu, S. S. and Bui, Q. N. (2016). Leak prediction model for water distribution networks created using a Bayesian Network learning approach. *Water Resource Management*, 30(8), 2719–2733.
39 Nguyen, L. D., Tran, D. Q. and Chandrawinata, M. P. (2016). Predicting safety risk of working at heights using Bayesian networks. *Journal of Construction Engineering and Management, ASCE*, 142(9), 04016041.
40 Khanzadi, M., Eshtehardian, E. and Esfahani, M. M. (2017). Cash flow forecasting with risk consideration using Bayesian Belief Networks (BBNS). *Journal of Civil Engineering and Management*, 23(8), 1045–1059.
41 Xia, N., Wang, X., Wang, Y., Yang, Q. and Liu, X. (2017). Lifecycle cost risk analysis for infrastructure projects with modified Bayesian networks. *Journal of Engineering, Design and Technology*, Vol. 15 Issue. 1, 79–103.
42 Yongho Ko and Seungwoo Han (2015). Development of construction performance monitoring methodology using the Bayesian probabilistic approach. *Journal of Asian Architecture and Building Engineering*, 14:1, 73–80.
43 Micán, C. A., Jiménez, V., Perez, V. and Borrero, J.A. (2013). Schedule risk analysis in construction project using RFMEA and Bayesian networks: The Cali-Colombia case study. 2013 IEEE International Conference on Industrial Engineering and Engineering Management, 10–13 Dec. 2013, Bangkok, Thailand, doi: 10.1109/IEEM.2013.6962445.
44 JokoWahyuAdi, T., Anwar, N. and Fahirah, F. (2016). Probabilistic prediction of time performance in building construction project using Bayesian Belief Networks – Markov chain. *ARPN Journal of Engineering and Applied Sciences*, 11(15), 9454–9460.
45 Sarasanty, D., Adi, T. J. W. and Wiguna, I. P. A. (2017). Probabilistic Model For Predicting Construction Worker Accident Based On Bayesian Belief Networks. The Third International Conference on Civil Engineering Research (ICCER), 1–2 August 2017, Surabaya – Indonesia, 452–458.
46 Zou, Y., Xu, B., Liu, S. and Zhou., J (2018). The simulation of schedule risk paths in submarine pipeline projects using Bayesian Networks. The International Conference on Mechanical, Electric and Industrial Engineering (MEIE2018), 26–28 May 2018, Hangzhou, China. Available from: https://iopscience.iop.org/article/10.1088/1742-6596/1074/1/012150/pdf.

47 Joseph, S. A., Adams, B. J., McCabe, B. (2010). Methodology for Bayesian belief network development to facilitate compliance with water quality regulations. *Journal of Infrastructure Systems, ASCE*, 16(1), 58–65.
48 Charles River Analytics (2004). About Bayesian belief network. *Charles River Analytics Inc.* www.cra.com/pdf/BNetBuilderBackground.pdf.
49 Bayers' theorem (2007). http://en.wikipedia.org/wiki/Bayes'_theorem.
50 Luu, T. V, Kim, S. Y, Nguyen, V. T and Ogunlana, S. O. (2009). Quantifying schedule risk in construction projects using Bayesian belief networks. *International Journal of Project Management*, 27(1), 39–50.
51 Nasir, D., McCabe, B. and Hartono, L. (2003). Evaluating risk in construction–schedule model (ERIC-S): construction schedule risk model. *Journal Construction Engineering Management, ASCE*, 129(5), 518–527.

6 The use of the analytical hierarchy and network process for risk management

Jonathan Nixon and Prasanta Kumar Dey

Introduction

The Analytical Hierarchy Process (AHP) was originally proposed by Thomas L. Saaty in the 1970s and has since gone on to become one of the most widely adopted multi-criteria decision-making methods used to date[1].

Decision-making problems continuously occur throughout all industries, sectors, businesses and people's lives in general. Many of these decisions that need to be made will have a whole host of surrounding issues and repercussions to be considered, thus ultimately they end up becoming very complex and difficult to manage. Decision-makers will typically use their judgement, experience or intuition to make choices, rather than using a holistic and structured approach to fully evaluate decision problems. And this can result in poor decisions being made. For this reason, Saaty developed AHP to provide decision-makers with a systematic method of decomposing a complex problem into several more manageable sub-problems. The AHP then uses science and psychology to evaluate and analyse all aspects of the problems. Applications of AHP originally demonstrated by Saaty included choosing a school for his son and transportation planning[1]. The AHP is now used all over the world in a vast range of applications including energy planning, marketing, information technology, education and policy making[2,3].

The AHP involves three main processes: (i) identifying the decision goal, selection criteria and alternatives, and organising them into a hierarchy; (ii) carrying out pairwise comparisons throughout all levels of the hierarchy and checking for consistency; and (iii) synthesising all the pairwise comparisons to determine each alternative's overall preference (see Figure 6.1)[4]. One major advantage of AHP is its ability to encompass both known and unknown data into the decision rationale. Real world data, statistics and probabilities can be used in relative form instead of making pairwise comparisons. However, where there is a gap in factual information, expert opinion and judgement can be used. This flexibility is particularly useful as there are often many uncertainties or subjective data in complex decision problems. To illustrate how AHP works, a simple example is now provided based on the goal of selecting the best car to purchase.

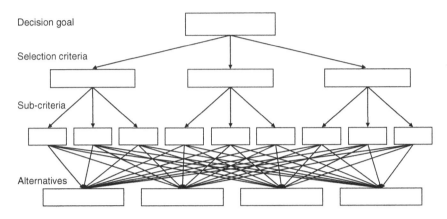

Decision goal

Selection criteria

Sub-criteria

Alternatives

Figure 6.1 Structure of the analytical hierarchy process

The analytical hierarchy process: an example for car selection

Suppose that a car has to be selected from three alternatives, Cars A, B and C. This decision is tackled by breaking down the problem into a hierarchy of four more manageable sub-problems, called criteria, which can be analysed individually. Further sub-criteria can be developed to give a more accurate selection. Selection of a car in this example is made based on the criteria of speed, comfort, cost and fuel economy (MPG). Figure 6.2 shows the resulting hierarchy.

The alternative cars are first compared against each criterion through a pairwise comparison matrix. Using factual data or judgement, each alternative is scored on a scale of 1–9 (1 = weak, 9 = strong) against the other alternatives to show their preference (see Table 6.1). If an alternative is worse, a reciprocal value is produced, e.g. 1/9. The matrix is then normalised by dividing a cell by its corresponding column total. The average of the row of the normalised table provides a priority vector (i.e. the preference in comparison to the other alternatives) for each alternative for the criterion analysed. An example is given for the criterion speed, showing that Car A is the fastest, and therefore has the highest priority vector (see Table 6.2).

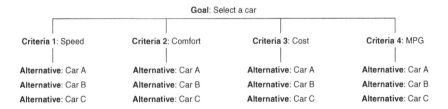

Goal: Select a car

Criteria 1: Speed	**Criteria 2**: Comfort	**Criteria 3**: Cost	**Criteria 4**: MPG
Alternative: Car A	**Alternative**: Car A	**Alternative**: Car A	**Alternative**: Car A
Alternative: Car B	**Alternative**: Car B	**Alternative**: Car B	**Alternative**: Car B
Alternative: Car C	**Alternative**: Car C	**Alternative**: Car C	**Alternative**: Car C

Figure 6.2 Simple hierarchy tree for the selection of a car

Table 6.1 Pairwise comparison scale values for the level of preference to be used in the pairwise comparison matrix

Verbal Judgement of Preference	Numerical Rating
Extremely Preferred	9
Very strongly to extremely preferred	8
Very strongly preferred	7
Strongly to very strongly preferred	6
Strongly preferred	5
Moderately to strongly preferred	4
Moderately preferred	3
Equally to moderately preferred	2
Equally preferred	1

Table 6.2 Normalised pairwise comparison matrix for alternatives preference against the speed criterion.

	Car A	Car B	Car C		Normalized			Priority Vector
Car A	1	8	5		0.75	0.67	0.79	0.74
Car B	1/8	1	1/3	=	0.09	0.08	0.05	0.08
Car C	1/5	3	1		0.15	0.25	0.16	0.19
Total	1.33	12	6.33					

In the pairwise comparison tables (Table 6.2), comparisons are made for each row from left to right against each column:

- Car A is equally preferred in comparison to Car A (1)
- Car A is very strongly to extremely preferred in comparison to Car B (8)
- Car A is strongly preferred in comparison to Car C (5)
- Car B is strongly to extremely unpreferred in comparison to Car A (1/8)

And so on . . .

The pairwise comparison process is repeated till a priority vector for each alternative is developed against every criterion. However, the importance of each criterion in relation to the other criteria is not specified. Thus, a final pairwise comparison matrix is completed to develop a weighting vector for each criterion (see Table 6.3). This stage of an AHP is often subjective and a study can be improved by obtaining the opinion of experts or stakeholders in a relevant field or project. The hierarchy tree can be updated to show all alternative priority vectors and criteria weighting vectors (see Figure 6.3).

The final overall value for how much each technology is preferred, is calculated by multiplying each alternative's priority vector by the corresponding criterion's weighting vector and totalling the values for each

Table 6.3 Pair-wise matrix for the weighting preference of each criterion

	Speed	Comfort	Cost	MPG		Normalized				Weighting Vector
Speed	1	1/4	1/7	1/5		0.06	0.08	0.03	0.10	0.07
Comfort	4	1	1/3	1/2	=	0.24	0.15	0.07	0.25	0.18
Cost	7	3	1	1/3		0.41	0.46	0.22	0.16	0.31
MPG	5	2	3	1		0.29	0.31	0.67	0.49	0.44
Total	17	6.5	4.48	2.03						

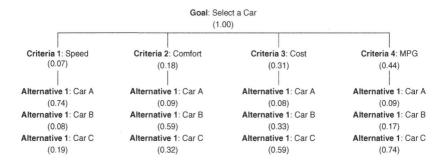

Figure 6.3 Updated hierarchy tree showing final priorities

alternative. The final ranking gives a relative value which can be expressed as a percentage of the preference. For the car selection example, alternative C is the preferred car with a preference of 58% (see Figure 6.4).

A consistency check can be performed to access the reliability of the results, this highlights any potential mistakes. For example, if Car A was ranked higher than Car B for speed and Car B was faster than Car C, Car C could not be preferred over Car A.

Saaty measured consistency by using the consistency ratio *CR*, which is calculated from a consistency index *CI* and random consistency index *RI*. Saaty determined that there is an acceptable level of inconsistency when the consistency ratio is less than or equal to 10%.

$$CI = \frac{\lambda_{max} - m}{m - 1}$$

(X.1)

Where λ_{max} is the sum of the priority vectors multiplied by the corresponding totalled value of the original pairwise matrix column and *m* is the size of the matrix. For the speed criterion (Table 6.2), λ_{max} is given by (1.33×0.74) $+ (0.08 \times 12) + (6.33 \times 0.19)$, which equals 3.08. The size of the matrix, *m*, is 3. The consistency ratio is finally calculated from,

$$CR = \frac{CI}{RI}$$

(X.2)

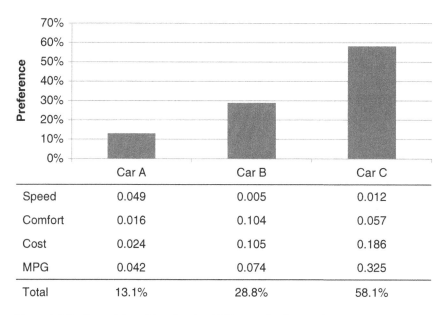

	Car A	Car B	Car C
Speed	0.049	0.005	0.012
Comfort	0.016	0.104	0.057
Cost	0.024	0.105	0.186
MPG	0.042	0.074	0.325
Total	13.1%	28.8%	58.1%

Figure 6.4 Preferential ranking for the AHP car selection study

Table 6.4 Random consistency index (RI)

m	1	2	3	4	5	6	7	8	9	10
RI	0	0	0.58	0.9	1.12	1.24	1.32	1.41	1.45	1.49

where the random consistency index, RI, is obtained from a standard table for the AHP (see Table 6.4). Therefore, for the speed criteria there is acceptable consistency of 0.0667 or 6.7%. Each criterion is evaluated and if all criteria have an acceptable consistency the AHP study is finished.

Managing risk using the AHP

Whilst there are numerous applications for the AHP, its use for managing risk and evaluating project uncertainties is particularly valuable. Thus, the uptake of the AHP by researchers and practitioners for managing risk in engineering and construction projects has started to increase significantly in recent years. Applications of the AHP include risk identification and assessment for highway construction projects in China[5], managing risk in supply chains[6] and analysing project risk in oil pipeline constructions in India[7]. Schoenherr et al. reports on the use of the AHP by an American manufacturing company to assess risk in their supply chain purchasing, and ultimately determine where to source components and parts [8].

There are numerous issues surrounding project planning and operations that result in varying degrees of risk: size, location, time, environment, availability, financing, etc. High risk projects will also often have high payoffs. The advantage of taking a holistic approach to decision-making using the AHP is that all these conflicting aspects can be simultaneously considered and incorporated into the decision rationale. Thus the AHP can be used to mitigate risk by fully evaluating and analysing the importance and contributions of various risk aspects in project management.

Mitigating risk in landfill site selection

Consider selecting a site for waste landfill. It is a project and decision problem surrounded in complex and abundant risks. Using the AHP we can aim to select a site that has the lowest associated risk and therefore the greatest chance for project success. To generate an AHP hierarchy (Figure 6.1), alternative sites are initially identified. In this example three sites are considered: Sites A, B and C. Next we identify the risk criteria and sub-criteria. For the sake of brevity the entire AHP analysis is not illustrated here, however, the process would be to use pairwise matrices to weight the importance of the individual risk criteria. Matrices, real world data, probabilities, etc. would then be used to score each site against each criterion.

Figure 6.5 delineates the hierarchy and potential risk factors for this site selection problem example. However, there are many further underlying issues relating to risk when selecting a landfill site. Assessment of the individual sites may reveal that a higher chance of public opposition may occur at a particular site due to the impact on human health and

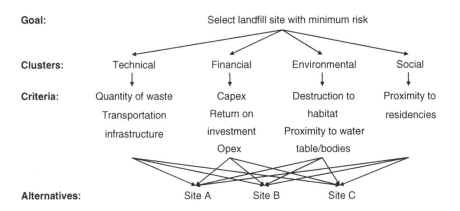

Figure 6.5 Hierarchy for landfill site selection using AHP

climate change. This could result in protests which would delay the project and result in bad publicity. If public opposition was considered, it is a criterion that is strongly linked to the other social and environmental criteria. A drawback of AHP is that it does not consider such dependencies among criteria.

The Analytical Network Process (ANP)

As criteria dependencies are not considered in the AHP, Saaty went on to develop the Analytical Network Process, which is an extension of the AHP to consider inner and outer dependencies between criteria and alternatives[9]. Instead of a hierarchy, in the ANP, a network of clusters with internal elements (criteria) is developed (see Figure 6.6). In comparison to the AHP, the ANP has additional layers of comparison matrices to establish the influence of criteria attributes on other clusters and elements. When completing the additional matrices, the question asked is: with respect to a parent element (and keeping in mind the goal of the analysis) which of two elements has the greatest influence on the parent element.

Going back to the landfill site selection problem, dependency links can now be established between the social and environmental risk criteria (elements). The majority of the financial risks are also dependent on the technical criteria. For example a higher quantity of waste to landfill will increase capital costs (large site required) and profits (payments to receive the waste, i.e. gate fees). However, if more waste is received, more land is required, thus there is a greater impact on climate change (increase in green house gas emissions) and destruction to habitat; and so the network

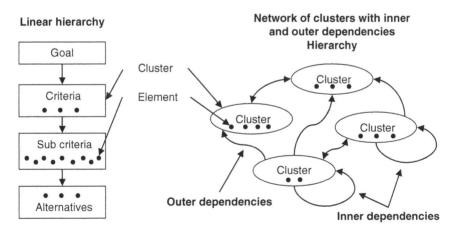

Figure 6.6 Comparison of AHP and ANP frameworks

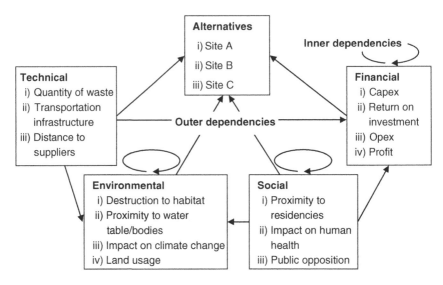

Figure 6.7 Network for landfill site selection using ANP

continues to grow and develop. An example of how the problem now looks like in a network is shown in Figure 6.7.

The solution algorithm for the ANP is different to the AHP. The ANP analysis works via the concept of a 'supermatrix', where the 'supermatrix' is a collection of all developed influence priority vectors established using pairwise comparisons. Next a weighted supermatix is developed that includes the importance weightings of the clusters with respect to the aim of the analysis. The final step is a 'limiting supermatrix' which provides the final preferential ranking of the alternatives by raising the matrix to powers until convergence is achieved. The mathematical derivation of this is shown in more detail in Ref.[9].

For simplicity, the AHP and the ANP are now compared considering only the environmental and social criteria for the landfill site selection problem. Table 6.5 tabulates all the priority vectors developed in an AHP analysis using pairwise comparisons for rating the sites' performance again each criterion and the importance of the criteria with respect to the goal. Table 6.6 shows how the supermatix for an ANP analysis now includes the dependency priority vectors among the criteria.

In a weighted supermatrix, priority vectors are multiplied by corresponding cluster weighting, if any have been specified. Whilst ANP allows cluster weightings to be specified, individual weightings for the criteria cannot be specified with respect to a goal. Some authors have therefore adopted a Hierarchical Analytical Network Process (HANP) method[10-13], which structures a network of clusters within a hierarchy

Table 6.5 Matrix of priority vectors for an AHP study (excludes weightings for clusters, i.e. the top level criteria)

	S. A	S. B	S. C	Destr. habitat	Impact on C.C.	Land usage	Prox. to water	Impact on h. h.	Prox. to residencies	Public opp.	Goal
Site A	0	0	0	0.58	0.33	0.55	0.10	0.23	0.12	0.33	0
Site B	0	0	0	0.18	0.33	0.21	0.36	0.65	0.61	0.41	0
Site C	0	0	0	0.23	0.33	0.24	0.54	0.12	0.27	0.26	0
Destruction to habitat	0	0	0	0	0	0	0	0	0	0	0.12
Impact on climate change	0	0	0	0	0	0	0	0	0	0	0.06
Land usage	0	0	0	0	0	0	0	0	0	0	0.25
Proximity to water table/bodies	0	0	0	0	0	0	0	0	0	0	0.57
Impact on human health	0	0	0	0	0	0	0	0	0	0	0.48
Proximity to residencies	0	0	0	0	0	0	0	0	0	0	0.11
Public opposition	0	0	0	0	0	0	0	0	0	0	0.41
Goal	0	0	0	0	0	0	0	0	0	0	0.00

Performance of each alternative against each criterion

Criteria weightings

Table 6.6 Unweighted ANP supermatix (excludes clusters weightings)

	S. A	S. B	S. C	Destr. habitat	Impact on c. c.	Land usage	Prox. to water	Impact on h. h.	Prox. to residencies	Public opp.
Site A	0	0	0	0.58	0.33	0.55	0.10	0.23	0.12	0.33
Site B	0	0	0	0.18	0.33	0.21	0.36	0.65	0.61	0.41
Site C	0	0	0	0.23	0.33	0.24	0.54	0.12	0.27	0.26
Destruction to habitat	0	0	0	0	0	0	0	0	0	0.43
Impact on climate change	0	0	0	0.09	0	0	0	0	0	0.13
Land usage	0	0	0	0.82	0	0	0	0	0	0.06
Proximity to water table/bodies	0	0	0	0.09	0	0	0	0	0	0.38
Impact on human health	0	0	0	0	0	0	0	0	0	0.8
Proximity to residencies	0	0	0	0	0	0	0	0	0	0.2
Public opposition	0	0	0	0	0	0	0	0	0	0

Dependencies among criteria

Table 6.7 Unweighted HANP supermatix showing priority vectors the alternative and criteria weightings and dependencies

	S. A	S. B	S. C	Destr. habitat	Impact on C.C.	Land usage	Prox. to water	Impact on h. h.	Prox. to residencies	Public opp.	Goal
Site A	0	0	0	0.58	0.33	0.55	0.10	0.23	0.12	0.33	0
Site B	0	0	0	0.18	0.33	0.21	0.36	0.65	0.61	0.41	0
Site C	0	0	0	0.23	0.33	0.24	0.54	0.12	0.27	0.26	0
Destruction to habitat	0	0	0	0	0	0	0	0	0	0.43	0.12
Impact on climate change	0	0	0	0.09	0	0	0	0	0	0.13	0.06
Land usage	0	0	0	0.82	0	0	0	0	0	0.06	0.25
Proximity to water table/bodies	0	0	0	0.09	0	0	0	0	0	0.38	0.57
Impact on human health	0	0	0	0	0	0	0	0	0	0.8	0.48
Proximity to residencies	0	0	0	0	0	0	0	0	0	0.2	0.11
Public opposition	0	0	0	0	0	0	0	0	0	0	0.41
Goal	0	0	0	0	0	0	0	0	0	0	0

Performance of each alternative against each criterion

Dependencies among criteria

Criteria weightings

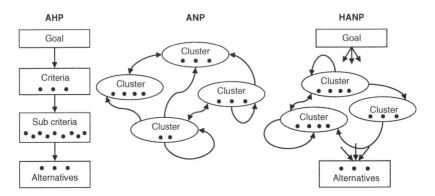

Figure 6.8 Comparison of an HANP framework with AHP and ANP

(see Figure 6.8). Whilst some flexibility of a full network is lost, HANP uses the advantages of AHP and ANP to enable both criteria dependencies and criteria weightings with respect to a goal to be modelled (see Table 6.7).

Conclusion

This chapter has introduced, outlined and demonstrated the use of three alternative multi-criteria decision-making methods: the analytical hierarchy process (AHP), the analytical network process (ANP) and the hierarchical analytical network process (HANP). Whilst each approach has its own advantages and disadvantages, all three of these methods are powerful tools that can be used by decision makers to mitigate risk in project planning. The AHP is a relatively simple method that enables subjective and objective information to be evaluated and can capture all aspects of a decision problem. It is this flexibility that has resulted in the AHP becoming one of the most widely adopted decision support tools and it has far reaching applications and uses across a wide range of sectors and disciplines. The ANP is an extension of the AHP to enable dependencies among different criteria that relate to the decision problem to be modelled. The ANP increases modelling complexity, however, the additional layers of analysis enable decision problems to be further evaluated. The HANP combines the attributes of the AHP and the ANP to further improve the reliability of results by performing extra evaluations.

Managing risk in engineering and construction projects is extremely critical as it helps to ensure successful and efficient operations. Surprisingly, key decisions that have to be made in industry are often not fully evaluated. Inappropriate decisions for site, capacity, supplier and technology selection are common problems which can result in unnecessary expense, future problems and even projects failing. With the growing size and complexity of engineering and construction projects, it is critical that decision makers fully evaluate problems to manage risk using systematic and scientific approaches like the AHP.

References

1 Saaty T.L. The analytic hierarchy process: Planning, priority setting, resource allocation. New York; London: McGraw-Hill International Book Co.; 1980.
2 Pohekar S.D., Ramachandran M. Application of multi-criteria decision making to sustainable energy planning–A review. Renewable and Sustainable Energy Reviews 2004; 8: 365.
3 Vaidya O.S., Kumar S. Analytic hierarchy process: An overview of applications. European Journal of Operational Research 2006; 169: 1.
4 Saaty T.L. What is the analytic hierarchy process?. New York: Springer; 1988.
5 Zayed T., Amer M., Pan J. Assessing risk and uncertainty inherent in Chinese highway projects using AHP. International Journal of Project Management 2008; 26: 408.
6 Gaudenzi B., Borghesi A. Managing risks in the supply chain using the AHP method. International Journal of Logistics Management 2006; 17: 114.
7 Dey P.K.. Managing project risk using combined analytic hierarchy process and risk map. *Applied Soft Computing* 2010; 10: 990.
8 Schoenherr T., Rao Tummala V.M., Harrison T.P. Assessing supply chain risks with the analytic hierarchy process: Providing decision support for the offshoring decision by a US manufacturing company. Journal of Purchasing and Supply Management 2008; 14: 100.
9 Saaty T. Fundamentals of the analytic network process – Dependence and feedback in decision-making with a single network. Journal of Systems Science and Systems Engineering 2004; 13: 129.
10 Momoh J.A., Zhu J. Optimal generation scheduling based on AHP/ANP. Systems, Man, and Cybernetics, Part B: Cybernetics, IEEE Transactions on 2003; 33: 531.
11 Yüksel İ.,I Dagdeviren M. Using the analytic network process (ANP) in a SWOT analysis – A case study for a textile firm. Information Sciences 2007; 177: 3364.
12 Liu K.F., Hsu C.-Y., Yeh K., Chen C.-W. Hierarchical analytic network process and its application in environmental impact evaluation. Civil Engineering and Environmental Systems 2011; 28: 1.
13 Nixon J.D., Dey P.K., Ghosh S.K., Davies P.A. Evaluation of options for energy recovery from municipal solid waste in India using the hierarchical analytical network process. Energy 2013; 59: 215–223.

7 Project risk management using fuzzy theory

Daniel Wright and Prasanta Kumar Dey

Project risk

Risk and uncertainty is an inherent and common characteristic of any project. The best possible way to mitigate the negative impact of an occurring project risk is to employ efficient and effective project risk management (PRM) practices. There are numerous benefits for adopting risk management techniques, with the overall goal being to ensure that project cost, schedule and safety are controlled. Despite the clear advantage, early work in the industry by Perry and Hayes (1985) highlighted that many projects, not just large capital projects, were still subject to cost and time overruns. A recent survey of construction companies and consultants in the UK still found that a large proportion still experience significant cost and time overruns despite having numerous control techniques at their disposal (Yakubu and Sun 2009).

The issue lies not with the PRM tools and techniques as the strengths and weaknesses of an approach are clearly known in advance of adoption, but likely reside with the decision-maker. Seminal work conducted by Akintoye and MacLeod (1997) highlighted that construction and project management practices, in their survey, overwhelmingly utilised intuition, judgement and experience over any other risk management technique. Moreover, the theory that not all risk management techniques are equally effective in increase in performance has been examined in a paper by Raz and Michael (2001). They surveyed software and high-tech Israeli project management companies to measure both the perceived value of PRM tools and a tools impact on overall performance. The findings showed that some of the 38 PRM tools analysed led to significantly greater risk management contribution and project management performance, tools such as risk impact analysis, risk classification, risk ranking, periodic document reviews and periodic trend reporting.

As there is not a truly objective way of analysing risk, and the subjectivity of risk means that this is unlikely to be possible in the near future, it is necessary to apply the most suitable risk management technique to a situation.

Project managers have long advocated decision-supporting models and methodologies to increase the reliability and accuracy of their analysis. However, a lack of familiarity and experience of the options available has greatly restricted the range of decision support model adoption, with the more 'complicated' techniques tending to be avoided (Akintoye and MacLeod 1997). As early as 1985, it was suggested that something needed to be done to increase the credibility of risk assessment techniques (Perry and Hayes 1985). This has also been echoed in later research with a call to explore new directions and more contemporary modelling techniques (Baloi and Price 2003). Of the contemporary new risk management methodologies, this chapter focuses on fuzzy logic: its application and advantages.

Fuzzy logic is fundamentally easy to understand, and its versatility has enabled it to be applied to a myriad of project risk management problems. As fuzzy logic is a contemporary technique, its application to the construction and project risk management industries has been concentrated on the past decade. Some of the early and most widely cited work in applying fuzzy logic to project management was by J. Tah and V. Carr (Tah, Thorpe et al. 1993; Tah and Carr 2000; Carr and Tah 2001). Their research 'enriched' earlier hierarchical risk breakdown structures, by integrating fuzzy linguistic variables to create a more formal reasoning process. Fuzzy risk models have been developed to assist in the reasoning and quality of contract form (Wong and So 1995), bid/no-bid (Lin and Chen 2004) and contractual risk allocation (Lam, Wang et al. 2007) decision-making. They have also been applied to project cash flow analysis (Boussabaine and Elhag 1999; Maravas and Pantouvakis 2012) and the CPM/PERT (Mon, Cheng et al. 1995; Shipley, de Korvin et al. 1997; Chanas and Zieliński 2002; Dubois, Fargier et al. 2003; Chen 2007). Furthermore, fuzzy logic has also been combined with other optimisation and decision-making methods; fuzzy logic with linear programming to suggest optimal corporate cash flow to potential projects (Lam, So et al. 2001), fuzzy multi-criteria decision making techniques (Zeng, An et al. 2007; Chen and Wang 2009) or with other artificially intelligent methods, such as adaptive fuzzy system (ANFIS) for construction risk assessment (Ebrat and Ghodsi 2011).

This chapter demystifies the application of fuzzy logic PRM by first reiterating the basics of the theory that are useful to practitioners and then applying the new method to an example case of bioenergy project risk management.

Risk management process

There are several British and international risk management standards that aim to support the project manager in the risk management process:

- BS (8444:1996) Risk Management: Guide to Risk Analysis of Technological Systems;
- ISO (31000:2009) and supporting ISO (31010:2009);
- HMGOV (2004) The Orange Book;
- IRM/AIRMIC/ALARM[1] (2002) Risk Management Standard; and
- PMBOK[2] (2009).

The process of risk management advocated by each standard largely remains the same, as shown in Table 7.1.

The standards reviewed in the table, have similar phases to risk management. The process to be taken forward throughout the paper is the ISO (31000:2009)

Table 7.1 PRM standards

Standard:	Risk definition:	Phases[1]:
BS 8444	Combination of the frequency, or probability, of occurrence and the consequence of a specified hazardous event[2]	Analysis: – scope def., – hazard identification, – risk estimation; Evaluation: – risk tolerability decisions – analysis of options Reduction/control: – decision making, – implementation, – monitoring
ISO 31000	Effect of uncertainty on objectives[3]	Identification Analysis Evaluation Treatment Monitor and review
HMGOV	Uncertainty of outcome	Identification Assess Address Review and reporting
PMBOK	Risk is an uncertain event or condition that, if it occurs, has an effect on at least one project objective[4]	Identification Analysis: – qualitative – quantitative Response Monitor and control
IRM/AIRMIC /ALARM[3]	The combination of probability of an event and its consequences[5]	Assessment: – Analysis: – identify, describe, estimate – Evaluation Reporting (threats and opportunities) Decision Treatment Residual risk reporting Monitoring

[1]Where present, context and communication phases have been removed; [2]Secondary cited from ISO/IEC Guide 51:1990; [3]Secondary cited from ISO/IEC Guide 73:2009; [4]Pg. 275; [5]Secondary cited from ISO/IEC Guide 73:2002.

method. It is worth noting that the supporting ISO (31010:2009) which covers the major risk assessment techniques does not refer to fuzzy or possibility theory as a method. However as similarities can be drawn with probability, which is widely cited, this should not be off-putting to utilising this technique.

Fuzzy logic and inferencing fundamentals

Fuzzy logic is a contemporary risk management method, but the underpinning theory has existed since Lotfi Zadeh's work in the 1960s. Fuzzy theory, sometimes referred to as possibility theory, is one of a restricted number of techniques well suited to risk analysis. This branch of modern mathematics has proven itself as an effective method of converting human perception and logic into intuitive control and decision-supporting tools. By mimicking human cognition and approximate reasoning with, often, partial information, this contemporary method represented a paradigm shift in the scientific community – a paradigm shift that they were initially not willing to accept. By utilising symbolic heuristic-based logic and moving away from the traditional numerical algorithmic logic, fuzzy logic is able to better accommodate particular problems. Zadeh (1973) argued that with this new approach " . . . we acquire a capability to deal with systems which are much too complex to be susceptible to analysis in conventional mathematical terms".

Crisp sets

A classical 'crisp' set is defined as a collection of objects or elements that belong to a group or 'set'. A good example of this is a simple two-set Venn diagram. At any point on the Venn diagram, an object or element $(x_1, x_2, x_3, \ldots, x_n)$ is a member of a universal set or discourse (X): $X = \{x_1, x_2, x_3, \ldots, x_n\}$ Furthermore, each object within the universal set belongs to either set A, set B or both sets.

For example, if an object, denoted as x is exclusively a member of set A, the following rule holds true: $A = \{x | x \in A\}$. It is also true that x does not belong to or is not a member of set B: $x \notin B$. An object that is exclusively a member of set B would be expressed in the same way, and an object at the intersection of A and B would either be $A \cap B$ or $A \cup B = \{x | x \in A \text{ or } x \in B$. Furthermore, the characteristic function tells one which objects in the universal set (X) are or are not a member of a set, for this example set A is shown:

$$X_A(x) = \begin{cases} 1 \text{ for } x \in A \\ 0 \text{ for } x \notin A \end{cases}$$

Where:

$X_A: X \rightarrow \{0, 1\}$

(Klir and Yuan 1995)

For each $x \in X$ it is possible to determine from the characteristic function whether it is or is not a member of a set, if $X_A(x) = 1$ it is a member and when $X_A(x) = 0$ it is not. In classic set theory this is dichotomous two-valued logic, as an object is only ever wholly a member of a set (1) or not a member of a set (0).

This orthodox approach to set theory is bound by logic and control, very clear lines that greatly differed from the 'shades of grey' applied in human cognition.

Fuzzy sets

Fuzzy sets differ from the orthodox approach as the characteristic function of a fuzzy set allows for a continuum of degrees of membership. Partial truths and information are rife in the 'real world' and crisp set theory was ill-designed for this problem. In fuzzy sets it is possible to accommodate these matters by allowing an object to be a member of a set to a matter of degree, denoted by μ. Where a fuzzy set A is a set of real numbers $\mathbb{R}$ is characterised by means of a membership level $\mu_A(x)$, $\mathbb{R} \rightarrow [0, 1]$. The extent to which an object within the universal set is a member of a particular set depends on the properties of the user defined membership function. Where a triangular fuzzy number $A = <a, b, c>$ has the following membership function:

$$A(x) = \begin{cases} 0, & x < a \\ \dfrac{x-a}{b-a} & a \leq x \leq b \\ \dfrac{c-x}{c-b}, & b \leq x \leq c \\ 0, & x > c \end{cases}$$

Fuzzy arithmetic (inc. α-cuts)

The extension principle (Zadeh 1965) underpins fuzzy theory by 'extending' the operations and definitions of ordinary 'crisp' mathematical concepts to fuzzy sets, of particular relevance to this paper is fuzzy arithmetic. This is often calculated with the use of α-cuts which are defined as a crisp set of elements belonging to a fuzzy set $\tilde{A}$ at least to the degree of α (Zimmerman 1990), and is denoted as:

$$^{\alpha}A = \{x \in X \mid \mu_{\tilde{A}}(x) \geq \alpha\}$$

Although all arithmetic operations are allowed for fuzzy sets, only the subtraction of sets is shown is utilised in this chapter. The subtraction of two fuzzy triangular sets A_1 and A_2 using the α-cut is given by:

$$A_1 - A_2 = \bigcup_{\alpha \in [0,1]} {}_\alpha (A_1 - A_2)$$

The functions for A_1 and A_2 are:

$$A_1(x) = \begin{cases} 0, & x < 5 \\ \dfrac{x-4}{1} & 4 \le x \le 5 \\ \dfrac{8-x}{3}, & 5 \le x \le 8 \\ 0, & x > 8 \end{cases}$$

$$A_2(x) = \begin{cases} 0, & x < 3 \\ x-3 & 3 \le x \le 4 \\ 5-x, & 4 \le x \le 5 \\ 0, & x > 5 \end{cases}$$

Giving their α-cuts as:

$${}^\alpha A_1(x) = [\alpha + 5, 8 - 3\alpha]$$

$${}^\alpha A_2(x) = [\alpha + 3, 5 - \alpha]$$

The solution is:

$${}^\alpha(A_1 - A_2) = [a, b] - [c, d] = [a - d, b - c] = [2\alpha, 5 - 2\alpha]$$ for $\alpha \in (0, 1]$.

As ${}^\alpha(A_1 - A_2)$ is a closed interval for each $\alpha \in (0, 1]$, $A_1 - A_2$ is also a fuzzy number (Klir and Yuan 1995) as shown in Figure 7.2.

Fuzzy inferencing (inc. risk associative model)

Fuzzy sets, although a paradigm shifting concept, only represent part of fuzzy logic and the human cognition process. Once one has accepted partial truths and shades of grey in the form of fuzzy sets, it is then necessary to mimic the way in which the human cognition process utilises intuition and heuristics to translate sensory inputs into useful outputs. This is often referred to as the fuzzy inferencing process and it requires three main components: fuzzy functions, fuzzy rules and logical operators to be configured. This is demonstrated with the Mamdani fuzzy method.

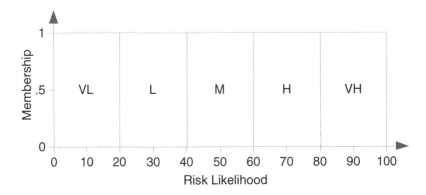

Figure 7.1 Crisp risk likelihood sets

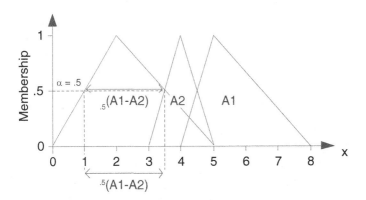

Figure 7.2 Fuzzy subtraction

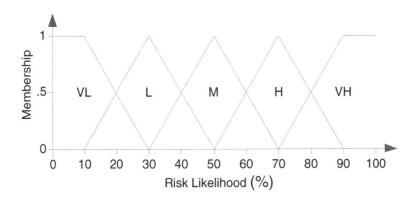

Figure 7.3 Fuzzy risk likelihood functions

The fuzzy functions in the figure are markedly different from the previous crisp functions (Figure 7.1). For this example, triangular fuzzy functions have been utilised to categorise each linguistic term. However, there are many possible linear and non-linear fuzzy functions that can be user defined to categorise any possible number of functions within a universal set or discourse.

Fuzzy functions commonly adopt linguistic terms to categorise characterise its membership in the universal discourse of 'risk likelihood'. By adopting approximate reasoning and taking a step back from the quest for absolute precision, Zadeh (1975) stated that it was possible to adopt linguistic variables "whose values are not numbers but words or sentences in a natural or artificial language". Linguistic variables are fully defined by a quintuple (v, T, X, g, m) where:

> v *is the variable name*
>
> T *is the set of linguistic terms of v*
>
> X *is the universal discourse*
>
> g *is a syntactic rule (a grammar) for generating linguistic terms*
>
> m *is semantic rule that assigns to each linguistics term* $t \in T$ *its meaning*
>
> (Klir and Yuan 1995)

Each linguistic variable (v) contains a given number of linguistic terms (T) generated by the syntactic rule (g), this is usually context-free grammar, such as: very low, low, moderate etc. The values of the linguistic terms (T) vary over the universal discourse (X). $T(v)$ defines the term set of the variable x, for instance:

T (*risk likelihood*) = {*very low, low, moderate, high, very high*}

The level of meaning or membership given to each linguistic term is determined by the semantic rule (m). These characteristics can be also displayed graphically:

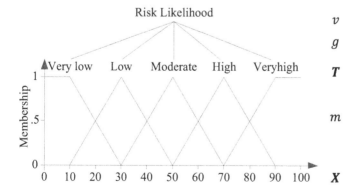

Figure 7.4 Linguistic variable definition

The base variable X is the percentage of risk likelihood. $m(t)$ is the rule that assigns a fuzzy set to each of the terms $t \in T$, for example:

$$m(low) = \{(x, \mu_{low}(x) | x \in [0, 100\%]\}$$

Analysis of risk events usually requires the evaluation of the likelihood of the event occurring as well as the severity or overall impact of the risk. A similar linguistic variable for risk severity could be generated but at present there is no way in which these two inputs can be brought together to infer an action to be taken by the decision-maker; this is where fuzzy inferences and heuristic rules are used.

Fuzzy rules are conditional *IF-THEN* statements that act as the brain of the fuzzy system by linking the model inputs to a useful output. As with the fuzzy functions, the fuzzy rules are designed to mimic the human cognition process: *IF* (condition) *THEN* (consequence), e.g. *IF risk severity is very high THEN reduce risk*. However, as Zadeh (1973) noted, this is poorly defined when used in communication between humans, something that needed to be remedied when mimicking this process. These heuristic rules are referred to as fuzzy rules, as the condition and consequence also happen to occur to a matter of degree, the extent to which a rule is activated or its strength depends on the formulated fuzzy membership functions. For longer inferencing chains, comprising of more than one condition v, it is necessary to define the operators (also known as Zadeh operators) that link the conditions together. Inference chains can contain as many conditions as required for the problem, although the larger the number of input conditions, the larger the number of heuristic rules required to fully map the model T'. At each point along the universal discourse, the number of active inference rules is S^v where S is the number of functions activated at a single point (Lam, Wang et al. 2007). Furthermore, a benefit of fuzzy inferencing is that not all rules are required for the model to function, but a positive relationship exists between inference numbers and model quality (Wong and So 1995).

As a risk associative model (RAM) is desired, it would be logical to produce a range of fuzzy functions for the most suitable risk action given the likelihood and severity of an event, where likelihood and severity are the conditions and risk response action is the consequence. These operators, similar to their Boolean ancestors, take the form of *AND, OR and NOT*, an example could be: *IF risk likelihood is high AND risk severity is very high THEN avoid risk*. As with the other parts of the fuzzy logic system, these operators also conform to the fuzzy degrees of truth or membership. So for any A and B fuzzy sets, the logical operators would be calculated as:

Negation $(not\ A) = \mu_{\neg A}(x) = 1 - \mu_A(x)$

Conjunction $(x\ AND\ y) = \mu_{x \wedge y}(x) = \min\{\mu_A(x), \mu_B(x)\}$

Disjunction $(x\ OR\ y) = \mu_{x \vee y}(x) = \max\{\mu_A(x), \mu_B(x)\}$

The min/max norm (shown above) fuzzy operators are the most commonly used, but there are a number of operator types to suit a fuzzy model's purpose. The final phase in the fuzzy logic inferencing process is to convert the fuzzy inferences into a useful output. This is often referred to as the 'defuzzification' stage.

Fuzzy defuzzification

The final stage is to defuzzify the triggered inferences, given the condition inputs. There are several methods for achieving this, but the most commonly employed is the centre of gravity approach. This can be approximately calculated without the requirement for integrals as:

$$y = \sum_{j=1}^{r} \mu_j . s_j \ / \ \sum_{j=1}^{r} \mu_j$$

Where:

S_j is the centre of gravity for the jth fuzzy set.

Fuzzy PRM methodology

The proposed fuzzy PRM methodology supports the decision-maker in the risk analysis, evaluation and treatment phases of the risk management. The identification and monitoring phases of the traditional approach remain unchanged. The proposed methodology is as follows:

1 *Define Fuzzy Inferences (Risk Associative Model)*
This the first stage of the process is to define the likelihood and severity functions as well as the RAM and the response output function for evaluating risks in the context of the project.

2 *Risk Identification*
The risk identification phase remains unchanged, risks are generated using the traditional methods: diagrammatic tools, checklists, and brainstorming etc.

3 *Risk Analysis*
Once the risks have been identified, it is necessary to define the risk with a fuzzy set and then to evaluate each risk with the fuzzy inferences from phase 1.

4 *Risk Treatment/Response*
The final of the covered phases is to apply a treatment or response to the risk event.

Numerical example: bioenergy project risk

An ever-increasing focus on the importance of low carbon and secure energy supply, the renewable energy industry has grown rapidly over the last decade. The European Union's legally binding 2020 targets require the UK to generate 15% of total energy consumption from renewable sources. In 2005, the UK generated only 1.3% of final energy consumption from renewable sources, meaning that the UK was largest percentage point increase of all the Member States (Lords 2008). Recent figures show that in the second quarter of 2011, the UK renewable energy sector generated 9.6% of total electricity demand – a 50% increase on the same period in 2010 (DECC 2011). However, 47% of the country's final energy consumption is for heating and only 1.5% is currently generated from renewable sources (DECC 2011).

Biomass energy (bioenergy) is vital to achieving the UK future renewable energy targets. As bioenergy is well suited to both small and large scales, able to provide a non-intermittent load of heat and power supply, and possible to be located next to heat and power demand sources, it is versatile. Yet, research has found that a magnitude of barriers, specific to bioenergy projects means that it is more difficult to deploy this technology type over the majority of renewable energy technologies. The barriers identified range from: regulatory (Costello and Finnell 1998) and policy issues (Hamilton 2009); financial (McCormick and Kaberger 2007); feedstock processing and logistics (Iakovou, Karagiannidis et al. 2010); and a wide range of barriers held by suppliers, developers and end-users (Adams, Hammond et al. 2011). These barriers have greatly impeded industry growth in this country.

Fuzzy PRM process

Example case

The example case is the development of a 1MWe biomass combined heat and power (bCHP) facility located in the West Midlands. The combustion plant burns virgin wood chip, with the produced electricity exported to the grid, and a proportion of the heat sold locally to a mixture of residential and commercial consumers. The project is financed from a combination of debt and equity sources.

Fuzzy inferences

The fuzzy inputs for the risk assessment model are the likelihood and the severity of a risk event. The output is the most appropriate action to be taken given the risk event's inputs. The likelihood linguistic set functions are the same as shown in Figure 7.4. However, the severity linguistic set functions differ as well as the output, as shown in Figures 7.5 and 7.6 respectively.

As the severity of the risk event is a more important factor in impeding the project's objectives, the linguistic terms increase more quickly than the likelihood terms over their respective universal discourse. The risk response discourse ranges from 0 to 6 with integers 1 to 5 representing the centre of one of five linguistic terms:

– **Retain (Ret)**

A suitable risk response for low likelihood and severity risks as the risk event is unlikely to impede the project objectives and there are likely to be numerous risks that fall into this category.

– **Monitor & Review (M&R)**

For risk events that could possibly lead to significant disruptions for the project, the most suitable strategy is to monitor and review the risk to

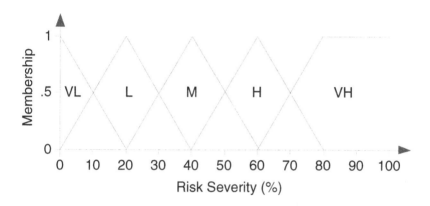

Figure 7.5 Fuzzy severity functions

Figure 7.6 Fuzzy response functions

ensure that it is controlled or re-evaluated if circumstances differ or more information is acquired.

– *Reduce (R)*

The risk event would lead to significant disruptions and proactive response to mitigate its effect before or upon occurrence is required.

– *Greatly Reduce (GR)*

For risk events with high impact on the project's objectives it is necessary to greatly reduce its effect. Similar to the previous category, mitigation strategies are required but to a greater extent as the threat is greater.

– *Avoid (A)*

Risk events with the highest levels or likelihood and severity require the most drastic action to avoid the occurrence of the risk as it is possible that this could lead to termination of the project. Risk reduction strategies are most likely insufficient to handle such critical risks or the cost of such measures would be inefficient or too costly.

The fuzzy associative matrix (Table 7.2) shows the heuristic rules required ($T^v = 5^2 = 25$) to fully map the possible inputs for the inferencing process.

At each condition input selected along the universal discourses of likelihood and severity, there are $S^v = 2^2 = 4$ active rules to inference. The RAM rules activated are true to a matter of degree along the universal discourse, this produces a surface of possible consequences for risk response actions (Figure 7.7).

Risk identification

The fuzzy process does not enhance the risk identification process so it would be necessary to generate the risk identification register by the usual

Table 7.2 Risk associative matrix

Conjunction AND		Severity				
		Very Low	*Low*	*Moderate*	*High*	*Very High*
Likelihood	**Very Low**	Retain	Retain	Monitor & Review	Reduce	Reduce
	Low	Retain	Monitor & Review	Monitor & Review	Reduce	Greatly Reduce
	Moderate	Monitor & Review	Monitor & Review	Reduce	Greatly Reduce	Avoid
	High	Reduce	Reduce	Greatly Reduce	Greatly Reduce	Avoid
	Very High	Reduce	Greatly Reduce	Greatly Reduce	Avoid	Avoid

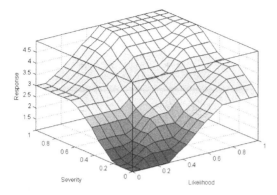

Figure 7.7 Inferencing Model Surface Map

means. Table 7.3 contains ten example risks relating to the development of a bioenergy facility in the UK.

Risk effects are a combination of time, cost and quality. Some risk events have multiple effects, such as the first and second risk event in the table. The changes to plant design would have direct time and cost impacts. Whereas any possible quality issues with the feedstock (7) could have a number of indirect consequences, such as blockage of feedstock feeding mechanisms and possible sub-optimal combustion in the boiler. These could have direct time and cost implications requiring plant shut-down and maintenance, but it may also qffect the operation of the plant over time which is harder

Table 7.3 Risk identification

No.	Description	Effect	Like.	Sev.	Impact (£,000)/days		
					a	b	c
1	Design changes to plant	Cost	60%	60%	5	30	90
2	Design changes to plant	Time	60%	60%	5	30	90
3	Feedstock price increase	Cost	45%	75%	0	15	40
4	Development cost increase	Cost	80%	40%	0	20	50
5	Maintenance cost increase	Cost	55%	20%	0	5	10
6	Grid connection delay	Time	30%	85%	0	60	120
7	Feedstock quality	Quality	60%	50%	–	–	–
8	Road widening cost increase	Cost	25%	20%	0	10	20
9	Breaching emissions regulations	Cost	50%	85%	–	–	–
10	Fall in gas prices makes required heat price uneconomical	–	35%	90%	–	–	–

Table 7.4 Fuzzy inferencing response table

	Likelihood		Severity		Inference Rules				Output					Score/Action
	$L\mu$	$U\mu$	$L\mu$	$U\mu$	1–3	1–4	2–3	2–4	1	2	3	4	5	
1	(M) .5	(H) .5	(M) 0	(H) 1	(R) 0	(GR) .5	(GR) 0	(GR) .5			0	.5		4 GR
2	(M) .5	(H) .5	(M) 0	(H) 1	(R) 0	(GR) .5	(GR) 0	(GR) .5			0	.5		4 GR
3	(L) .25	(M) .75	(H) .25	(VH) .75	(R) .25	(GR) .25	(GR) .75	(A) .75			.25	.25	.75	4.4 GR
4	(H) .5	(VH) .5	(L) 0	(M) 1	(R) 0	(GR) .5	(GR) .0	(GR) .5			0	.5		4 GR
5	(M) .75	(H) .25	(VL) 0	(L) 1	(MR) 0	(MR) .75	(R) 0	(R) .25		.75	0			2 M&R
6	(VL) 0	(L) 1	(H) 0	(VH) 1	(R) 0	(R) 0	(R) 0	(GR) 1			0	1		4 GR
7	(M) .5	(H) .5	(M) .5	(H) .5	(R) .5	(GR) .5	(GR) .5	(GR) .5			.5	.5		3.5 GR
8	(VL) .25	(L) .75	(VL) 0	(L) 1	(Ret) 0	(Ret) .25	(Ret) 0	(MR) .75	0	.75				2 M&R
9	(L) 0	(M) 1	(H) 0	(VH) 1	(R) 0	(GR) .0	(GR) .0	(A) 1			0	0	1	5 A
10	(L) .75	(M) .25	(H) 0	(VH) 1	(R) 0	(GR) .0	(GR) .0	(A) .25			0	0	.25	5 A

to quantify. Risk event 9, breaching emission regulations under the appropriate act for small-scale biomass power stations (Clean Air Act) could go unnoticed if for a short period, but if the Local Authority becomes aware of this, a fine or shutdown of the plant may be required until the problem is resolved. The final risk, risk 10, concerns a drop in gas prices, leading to the minimum price required by the bCHP no longer being competitive with gas equivalent. Consumers would not choose the more expensive option, meaning that the plant is possibly uneconomical to continue operating. The last two risks in particular are very severe and the implications of these risk events could lead to the end of the project.

Risk analysis and response

The application of the fuzzy inferencing process, utilising the RAM methodology, yields the following results:

The inferencing method is able to accommodate qualitative as well as quantitative risk events. It also clearly states the recommended risk response action given the likelihood and severity of an occurring risk. However, it is then necessary for the decision-maker to formulate suitable risk response actions from the suggest measure. Particularly in the case of risk 10, it may be difficult to fully avoid the risk event as it is exogenous to the project's control and cannot simply be engineered out of the process.

Quantitative response

Where it is possible to quantify the impact of the risk event on occurrence, fuzzy set arithmetic can be used to illustrate the effectiveness of risk response actions.

The quantifiable risk events in Table 7.5 have a risk response with a fuzzy effectiveness range. By applying the extension and α-cut techniques, it is possible to subtract the effectiveness from the inherent risk event to give an improved residual risk event impact.

Conclusions

As demonstrated, the fuzzy PRM methodology can support the decision-maker in handling decisions regarding possible quantitative and qualitative risk events in a unique and potentially powerful way. The method is conceptually easy to understand and apply. A fuzzy inferencing model can be tailored to any PRM decisions, but is most suited to unstructured decision processes. Heuristic rules can be developed to ensure that the risk management process is methodical and potentially less subjective than an individual decision-maker's judgement alone. However, as stated, linguistic inputs grow in size as do the number of rules to fully map all inference chains. This does require expert input and the inferencing process is only as good as the rules.

Table 7.5 Fuzzy response table

No.	Description	Effect	Impact (£,000)/ days			Response	Effectiveness (%)			Residual (£,000)/days		
			a	b	c		a	b	c	a	b	c
1	Design changes to plant	Cost	5	30	90	Contractual clause/restrict	70	90	95	*0.25*	*3*	*27*
2	Design changes to plant	Time	5	30	90	Contractual clause/restrict	60	90	95	*0.25*	*3*	*27*
3	Feedstock price increase	Cost	0	15	40	Supplier contractual clause	75	80	80	*0*	*3*	*10*
4	Development cost increase	Cost	*0*	*20*	*50*	Enforce budgets and controls	*40*	*65*	*75*	*0*	*7*	*30*
6	Grid connection delay	*Time*	0	60	120	Ensure compliance	10	30	35	*0*	*42*	*108*

The quantitative response to risk events is particularly suited to risk forecasting with limited information. It provides a useful output where there is uncertainty or vagueness present. It is also well placed to produce risk management information that is better than 'worst case' scenario alone. Traditional sensitivity analysis and other financial metrics can be infused with fuzzy sets to incorporate uncertainty in forecasting, without having to rely on stochastic simulation. However, there are some weaknesses to the approach. The use of triangular $A = <a, b, c>$ fuzzy distributions or similar can be overly simplistic if there is sufficient information to assemble probability distributions for risk event impact or interdependence between risk events, this would be better suited to stochastic methods.

Notes

1 Institute of Risk Management (IRM); The Association of Insurance and Risk Managers (AIRMIC); The National Forum for Risk Management in the Public Sector (ALARM).
2 Project Management Institute – Project Management Body of Knowledge.

References

Adams, P. W., G. P. Hammond, et al. (2011). "Barriers to and drivers for UK bioenergy development." *Renewable and Sustainable Energy Reviews* 15(2): 1217–1227.
Akintoye, A. S. and M. J. MacLeod (1997). "Risk analysis and management in construction." *International Journal of Project Management* 15(1): 31–38.

Baloi, D. and A. D. F. Price (2003). "Modelling global risk factors affecting construction cost performance." *International Journal of Project Management* 21(4): 261–269.

Boussabaine, A. H. and T. Elhag (1999). "Applying fuzzy techniques to cash flow analysis." *Construction Management and Economics* 17(6): 745–755.

BS (8444:1996). Risk management – Guide to Risk Analysis of Technological Systems.

Carr, V. and J. H. M. Tah (2001). "A fuzzy approach to construction project risk assessment and analysis: construction project risk management system." *Advances in Engineering Software* 32(10–11): 847–857.

Chanas, S. and P. Zieliński (2002). "The computational complexity of the criticality problems in a network with interval activity times." *European Journal of Operational Research* 136(3): 541–550.

Chen, P. and J. Wang (2009). *Application of a Fuzzy AHP Method to Risk Assessment of International Construction Projects*. Electronic Commerce and Business Intelligence, 2009. ECBI 2009. International Conference on.

Chen, S.-P. (2007). "Analysis of critical paths in a project network with fuzzy activity times." *European Journal of Operational Research* 183(1): 442–459.

Costello, R. and J. Finnell (1998). "Institutional opportunities and constraints to biomass development." *Biomass and Bioenergy* 15(3): 201–204.

DECC (2011). Renewable Heat Incentive. D. o. E. a. C. Change. London, Department of Energy and Climate Change.

DECC (2011). Statistical Press Release. D. o. E. a. C. Change. London, Department of Energy and Climate Change.

Dubois, D., H. Fargier, et al. (2003). "On latest starting times and floats in activity networks with ill-known durations." *European Journal of Operational Research* 147(2): 266–280.

Ebrat, M. and R. Ghodsi (2011). "Risk assessment of construction projects using network based adaptive fuzzy systems." *International Journal of Academic Research* 3(1): 411–417.

Hamilton, K. (2009). *Unlocking Finance for Clean Energy: The Need for 'Investment' Grade Policy*. London, Chatham House.

HMGOV (2004). *The Orange Book: Management of Risk – Principles and Concepts*. London.

Iakovou, E., A. Karagiannidis, et al. (2010). "Waste biomass-to-energy supply chain management: A critical synthesis." *Waste Management* 30(10): 1860–1870.

IRM/AIRMIC/ALARM (2002). A Risk Management Standard. London, Institute of Risk Management (IRM); The Association of Insurance and Risk Managers (AIRMIC); The National Forum for Risk Management in the Public Sector (ALARM).

ISO (31000:2009). Risk management – Principles and guidelines.

ISO (31010:2009). Risk management – Risk assessment techniques.

Klir, G. J. and B. Yuan (1995). *Fuzzy Sets and Fuzzy Logic*. Upper Saddle River, Prentice Hall Inc.

Lam, K. C., A. T. P. So, et al. (2001). "An integration of the fuzzy reasoning technique and the fuzzy optimization method in construction project management decision-making." *Construction Management and Economics* 19(1): 63–76.

Lam, K. C., D. Wang, et al. (2007). "Modelling risk allocation decision in construction contracts." *International Journal of Project Management* 25(5): 485–493.

Lin, C.-T. and Y.-T. Chen (2004). "Bid/no-bid decision-making – a fuzzy linguistic approach." *International Journal of Project Management* 22(7): 585–593.

Lords, H. o. (2008). *The EU's Target for Renewable Energy: 20% by 2020. H. o. Lords*. London, Authority of the House of Lords.

Maravas, A. and J.-P. Pantouvakis (2012). "Project cash flow analysis in the presence of uncertainty in activity duration and cost." *International Journal of Project Management* 30(3): 374–384.

McCormick, K. and T. Kaberger (2007). "Key barriers for bioenergy in Europe: Economic conditions, know-how and institutional capacity, and supply chain coordination." *Biomass and Bioenergy* 31(7): 443–452.

Mon, D.-L., C.-H. Cheng, et al. (1995). "Application of fuzzy distributions on project management." *Fuzzy Sets and Systems* 73(2): 227–234.

Perry, J. G. and R. W. Hayes (1985). "Risk and its management in construction projects." *Proceedings of Institution of Civil Engineers* **78**(Part 1): 499–521.

PMBOK (2009). Newton Square, Project Management Institute.

Raz, T. and E. Michael (2001). "Use and benefits of tools for project risk management." *International Journal of Project Management* 19(1): 9–17.

Shipley, M. F., A. de Korvin, et al. (1997). "BIFPET methodology versus PERT in project management: Fuzzy probability instead of the beta distribution." *Journal of Engineering and Technology Management* 14(1): 49–65.

Tah, J. H. M. and V. Carr (2000). "A proposal for construction project risk assessment using fuzzy logic." *Construction Management and Economics* 18(4): 491–500.

Tah, J. H. M., A. Thorpe, et al. (1993). "Contractor project risks contingency allocation using linguistic approximation." *Computing Systems in Engineering* 4(2–3): 281–293.

Wong, K. C. and A. T. P. So (1995). "A fuzzy expert system for contract decision making." *Construction Management & Economics* 13(2): 95.

Yakubu, A. O. and M. Sun (2009). "Cost time control construction projects: A survey of contractors and consultants in the United Kingdom (UK)." *Construction Information Quarterly* 11(2): 52–59.

Zadeh, L. A. (1965). "Fuzzy sets." *Information and Control* 8(3): 338–353.

Zadeh, L. A. (1973). "Outline of a new approach to the analysis of complex systems and decision processes." *Systems, Man and Cybernetics, IEEE Transactions on* **SMC-3**(1): 28–44.

Zadeh, L. A. (1975). "The concept of a linguistic variable and its application to approximate reasoning." *Information Sciences* 8(3): 199–249.

Zeng, J., M. An, et al. (2007). "Application of a fuzzy based decision making methodology to construction project risk assessment." *International Journal of Project Management* 25(6): 589–600.

Zimmerman, H.-J. (1990). *Fuzzy Set Theory and Its Applications.* Dordrecht, Kluwer Academic Publishers.

8 Development of influence diagrams for assessing risks

Djoen San Santoso

Introduction

Risk has been part of daily life activities in almost every sector regardless whether its existence is considered in the activity or not. The construction industry with its complexity and the involvement of many stakeholders is subject to more risk and uncertainty than other industries (Flanagan and Norman 1993). Managing risks effectively and systematically in a construction project has become one of the important factors that contributes to the success of a project. Nevertheless, Laryea (2008) asserted that risk in the construction industry is poorly analyzed in comparison to other industries such as finance or insurance. A focus on and improvement in risk assessment is therefore imperative to shift the perception on risks from threats to project success to opportunities to enhance the chances of project success (Project Management Institute, 2004).

It is well understood that not every risk involved in a project can all the time be anticipated. However, it is essential that any possible risks, as many as possible including improbable yet potential risks that may occur in the project, are identified (Papageorge, 1988). Identification of risks is the initiation process before moving to the assessment stage and this risk identification is typically not a big issue. When the process moves to assessment, traditionally, quantitative risk assessment has been widely used (Tah and Carr 2001). The most popular way of quantifying risk is based on the frequency or probability that the risk may occur in the project and the impact of that risk to the project. Subjective valuation for these two measurements is usually used in this assessment, which depends on the experience of the engineer in the construction industry, the experience on the specific project that is under consideration, the experience in facing and managing the risk in question and many other elements. This has generated a lot of debate on the reliability and objectivity of the assessment, which has made risk assessment a controversial issue (Baloi and Price 2003). Many approaches have been developed including methods to reduce subjectivity in the assessment, such as Fuzzy Sets Theory and Analytical Hierarchy Process. There is not yet common ground to decide which approach or method is the best one.

Taroun (2014) from his review of research on construction risk in the last five decades indicated that a comprehensive assessment approach to capture risk impact on different project objectives has to be improved by integrating professional experience in the process to bridge the gap between the theory and practice of risk assessment. He identified that interdependencies between project risks are one of the main factors that need to be improved. In developing interdependencies between risks, experience in construction projects is necessary to understand the interaction of one risk with other risks: how one risk affects other risks and is influenced by others. Influence diagrams can be considered as the appropriate method to model the interaction and interdependencies between risks that are involved in a project, where the cause-and-effect flows can be identified and traced to monitor the sources and consequences. This chapter focuses on the development process of influence diagrams by providing step-by-step guidance to build a comprehensive network covering all risks and factors in assessing risks for better management of construction projects.

Influence diagrams in risk assessment

The influence diagram has been a part of early development in modeling and assessing project risks. Methods using probability theory, Monte Carlo Simulation and the fuzzy approach have been used earlier in assessing project risks. While these quantitative approaches provide numbers as their results in analyzing the interactions of risks, there is a need to have a graphical illustration explaining the relationship between risks and the evolvement of risks to consequences that can be understood without difficulty by general audiences. With this consideration, the influence diagram was introduced to help display and explain the logic of the quantitative assessment of risks.

One of the earliest research studies that considered the application of influence diagram was conducted by Al-Bahar (1988, as cited in Al-Bahar and Crandall 1990). He considered the influence diagram as one of the advanced methods in analyzing risks, in which the interaction between risks can be easily observed through a graphical representation. This is the main advantage of applying the method in analyzing risks. He used the influence diagram together with Monte Carlo Simulation as tools to assess project risks in developing systematic risk management approach. Niwa (1989) used influence diagrams to explain the risk mechanism models that he generated by analyzing cases from large construction projects, where knowledge-based systems were also developed and implemented to help manage project risks.

The popularity of influence diagrams in assessing risks is still growing. The approach was used in integration with other quantitative methods to model the interactions and interrelationships of risks for the quantification process. On the qualitative method, influence diagrams can be developed thoroughly and holistically from major or critical risks identified from the project or case. Comprehensive discussion can follow where detailed aspects of risks can be detected. With the usefulness and easy-to-understand approach, the method is still used in improving the assessment and management of risks even after a decade from the inception of the use of the method. The method was used in the assessment of risks in high-rise building construction in Indonesia (Santoso, et al. 2003). Poh and Tah (2006) developed influence diagrams to help analyze the interdependencies between duration and cost parameters in modelling risk impacts on construction tasks. The influence diagram was used by Dikmen et al. (2007) to construct a risk model, which was then integrated with the fuzzy approach to estimate a cost overrun risk rating for international construction projects. Santoso et al. (2012) also used influence diagrams to understand and explain the interaction and interrelationship of investors' risks in public-private partnerships for tollway construction and operation. Recently, in 2014, the influence diagram was used for mapping risk using a knowledge-based approach where vulnerability was integrated in the assessment process for cost estimation of international construction projects (Yildiz et al. 2014).

Framework in developing influence diagrams

Influence diagrams have been used in assessing risks for the last 20 years. However, the question still remains how to properly develop a solid influence diagram as previous studies mostly, if not all, never provide clear guidance in their development. This happens maybe with the assumption that readers are expected to know how to develop an influence diagram, which may not be the case. With this consideration, a framework is developed to provide step-by-step guidance in constructing an influence diagram as shown in Figure 8.1. Each step is discussed in detail after this and a case study is provided to help understand the application of the framework in a research study in the next session.

Step 1: identify risk events related to the project

This process requires investigation and collection of risk events related to the project from literature review and experience from construction sites. Initially, risk events can be obtained from research studies already conducted in the context of the project under study. It is recommended to group the risk events by the source of risk or the party responsible for the risks. This will help to provide an organized and systematic list of risk events. Experienced researchers may add more events to the list.

However, collection of risk at this stage may not be sufficient. As any sound research study, discussions with experts who are knowledgeable

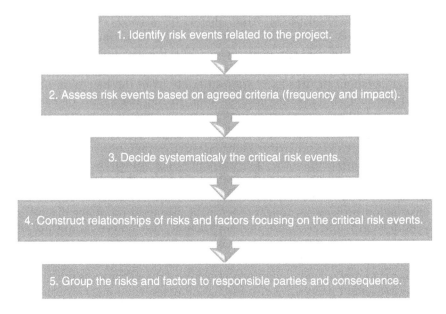

Figure 8.1 Framework for developing influence diagram

about or have direct experience with the nature of the project and are familiar with the context of the project (location specific) are essential to validate the list. A developing country may have its own unique risk events that exist only in the developing country. At the same time, not all risk events from a developed country are applicable or exist in a developing country. Each developing country may also have its own unique risk events due to its specific environment or political situation or other reasons. Therefore, suggestions from experts can help making sure that most, if not all, important risk events related to the project under study have been covered in the list.

Step 2: assess risk events

Typically, assessment of risk events is based on two aspects: frequency or probability of occurrence and degree of impact. These two aspects are well recognized and accepted by many practitioners and researchers (Santoso et al. 2012; Roumboutsos and Anagnostopoulos 2008; Santoso et al. 2003; Shen et al. 2001, Kumaraswamy 1997; Chapman and Ward 1997; Zhi 1995; Edwards 1995; Al-Bahar and Crandall 1990). The Likert scale can be used for assessing these two measurements in the questionnaire. Identification of suitable respondents (position, years of experience, company type, etc.) eligible for the survey should be done before conducting the survey. Respondents will be asked to give their assessment for each risk event on the provided scale. After that, the mean value of each measurement can be computed for each risk event by averaging the measurement based on the number of respondents.

Step 3: decide systematically the critical risk events

Mean values for frequency of occurrence and degree of impact can be used as references in deciding the critical risk events. Two approaches can be used to systematically decide the critical risk events. The first one is by using an index that takes into account both measurements. A risk index (RI) can be calculated based on the two measurement values by multiplying the mean values of the occurrence of risk with the mean value of the impact of risk.

As the assessment is typically collected from many respondents, it is important to understand the proper way of calculating the risk index. To avoid underestimation of the value of risk index, Tsu (2002) suggested that the average values of occurrence and impact for each risk event should be counted first from all respondents before multiplying the both values. Therefore, the risk index should not be calculated by averaging risk index counted from each respondent.

RI = Mean of occurrence x Mean of impact

$$RI^i = \left(\sum_{j=1}^{n} \frac{\alpha_j^i}{n} \right) \left(\sum_{j=1}^{n} \frac{\beta_j^i}{n} \right)$$

RI^i is the risk index for risk i.
n is the number of respondents.
α_j^i is the probability of occurrence of risk i, assessed by respondent j.
β_j^i is the degree of impact of risk i, assessed by respondent j.

In selecting the critical risk events, the easiest and most convenient way is to take the top ten risk events. However, this approach does not have strong rationalization to defend the selections. Other risk events that significantly affect the project may not only be in the first top ten and, in this case, they will be excluded from further analysis. It is always good to have a specific value as a reference in defining which risk events should be considered as the critical risk events. Assuming that the Likert scale is used in providing options of answers for respondents in assessing the risk events, this cut-off value will depend on the scale. A five-point Likert scale used in the questionnaire will have a different cut-off value than a seven-point Likert scale used as options of answer in assessing the risk event. Justification in defining this cut-off value should be provided as well.

For this approach, it is important that both assessments use the same number of points of Likert scale in the options for answers so the multiplication of the measurement values can have the same logic as the measurements. Therefore, if degree of impact uses a five-point Likert scale in the measurement, frequency of occurrence also needs to have a five-point Likert scale in the measurement.

Another approach in selecting the critical risk events is by using cross-tabulation considering probability or frequency of occurrence and degree of impact. In this approach, the occurrence of risk can be categorized into three levels, i.e. seldom, sometimes, and often. The range of mean values for each category should be logically defined by considering the scale used in the questionnaire. Likewise, the consequence of risk to project can be categorized into low, medium and high. Again, the range of mean values for each category need to be defined for allocating risk events into proper category. The categories of occurrence and impact are then tabulated to x-axis and y-axis as shown in Figure 8.2. Critical risk events can be defined using this tabulation.

As shown in Figure 8.2, critical risk events are those events that fall within the CR area, which are risk events that have at least medium level for both frequency and impact. Other risk events outside of this area will not be considered for further analysis. One may decide not to consider risk events that sometimes occur with medium impact as critical. This should be decided based on project characteristics and environment. The decision is all yours with proper justification.

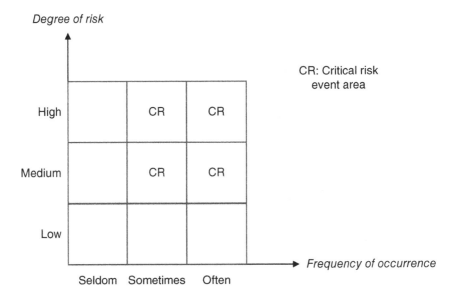

Figure 8.2 Cross tabulation in selecting critical risk events

*Step 4: construct relationships of risks and factors focusing on
the critical risk events*

In this step, the influence diagram is developed by considering the critical
risk events as the foundation. Among the critical risk events, a few of them
are usually linked to each other directly or indirectly through other risk
events or factors. Important factors that contribute to the generation of
the risk events are necessary to be included in the diagram so the overall
view of the mechanism can be generated. The relationship can be drawn
by connecting the two risk events or factors to explain the cause-and-effect
relationship. Typically, in this process, one or two main risk events are the
main risk events that other risks or factors are directly or indirectly linked
to this/these risk(s). There is no fixed number to define how many influence diagrams should be developed in this process. The important part
is that all critical risk events should be explained in the risk mechanism
diagrams to capture and understand their cause-and-effect relationships. It
is recommended to have consultations with experts in the field to develop
solid risk mechanism diagrams. Certain diagrams may have complex and
many factors involved in the mechanism, while other diagrams may look
simple with only one risk event as the focus. This is fine as long as experts
have been consulted to review the diagrams. The complexity of the network developed for the influence diagram should also consider whether
the quantitative method is a part of the assessment or not. If so, risks or
factors included in the diagram should be able to be measured for the computation process. Otherwise, the quantitative method could not fully cover
all aspects reflected in the qualitative method (influence diagram).

Step 5: group the risks and factors to responsible parties and consequences

Finally, the risk events and factors in the influence diagram can be grouped
according to those responsible for the risks and which risks or factors are
the consequences. This will help to provide a clear picture on how the
risks should be properly managed. In this regard, the parties involved in
the project can be aware of their responsibilities, and which risks they are
responsible for. Therefore, suitable actions can then be prepared to anticipate the occurrence of the risks by the concerned party.

As has been previously explained, the influence diagram method can
be used in integration with a quantitative method to model the interactions and interrelationships of risks and factors before the computation
process. With the quantification aspect as a part of the assessment, it can
be expected that the risks and factors included in the influence diagram
are those that are measurable. This can be a minor drawback in developing a comprehensive influence diagram. From a different perspective,
when an influence diagram is developed more to provide a qualitative
explanation or discussion on critical risk events, more details and a

comprehensive network can be developed by involving more aspects in constructing interactions and interrelationships for the model without concerning whether the aspects are measurable or not. Each perspective has its own strength points which highlight the importance of influence diagrams in the assessment of risks.

The development of influence diagrams provided in this chapter focuses more on a certain topic than individual risk. This topic orientation is expected to provide an answer to one of the concerns in the assessment of risks in the project raised by researchers and professionals as it is expected that the assessment is supposed to provide or can be directed to provide clear information on the impact to the project level or objectives of the project of time, cost and quality. With this in consideration, an influence diagram provides flexibility as its main purpose is to provide graphical representation in explaining the relationships among risks and factors involved in the process. Therefore, when it is necessary, the diagram can be expanded or integrated with other diagrams to have a more comprehensive and overall perspective of risks at a higher level, i.e. project level, where time, cost and quality of the project can be the ultimate consequences. This, however, does not necessarily mean that the quantitative method will follow. As long as the concern is only about influence diagrams, the method can be used to serve the needs in assessing the risks to the project level, not only on the activity level. This is another selling point for influence diagrams.

Case study

A case study, using data collected from high-rise building contractors in Jakarta, Indonesia, is presented to provide better understanding of the framework. In total, 130 risk events were used for the survey as a a result of literature review and after filtering and modification to suit to conditions in Jakarta (step 1). These risks were classified into nine categories and 12 subcategories. Based on this list, a questionnaire was designed to solicit assessment of respondents from contractors of high-rise building construction industry in terms of frequency of occurrence and degree of risk impact.

In step 2 for the case study, different scales were used in the assessments of the risk events. A three-point scale was used for assessing degree of risk impact with options low, medium and high and weights of 1, 2 and 3 were assigned respectively to those choices. Meanwhile, frequency of occurrence of never, low, intermediate and high were used with the assigned weights of 1, 2, 3 and 4, respectively. Besides soliciting the assessment of respondents regarding frequency of occurrence and degree of risk impact for high-rise building construction, general information about respondents was also obtained.

From the assessment of frequency of occurrence and degree of risk impact, cross-tabulation was used to identify critical risk events (step 3).

The mean value of degree of risk impact was used to classify the risks into low, medium and high impact. Risk events were also classified into three categories, i.e. seldom, sometimes and often based on the mean value of frequency of occurrence. The schemes for the classifications are as below:

Degree of impact:

$2.50 \leq$ mean value $\leq 3.00 \rightarrow$ high impact

$2.00 \leq$ mean value $< 2.50 \rightarrow$ medium impact

$1.00 \leq$ mean value $< 2.00 \rightarrow$ low impact

Frequency of occurrence:

$3.25 \leq$ mean value $\leq 4.00 \rightarrow$ often occur

$2.50 \leq$ mean value $< 3.25 \rightarrow$ sometimes occur

$1.00 \leq$ mean value $< 2.50 \rightarrow$ seldom occur

The critical risk events are defined as risk events that have at least medium level for both frequency of occurrence (sometimes) and degree of risk impact (medium). This criterion is the same as shown in Figure 8.2. Based on this, 16 risk events were identified as critical out of 130 risk events, as presented in Table 8.1 in the order of the risk code.

In developing a risk mechanism diagram for the fourth step of the framework, only one diagram is developed from the critical risk events as an example. Niwa (1981, cited in Niwa 1989) proposed a risk mechanism model to analyze and manage recurrence of similar risks in large construction projects in the USA and Europe. This study adopted and modified his model in developing risk mechanism into influence diagram. For the example from the case study, the focus is on subcontractor. There are four risk events related to subcontractors in the critical list: subcontractors unable to finish work on time (R023), low quality work done by subcontractors (R024), ;ow productivity of subcontractors (R026), problems in coordination of subcontractors (R027). These risks are used as a base to generate relationships with other risk events or other factors in developing a risk mechanism model.

In Figure 8.3, the critical risk events can be identified from their bold font, while others are in normal font. Three non-critical risk events and four factors are included in the influence diagram to explain the risk mechanism model. As the perspective of the research is from the contractor side, some risks and factors can be identified to be under their control. The risks and factors are categorized as managerial and operational error. In this category, interactions and interrelationships can be observed for two risk events: lack of experienced engineers (R038) and contractor is not capable to coordinate properly (R066), and two factors: schedule not being updated and work progress is not on schedule. The risks and factors

Table 8.1 Critical risk events

Risk code	Risk Events	Mean value	
		Frequency of occurrence	*Degree of Risk Impact*
R023	Subcontractors unable to finish work on time	2.660	2.057
R024	Low quality work done by subcontractors	2.566	2.075
R026	Low productivity of subcontractors	2.547	2.000
R027	Problems in coordination of subcontractors	3.019	2.302
R041	Delay in material and shop drawing approval	2.868	2.377
R042	Communication and coordination problems by consultant	2.792	2.226
R045	Interference	2.769	2.077
R046	Change orders	3.058	2.288
R048	Quality expected beyond standard and specs	2.769	2.154
R052	Late in material delivery	2.774	2.358
R056	Low productivity and efficiency of equipment	2.528	2.170
R066	Coordination problems during construction	2.811	2.132
R074	Delay of information from designers	2.679	2.113
R092	Errors in drawings	2.623	2.264
R093	Incomplete design scope	2.868	2.132
R108	Slow payment by client	2.962	2.264

that are not controllable by contractor are grouped as environmental factors. One risk event, subcontractor lacks required technical skills (R022), and two factors, family/relative business and subcontractor is not competent, are included in the environmental factor with none of the critical risk events present in Figure 8.3. Three risk events are considered as consequence risks in the model and all of them are critical risks: subcontractor unable to finish work on time (R023), low quality work done by subcontractor (R024), and low productivity of subcontractor (R024). In this model, problems in coordination and supervision of subcontractors (R027) works as a mediator.

It should be kept in mind that the risk mechanism model was developed based on the critical risk events as the core. These critical risk events were retrieved after analyzing data collected from the survey. Therefore, the model reflected in the influence diagram should represent the issues of a specific case (study area), which may also be possible to be generalized. With this in mind, explaining the risk mechanism model directly to the practical aspects of the context or the study area is necessary. This is done in the following paragraphs for the context of high-rise building contractors in Jakarta. Six experienced engineers were consulted in producing the explanation.

Indonesian people have a strong tradition of helping each other, especially their relatives or extended family. Koentjaraningrat (1997) in his research about Indonesian culture provides evidence on this kinship spirit in Indonesian life where he observed that Indonesian people always try to be involved with their community, especially their relatives. The culture also works in the construction industry. Some clients insist on using certain subcontractors for their projects because they are relatives even though those subcontractors may not have any necessary skill or competency to do the job. With this kind of practice, oftentimes, subcontractors create problems to the main contractor in the execution of work.

Coordination problems can happen to contractors. It was not unusual to find on the site that site engineers do not update project schedules either because they are not able to do so or they are too busy. In addition, it is also quite common in construction projects to find that work is not progressing according to schedule. The occurrence of these two issues have made the situation not conducive for the coordination of site work because contractors cannot know precisely about the progress and development of the project. The involvement of less experienced engineers may worsen the situation where contractors will later encounter difficulties in coordination and supervision of work and manpower, including subcontractors.

The consequences for this model are all critical risk events, which cover subcontractor inability to finish on time (R023), subcontractors experience low productivity in their work (R026), and low quality of work performed by subcontractor (R024). This gives a strong signal that contractors need to seriously manage all risks and factors involved in the model as much as possible to prevent or at least minimize the risks to the project, especially those under the control of contractors.

Conclusions

The influence diagram with its graphical presentation is a useful method in assessing risks. This chapter provides step-by-step guidance in developing a solid influence diagram, starting from listing and justifying risk events related to projects under study to constructing the relationships among the risks and factors which afterward can be grouped into responsible parties and consequences. The involvement of experts in the topic and study area during the development process can help produce a solid and reliable influence diagram and explanation. A case study is provided to help understand the process.

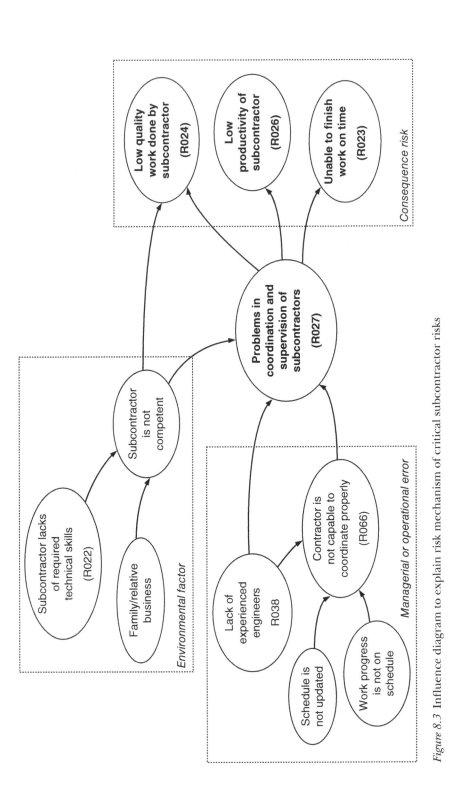

Figure 8.3 Influence diagram to explain risk mechanism of critical subcontractor risks

References

Al-Bahar, J.F. and Crandall, K.C. (1990) Systematic risk management approach for construction project. *Journal of Construction Engineering and Management*, 116(3), 533–546.

Baloi, D., Price, A.D.F. (2003) Modelling global risk factors affecting construction cost performance. *International Journal of Project Management*, 21, 261–269.

Chapman, C. and Ward, S. (1997) *Project Risk Management: Process, Techniques and Insights*. Chichester: John Wiley & Sons Ltd.

Dikmen, I., Birgonul, M.T. and Han, S. (2007) Using fuzzy risk assessment to rate cost overrun risk in international construction projects. *International Journal of Project Management*, 25, 494–505.

Edwards, L. (1995) *Practical Risk Management in the Construction Industry*. London: Thomas Telford.

Flanagan, R. and Norman, G. (1993) *Risk Management and Construction*. Oxford: Blackwell Scientific Publications.

Koentjaraningrat (1997) *Kebudayaan, Mentalitas dan Pembangunan*, 18th edition. PT. Gramedia Pustaka Utama, Jakarta (in Indonesian).

Kumaraswamy, M.M. (1997) Conflicts, claims and disputes in construction. *Engineering, Construction and Architectural Management*, 4(2), 95–111.

Laryea, S. (2008) Risk pricing practices in finance, insurance, and construction, COBRA 2008. The construction and building research conference of the Royal Institution of Chartered Surveyors, Dublin.

Niwa, K. (1989) *Knowledge-based Risk Management in Engineering: A Case Study in Human Computer Cooperative System*. Toronto: John Wiley & Sons, Inc.

Papageorge, T.E. (1988) *Risk Management for Building Professionals*. Kingston: R.S. Means Company, Inc.

Poh, Y. and Tah, J. (2006) Integrated duration–cost influence network for modelling risk impacts on construction tasks. *Construction Management & Economics*, 24, 861–868.

Project Management Institute (2004) *A Guide to the Project Management Body of Knowledge: PMBOK Guide*. 3rd Edition. Pennsylvania: Project Management Institute.

Roumboutsos, A. and Anagnostopoulos, K.P. (2008) Public-private partnership projects in Greece: risk ranking and preferred risk allocation. *Construction Management and Economics*, 26(7), 751–763.

Santoso, D.S., Ogunlana, S.O. and Minato, T. (2003) Assessment of risks in high rise building construction in Jakarta. Engineering, *Construction and Architectural Management*, 10(1), 43–55.

Santoso, D.S., Joewono, T.B., Wibowo, A., Sinaga, H.P.A. and Santosa, W. (2012) Public-private partnerships for tollway construction and operation: risk assessment and allocation from the perspective of investors. *Journal of Construction in Developing Countries*, 17(2), 45–66.

Shen, L.Y., Wu, G.W.C. and Ng, C.S.K. (2001) Risk assessment for construction joint ventures in China. *Journal of Construction Engineering and Management*, 127(1), 76–81.

Tah, J. and Carr, V. (2001) Knowledge-based approach to construction project risk management. *Journal of Computing in Civil Engineering*, 15, 170–177.

Taroun, A. (2014) Towards a better modelling and assessment of construction risk: Insights from a literature review. *International Journal of Project Management*, 32, 101–115.

Tsu, R.Y.C. (2002) Discussion of "Risk assessment for construction joint ventures in China" by L.Y. Shen, George W.C. Wu, and Catherine S.K. Ng. *Journal of Construction Engineering and Management*, 128(2), 194–194.

Yildiz, A.E., Dikmen, I., Birgonul, M.T., Ercoskun, K. and Alten, S. (2014) A knowledge-based risk mapping tool for cost estimation of international construction projects. *Automation in Construction*, 43, 144–155.

Zhi, H. (1995) Risk management for overseas construction projects. *International Journal of Project Management*, 13(4), 231–237.

9 Risk assessment in construction and engineering projects with SDANP plus

Prince Boateng

Introduction

No construction project is risk free. To make decisions for efficient coordination, monitoring and management of projects, there is a need to set priorities, rank the priorities and allocate resources among available options ("alternatives"). To achieve this, the Project Manager (PM) and his team can derive science-based measurements and methods to develop criteria and weight them, and then evaluate the alternatives against these criteria. Using the *Analytic Network Process* (ANP) project managers can derive such weights by conducting pairwise comparisons of criteria with respect to their importance, likelihood or preference. With the AHP computations, PMs can integrate results into a system dynamics (SD) methodology to understand the dynamics of risks that impact on the objectives of these projects over time. This combined approach is crucial in interpreting evidence of qualitative factors and designing measures to deal with the likely impact on the projects. Furthermore, PMs can use this unambiguous combined method also known as SDANP methodology to assess and address value trade-offs.

The purpose of this chapter is to introduce the SDANP method to risk management in engineering and construction projects and to explain how to apply it for deriving criteria weights and assessing risks overtime during project execution.

Methods

The Analytical Network Process (ANP)

The ANP is a methodological tool developed by Thomas Saaty. This tool is chosen for this book chapter because of its significance in multi-criteria decision making (MCDM) when an extensive number of factors are involved. The ANP is a general hierarchical process for ranking alternatives based on some set of criteria. Unlike the Analytical Hierarchy Process (AHP) where decision problems are formed in hierarchy linear structure, the ANP has clusters organized and arranged in a particular

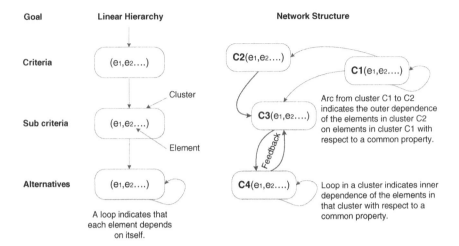

Figure 9.1 Comparison of Hierarchy (AHP) and Network (ANP) Structure
Source: (Boateng et al. 2015)

order and directions that allow interdependency (inner dependence, outer dependence and feedback) among the decision clusters and even among elements within the same cluster (Saaty 2000 and 2005).

Figure 9.1 further illustrates the differences in relationships and links between AHP hierarchy and ANP network. Inner dependencies indicate the dependences of elements of a cluster on each other while the outer dependencies designate feedback between clusters from one level to another level and can be expressed either from one cluster directly to another one or by transiting influence through intermediate clusters along a path that sometimes returns to the original cluster, forming a cycle.

Based on the above explanation, one can conclude that the

- *AHP decomposes a problem into several levels making a hierarchy in which each decision element is considered to be independent.*
- *AHP allows groups or individuals to combine qualitative and quantitative factors in decision making processes.*
- *AHP and ANP are both Multi Criteria Decision Making methods for complicated and unstructured problems.*
- *ANP extends AHP to problems with dependence and feedback.*
- *ANP allows for more complex interrelationships among decision elements by replacing the hierarchy in AHP with a network.*

As an MCDM method, ANP is used in this chapter to synthetize expert judgements into numerical values given their specific subjectivity inputs. Its purpose is to prioritize lists of construction and engineering risks based on their relative importance in an organization.

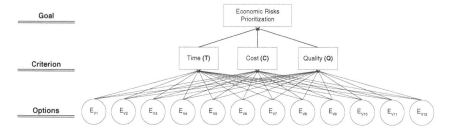

Figure 9.2 Generic Analytical Network Process Model for Economic Risks
        Prioritization

In this chapter, we shall consider four major steps involve in using the
ANP for risk prioritization:

   Step 1: Model construction

   Step 2: Pairwise comparisons and local priority vector

   Step 3: Normalized priorities

   Step 4: Final priorities.

Step 1: Model construction

In this step, a hierarchical system is set up by decomposing the problem
into a hierarchy of interrelated elements. Figure 9.2 illustrates an ANP network structure for economic risks prioritization for a construction project.
The structure consists of 12 decision elements grouped into one decision
cluster. Note that in the case of risks influences, the elements in a cluster
can influence some or all of the elements of any clusters.

The 'Goal' Cluster

   *This contains only one element as the statement of the purpose for risk
   prioritization within which the category of 'high risks' are listed according
   to the results from the pairwise comparison calculation.*

The 'Criterion' Cluster

   *This cluster consists of potential consequences of elements of potential risks
   on project cost, time and quality.*

The 'Options' Cluster

   *This cluster contains potential economic risk variables. The arrows indicate
   relationships between elements in one cluster against elements in other clusters.*

Step 2: Pairwise comparisons

We compared the comparative weights between the attributes of the decision elements to form the reciprocal matrix. In ANP, pairwise comparisons of the elements in each level are conducted with respect to their relative importance to their control criterion. The correlation matrices are prepared on a 1–9 ratio scale to determine the relative preferences for two elements of the network in the matrix. A score of 1 indicates that the two options have equal importance whereas a score of 9 indicates dominance of the component under consideration over the comparison component matrices.

Considering the example in figure 9.2, elements of each cluster (time, cost and quality) are compared pairwisely as follows:

- Criterion T vs. Criterion C
- Criterion T vs. Criterion Q
- Criterion C vs. Criterion Q

For each pairwise, we ask if one criterion is more important (or risky, likely, preferred, etc.) than the other? If yes, by how much more? To determine the relative *"how much more,"* we use the fundamental scale measurement tool for pairwise comparisons for data transformation (see Figure 9.3). A score of 1 indicates that the two options have equal importance whereas a score of 9 indicates dominance of the component under consideration over the comparison component matrices. Note that the judgement of *"how much more"* is just a qualitative strength of importance associated values $(1, 2 \ldots 9)$. It is based only on the qualitative description such as moderate, strong, very strong and extreme. It is our interpretation of the relative importance of one criterion compared to another.

Based on the measurement tool for the pairwise comparison represented in figure 9.3 a decision maker can conclude that:

Criterion A, **Criterion B** and **Criterion C** are all equally important pairwise.

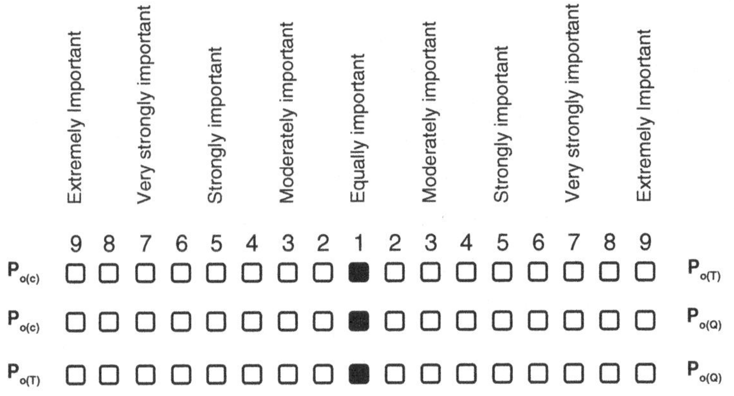

Figure 9.3 Fundamental scale measurement tool for pairwise comparisons

Next, we create a comparison matrix using the results from Figure 9.3. We will do this incrementally starting with the diagonal values which are always **1s**. Per definition, pairwise comparison for criteria T–T = C–C = Q–Q = 1. This is so because each of the criteria represents the rating of an option compared to itself.

	Criterion T	*Criterion C*	*Criterion Q*
Criterion T	1		
Criterion C		1	
Criterion Q			1

1

To complete the matrix, we read across the rows and columns. But since all the **CRITERION are equally important** when compared with **one another,** we insert 1s into the matrix as indicated below.

	Criterion T	*Criterion C*	*Criterion Q*
Criterion T	1	1	1
Criterion C	1	1	1
Criterion Q	1	1	1

2

The process continues and next is to calculate the geometric mean $(Gm_{po(C)})$ value of the rating results with respect of the project objectives per criterion. The following formula is applied to the relative ratings in the criterion rating matrix (2), where Gm represents the geometric mean of the rating results of criteria $(C_T, C_C$ and $C_Q)$ or of any 'n' criteria:

$$Gm_{po(C)} = (1 \times Rc_T c_C \times Rc_T c_Q \times \dots \dots \dots \times Rc_T c_n)^{1/n}$$

For each of the chosen three criterions, the resultant formula above is used to calculate the Gm for their respective geometric means. The geometric mean of the rating results of criterion CT, and noting that $Rc_T c_T = 1$ per definition, is:

$$Gm_T = (1 \times Rc_T c_c \times Rc_T c_Q)^{1/3}$$

A column can be added to the criterion rating matrix to show this calculation result, producing the following matrix below:

	C_T	C_C	C_Q	$Gm = \sqrt[3]{\left(1 \times R_{cTcC} \times R_{cTcQ}\right)}$
C_T	1	1	1	1.00
C_C	1	1	1	1.00
C_Q	1	1	1	1.00
		$\Sigma Gm =$		**3.00**

Step 3: Normalized priority value

In this step, the subjective geometric weights are normalized to obtain the priority value (P_v) of each criterion. The (P_v) is calculated by dividing each of the geometric mean values by the sum of the geometric mean values using the formula below:

$$P_V = Gm \: / \: (Gm_T + Gm_c + \ldots \ldots + Gm_n)$$

A column can be added to the criterion rating matrix to show this calculation result, producing the following matrix:

Matrix(Po)	C_T	C_C	C_Q	$Gm = \sqrt[3]{\left(1 \times R_{cTcC} \times R_{cTcQ}\right)}$	$P_V = (Gm / (\Sigma Gm))$
C_T	1	1	1	1.00	0.333
C_C	1	1	1	1.00	0.333
C_Q	1	1	1	1.00	0.333
			$\Sigma Gm =$	3.00	

Similarly, we created comparison matrices for the option clusters with respect to project time, cost and quality using the above procedures and table 9.1.

We used Table 9.1 to transform data when two or more decision makers have different relative judgements during a decision making process. For example, in Table 9.2, two independent Decision makers (Dm) judged that:

Dm₁: E_{V1} is 'Very strongly risky' than E_{V2} and arrived at a judgement scale of 7

Dm₂: E_{V2} is 'Equally to moderately risky' than E_{V1} and arrived at a judgement scale of 2

Table 9.1 Relative importance and data transformation in pairwise comparison

Intensity of importance	Definition (Saaty 1996)	Data transformation mechanism (Chen, et al. 2011)
1	Equal	1:1
2	Equally to moderately dominant	2:1, 3:2, 4:3, 5:4, 6:5, 7:6, 8:7, 9:8
3	Moderately dominant	3:1, 4:2, 5:3, 6:4, 7:5, 8:6, 9:7
4	Moderately to strongly dominant	4:1, 5:2, 6:3, 7:4, 8:5, 9:6
5	Strongly dominant	5:1, 6:2, 7:3, 8:4, 9:5
6	**Strongly to very strongly dominant**	**6:1, 7:2, 8:3, 9:4**
7	Very strongly dominant	7:1, 8:2, 9:3
8	Very strongly to extremely dominant	8:1, 9:2
9	Extremely dominant	9:1

Source: (Boateng 2014)

The group decision can then be said to be in the ratio of 7:2. Tracing the 7:2 ratio on column 3, row 6 of table 9.1, the combined decision can then be said to be *'Strongly to very strongly dominant'* with a judgement value of 6. This value is then inserted into row 1, column 2 of the pairwise comparison matrix (see table 9.2). The process is repeated to complete the entire matrix and any other matrices with group decision inputs are completed.

Table 9.2 Pairwise comparison for economic risk variables with respect to time

Cost	RJ*	Risk Code	E_{V1}	E_{V2}	E_{V3}	E_{V4}	E_{V5}	E_{V6}	E_{V7}	E_{V8}	E_{V9}	E_{V10}	E_{V11}	E_{V12}	Priorities
Time	7	E_{V1}	1	6	1	6	6	6	3	5	4	1	1	1/2	0.14
	2	E_{V2}	1/6	1	1/6	1	1	1	1/4	1/2	1/3	1/6	1/6	1/7	0.02
	7	E_{V3}	1	6	1	6	6	6	3	5	4	1	1	1/2	0.14
	2	E_{V4}	1/6	1	1/6	1	1	1	1/4	1/2	1/3	1/6	1/6	1/7	0.02
	2	E_{V5}	1/6	1	1/6	1	1	1	1/4	1/2	1/3	1/6	1/6	1/7	0.02
	2	E_{V6}	1/6	1	1/6	1	1	1	1/4	1/2	1/3	1/6	1/6	1/7	0.02
	5	E_{V7}	1/3	4	1/6	4	4	4	1	3	2	1/5	1/5	1/4	0.06
	3	E_{V8}	1/5	2	1/5	2	2	2	1/3	1	1/2	1/5	1/5	1/6	0.03
	4	E_{V9}	1/4	3	1/4	3	3	3	1/2	2	1	1/4	1/4	1/5	0.05
	7	E_{V10}	1	6	1	6	6	6	5	5	4	1	1	1/2	0.15
	7	E_{V11}	1	6	1	6	6	6	5	5	4	1	1	1/2	0.15
	8	E_{V12}	2	7	2	7	7	7	4	6	5	2	2	1	0.21

Note:

1 P_O refers to project objectives
2 * refers to relative judgement
3 E_V Economic risk variable

Table 9.3 Pairwise comparison for economic risk variables with respect to cost

P_O	RJ*	Risk Code	E_{V1}	E_{V2}	E_{V3}	E_{V4}	E_{V5}	E_{V6}	E_{V7}	E_{V8}	E_{V9}	E_{V10}	E_{V11}	E_{V12}	Priorities
Cost	9	E_{V1}	1	6	3	7	7	7	3	5	4	3	2	1	0.20
	4	E_{V2}	1/6	1	1/4	2	2	2	1/4	1/2	1/3	1/4	1/5	1/6	0.03
	7	E_{V3}	1/3	4	1	5	5	5	1	3	2	1	1/2	1/3	0.09
	3	E_{V4}	1/7	1/2	1/5	1	1	1	1/5	1/3	1/4	1/5	1/6	1/7	0.02
	3	E_{V5}	1/7	1/2	1/5	1	1	1	1/5	1/3	1/4	1/5	1/6	1/7	0.02
	3	E_{V6}	1/7	1/2	1/5	1	1	1	1/5	1/3	1/4	1/5	1/6	1/7	0.02
	7	E_{V7}	1/3	4	4	5	5	5	1	3	2	1	1/2	1/3	0.09
	5	E_{V8}	1/5	2	2	3	3	3	1/3	1	1/2	1/3	1/4	1/5	0.04
	6	E_{V9}	1/4	3	3	4	4	4	1/2	2	1	1/2	1/3	1/4	0.06
	7	E_{V10}	1/3	4	4	5	5	5	1	3	2	1	1/2	1/3	0.09
	8	E_{V11}	1/2	5	5	6	6	6	2	4	3	2	1	1/2	0.14
	9	E_{V12}	1	6	6	7	7	7	3	5	4	3	2	1	0.20

Note:

1 P_O refers to project objectives
2 * refers to relative judgement
3 E_V Economic risk variable

Table 9.4 Pairwise comparison for economic risk variables with respect to quality

P_O	RJ*	Risks Code	E_{V1}	E_{V2}	E_{V3}	E_{V4}	E_{V5}	E_{V6}	E_{V7}	E_{V8}	E_{V9}	E_{V10}	E_{V11}	E_{V12}	Priorities
Quality	6	E_{V1}	1	5	1	5	5	5	2	4	4	1	1	1	0.14
	2	E_{V2}	1/5	1	1/5	1	1	1	1/4	1/2	1/2	1/5	1/5	1/5	0.03
	6	E_{V3}	1	5	1	5	5	5	2	4	4	1	1	1	0.14
	2	E_{V4}	1/5	1	1/5	1	1	1	1/4	1/2	1/2	1/5	1/5	1/5	0.03
	2	E_{V5}	1/5	1	1/5	1	1	1	1/4	1/2	1/2	1/5	1/5	1/5	0.03
	2	E_{V6}	1/5	1	1/5	1	1	1	1/4	1/2	1/2	1/5	1/5	1/5	0.03
	5	E_{V7}	1/2	4	1/2	4	4	4	1	3	3	1/2	1/2	1/2	0.09
	3	E_{V8}	1/4	2	1/4	2	2	2	1/3	1	1	1/4	1/4	1/4	0.04
	3	E_{V9}	1/4	2	1/4	2	2	2	1/3	1	1	1/4	1/4	1/4	0.04
	6	E_{V10}	1	5	1	5	5	5	2	4	4	1	1	1	0.14
	6	E_{V11}	1	5	1	5	5	5	2	4	4	1	1	1	0.14
	6	E_{V12}	1	5	1	5	5	5	2	4	4	1	1	1	0.14

Note:

1   P_O refers to project objectives
2   * refers to relative judgement
3   E_V Economic risk variable

Step 4: Final priorities

This is the final stage where the relative weights of the decision elements are aggregated to determine the most risky economic risk variables (alternatives).

The next step is to check consistencies of the results for the matrices consistency ratio of 0.1 to judge whether the comparison is consistent. The Consistency Ratio (CR) is a widely used consistency test method in both AHP and ANP. The CR is used to check the consistencies of the values obtained according to the pairwise comparison. In the AHP method, participants and decision makers or experts who make judgements or preferences must go through the consistency test. The reasons are because the final risk assessment and decision analysis could be inaccurate if the priority values are calculated from the inconsistent comparison matrix. Therefore, the consistency of each comparison matrix has to be tested before the comparison matrices are used to assess risk and analyze a decision. If the consistency test for the comparison matrix failed, the inconsistent elements in the comparison matrix would have to be identified and revised; otherwise, the result of risk assessment and decision analysis would be unreliable.

To determine the consistency of participants' judgement on the level of economic risks impacts on a project, a CR of the comparison matrices are calculated using the process below:

Step 1: Calculate the maximum eigenvalue ($\lambda_{max.}$) of one comparison matrix

Step 2: Calculate the value of Consistency Index (*CI*)

Step 3: Calculate the CR using the formula $CR = CI/RI$

Step 4: Compare the value of CR with the consistency threshold 0.1 to judge whether the comparison is consistent.

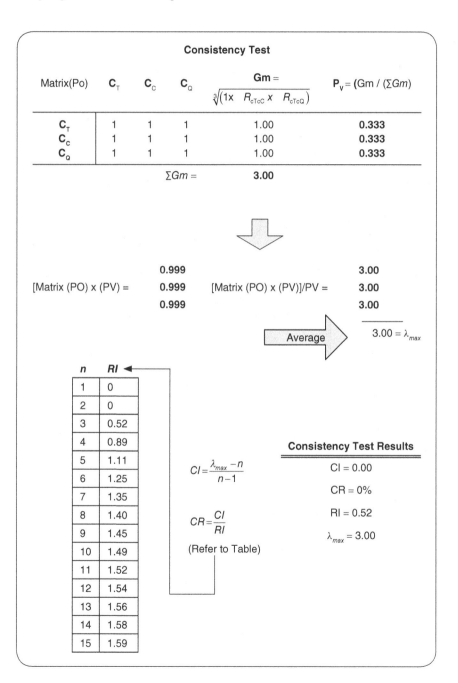

Table 9.5 Final ANP decision making priority results for economic risks variables

Potential risk	Risk code	Risk priorities			Synthesized priorities				Ranking
		Local priorities							
		Cost (0.33) $w(c)$	Time (0.33) $w(t)$	Quality (0.33) $w(q)$	Cost $(w(C))$ $0.33 \times w(c)$	Time $(w(T))$ $0.33 \times w(t)$	Quality $(w(Q))$ $0.33 \times w(q)$	Total (R_{PV}) $\Sigma w_{(C,T,Q)}$	
Economic risks (P_R)	E_{V1}	0.20	0.14	0.14	0.07	0.05	0.05	0.17	2
	E_{V2}	0.03	0.02	0.03	0.01	0.01	0.01	0.03	8
	E_{V3}	0.09	0.14	0.14	0.03	0.05	0.05	0.13	4
	E_{V4}	0.02	0.02	0.03	0.01	0.01	0.01	0.03	8
	E_{V5}	0.02	0.02	0.03	0.01	0.01	0.01	0.03	8
	E_{V6}	0.02	0.02	0.03	0.01	0.01	0.01	0.03	8
	E_{V7}	0.09	0.06	0.09	0.03	0.02	0.03	0.08	6
	E_{V8}	0.04	0.03	0.04	0.01	0.01	0.01	0.03	8
	E_{V9}	0.06	0.05	0.04	0.02	0.02	0.01	0.05	7
	E_{V10}	0.09	0.15	0.14	0.03	0.05	0.05	0.13	4
	E_{V11}	0.14	0.15	0.14	0.05	0.05	0.05	0.15	3
	E_{V12}	0.20	0.21	0.14	0.07	0.07	0.05	0.19	1

Values of CI, RI, CR and inconsistency for all the pairwise comparison matrices

Criteria	Values λmax	CI	RI	CR	Inconsistency
Cost	12.49	0.040	1.540	0.030	0.000
Time	12.35	0.030	1.540	0.020	0.000
Quality	12.20	0.020	1.540	0.010	0.000

Note:

1 E_V refers to economic risk variable   4 CI refers to consistency index
2 R_{PV} refers to final risk priority value   5 RI refers to relative index
3 λ_{max} refers to maximum eigenvalue   6 CR refers to consistency ratio

Where CR ≤ 0, it meant that decision makers' judgements satisfy this consistency. If not, the experts had conflicting judgements and therefore, the inconsistent elements in the comparison matrix have to be identified and revised. The consistency test process above presents the maximum Eigenvector (λ_{max}), CI, Relative Index (RI) and CR achieved for the economic risk factors in this example. Table 9.5 indicates the full consistency test results.

Also, analysis of the results in Table 9.5 indicates that decision makers' answers to the prioritization on project objectives are consistent. The final Risk Priority Value (R_{PV}) for the economic risks is the addition of the synthesized weights of cost (W_C) Time (W_T) and quality (W_Q). As indicated in Table 9.5, the local priority values suggest equal importance of the project objectives to decision makers during evaluation.

Consequently, the R_{PV} for E_{V1} is 0.17 (17%). The R_{PV} for each risk variable is the final risk priority index that could be used as an indicator to attract a developer's attention to potential risks that have the highest level of impacts on project objectives. The values could also be imported into the system dynamic methodology to simulate the behaviour of such risks over time so that appropriate mitigation procedures could be initiated.

Risk rating

In this section, the process of rating categories for each economic risk was established. The results obtained from the pairwise comparison are used to provide a verbal risk rating for a better representation of the level of impacts of identified risks on the project objectives. Potential risks as alternatives were evaluated by selecting the appropriate rating category on their level of impacts on each criterion (objectives) as Very High (5), High (4), Medium or Moderate (3), Low (2) and Very Low (1). They were compared for preference using a pairwise comparison matrix in the usual way as discussed earlier to obtain the total priorities.

To obtain idealized priorities, the total priorities are further normalized by dividing each total priority by the largest of the priorities. For example, in Table 9.6, the total priority for a 'High' rating is 0.26. By dividing the 0.26 by 0.42 which is the largest rating among the total priorities, the result becomes 0.62 as an idealized priority value. The idealized priorities are therefore used in ranges for risks rating. For example, a priority value greater than 0.62 is classified as having a very high risk impact on the project objectives and so on. The rating categories for the five scales are established in Table 9.6.

Finally, Table 9.7 provides the verbal ratings of how the 12 potential economic risks as alternatives on each covering criterion (Time, Cost and Quality) and their corresponding numerical ratings are rated.

Table 9.6 Deriving priorities for risks ratings

Risks rating for mega construction projects		VH	H	M	L	VL	Total priorities	Idealized priorities	Numerical risks rating (%)
Very High	VH	1	2	3	4	5	0.42	1.00	> 62
High	H	1/2	1	2	3	4	0.26	0.62	38–61
Moderate	M	1/3	1/2	1	2	3	0.16	0.38	24–37
Low	L	1/4	1/3	½	1	2	0.10	0.24	14–23
Very Low	VL	1/5	1/4	1/3	1/2	1	0.06	0.14	< 14
Total priorities							**1.00**		

Table 9.7 Verbal ratings for economic risk variables

Risk code	Risks	Ideal synthesized risk priority values (%)	Verbal ratings
E_{V1}	Change in government funding policy	100	Very high
E_{V2}	Taxation changes	18	Low
E_{V3}	Change in government	76	Very high
E_{V4}	Wage inflation	18	Low
E_{V5}	Local inflation change	18	Low
E_{V6}	Foreign exchange rate	18	Low
E_{V7}	Material price changes	47	High
E_{V8}	Economic recession	18	Low
E_{V9}	Energy price changes	29	Medium
E_{V10}	Catastrophic environmental effects	76	Very high
E_{V11}	Project technical difficulties	88	Very high
E_{V12}	Project delays of all forms	100	Very high

Notes: Very High = (>62%), High = (38–61%), Medium = (24–37%), Low = (14–37%), and Very Low = (<14%).

The System Dynamics

SD is a powerful tool that improves learning about a business or a company, market and competitors; portrays the cognitive limitations on the information gathering and processing power of human mind; facilitates the practice of considering opinions; and supports building of "what if" scenarios. Although researches on system dynamics modelling is very rich with applications in many fields, not many such system dynamics models were applied so far on real case projects. In this chapter we portray current approaches to the development of system dynamics models. These are

1   Problem identification, definition and initial model development based on influence diagram
2   Model verification (using expert opinion)

3   Final model development and simulation (analysis of model behaviour)
4   Model validation using software tools and a case study
5   Policy analysis, model use or implementation.

The concept is to understand how the parts in a system interact with one another. Also, it is to show how a change in one variable can affect the other over time and in turn affects the original variable (See Figure 9.4).

SD can be used to model systems in both qualitative and quantitative manner and can be constructed from three basic building blocks: positive feedback or reinforcing loops, negative feedback or balancing loops, and delays. Positive loops (called reinforcing loops) are self-reinforcing while negative loops (called balancing loops) tend to counteract change. Delays in SD models indicate potential instability in the system. Figure 9.4a shows how a reinforcing loop feeds on itself to produce a growth in a system to correspond to positive feedback loops in control theory. For example, in Figure 9.4a, an increase in entity (1) leads to an increase in entity (2) (as indicated by the "+" sign) and that in turn leads to additional increase in variable (A) and so on. The "+" sign indicates on the head of the arrow does not necessarily mean that the values produced in the system will increase. It is just that entity (1) and entity (2) will change in the same direction of polarity. If entity (1) decreases, then entity (2) will decrease. In the absence of external influences, both entities (1 and 2) will clearly grow or decline exponentially. Reinforcing loops generate growth, amplify deviations, and reinforce change. A balancing loop indicated in Figure 9.4b is a structure that changes the current value of a system variable or a desired or reference variable through some action. It corresponds to a negative feedback loop in control theory. A (-) sign indicates that the values of the variables

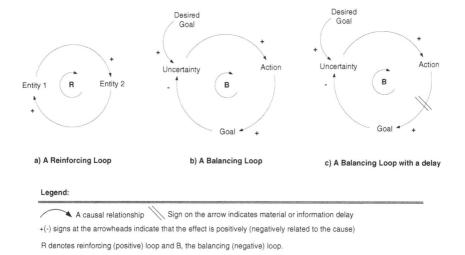

a) A Reinforcing Loop   b) A Balancing Loop   c) A Balancing Loop with a delay

Legend:

A causal relationship   Sign on the arrow indicates material or information delay

+(-) signs at the arrowheads indicate that the effect is positively (negatively related to the cause)

R denotes reinforcing (positive) loop and B, the balancing (negative) loop.

Figure 9.4 The three components of system dynamics models

change in opposite directions. The difference between the current value and the desired value is perceived as an error. An action proportional to the error is taken to decrease the error so that, over time, the current value approaches the desired value. The third basic element is a delay, which is used to model the time that elapses between cause and effect. A delay is indicated by a double line, as shown in Figure 9.4c. Delays make it difficult to link cause and effect (dynamic complexity) and may result in unstable system behaviour.

Based on a verified Causal Loop Diagram, a stock and flow diagram indicated in Figure 9.5 can be developed using the priority values of risks derived from the ANP computation and the inputs which the experts provided to facilitate in-depth stock and flow modelling and risk simulation overtime.

The governing equations used to calculate the entire system parameters can also be formulated at this point.

To understand accumulation process of inflow of uncertainties, it is important to know the mathematical meaning used to integrate the flow of risk influences into the system. Based on a mathematical definition of the integral, the level of risk impacts inside a stock will be the integration of total flows of uncertainties on the stock as indicates in the equation below.

$$\text{Stock } (t) = \int_0^t [\text{flows}_{total}(s)] ds$$

Where,

- $\int_0^t [\text{flows}_{total}(s)$ is a function of the total flow in the system.

Inflow (economic uncertainty) indicates the increasing amount of risk level in the stock (risk accumulation container). In the other hand, outflow (certainty) decreases the level of risk impacts in the stock.

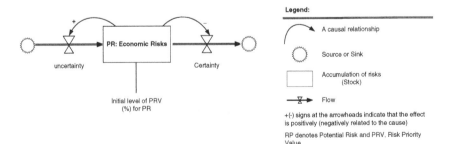

Figure 9.5 A simple stock and flow model

Using AHP's R_{PV} as the quantity of risk impact level in the stock at the initial time, the equation stated above becomes the equation below as follows:

$$Stock(t) = \int_0^t [flows_{total}(s) - Outflow(s)]ds + Stock(0)$$

Where,

– Stock(0) is the stock of risk impact level (RPI) at the initial time, $(t = 0)$.

In systems dynamics, verbal descriptions and causal loop diagrams are more qualitative; stock and flow diagrams and as well as model equations are more quantitative ways to describe a dynamic situation. Since system dynamics is largely based on the soft systems thinking (learning paradigm), it is well suited to be applied on those managerial problems which are ambiguous and require better conceptualization and insight (Madachy 2007).

In this chapter, the SD approach will be demonstrated on the example of development of a simple economic risk model to express the impacts of such risks on a construction and engineering project in the UK.

SDANP risk modelling

This session will provide technical details on the development of SD models for risk simulation on construction and engineering projects and will utilize the data from the ANP computation to demonstrate the procedure and effectiveness of using SD in risk simulation overtime.

Economic risk system model

Following the development of the AHP models, a high level causal loop diagram (CLD) for the entire economic risk system is constructed (see Figure 9.6). Variables in the high level causal diagram are causally related in the form of loops and may either have no influence or have a positive or negative influence on another factor. As a result of many system variables (entities) and loops in Figure 9.6, a construction and engineering project can be said to belong to the class of complex dynamic systems that, according to Sterman (1992), exhibits five major characteristics:

- They are complex and consist of multiple components
- They are highly dynamic
- They involve multiple feedback processes
- They involve non-linear relationships and
- They involve both "hard" and "soft" data.

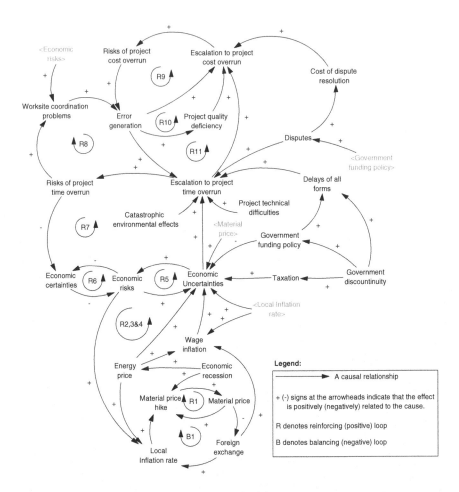

Figure 9.6 Causal loop diagram for economic risks system

Causalities for the economic risks stock variables

For each entity in the economic risks causal loop model, two different causality trees can be drawn. The first, called *"causes tree"*, represents the entity in question as the end of the tree and includes all the variables (entities) that influence it. The second tree-like diagram, called *"uses tree"*, has the entity in question at its head, and shows all other entities influenced by it.

For example, in Figure 9.7 the "causes" trees for the economic risks entity have the risk variables in question at the head, while in Figure 9.8, the "uses" tree shows all other risk variables influenced by it for the economic risks entity.

Considering the relation of *economic uncertainties* variable in Figure 9.6, by pure logic, project participants can associate and represent both the *causes* and *uses* tree diagrams in Figures 9.7 and 9.8 as shown in Figure 9.9 and Table 9.8 respectively.

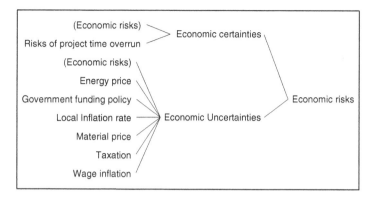

Figure 9.7 Causes tree diagram for the economic risks entity

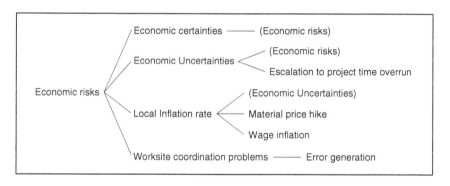

Figure 9.8 Uses tree diagram for the economic risks entity

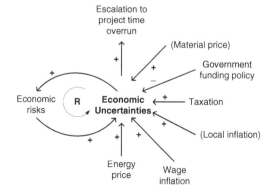

Figure 9.9 Causality of economic uncertainties

Table 9.8 Influence of economic uncertainties

Economic uncertainties			
Influence by		Influence on	
Positive	Negative	Positive	Negative

– Local inflation
– Wage inflation
– Energy price
– Material price
– Taxation
– Economic risks
	– Government
	  funding policy
		– Escalation to
		  project time
		  overrun
			——

In all causal tree diagrams entities in parentheses, that is (local inflation and material price) indicated in Figure 9.9, denote that these entities appears at least twice in the tree, and are therefore contained in a loop, as indeed can be seen in the causal loop model presented in Figure 9.6. It can also be observed that Figure 9.6 contains several risk variables and loops which indicate cause-effect interrelationships for the entire economic risks causal loop model.

Initial Model Development

The initial model defines the dynamic hypothesis of the economic risks system model considered in this chapter. The dynamic hypothesis is a statement that can be proved right or wrong after a thorough research based investigation. Ranganath and Rodrigues (2013) stated in system dynamics theory and case studies that a dynamic hypothesis can be proposed in the form of a statement, a causal loop diagram, or stock and flow diagram. According to Ranganath and Rodrigues (2013), it basically expresses the model in a systematic term for conducting simulation by deciding on the factors of study. Refinement and revision can be made on this dynamic hypothesis, as and when required, because no model can be perfect in all respects.

Elements of the dynamic hypothesis in this book chapter are derived from literature review, interviews conducted with academic staff belonging to higher education academy in the UK, the Netherlands, Germany, Italy and other EU member countries, consultants, project managers, engineers, top level management staff involved in megaproject development and through the modeler's interaction with industry practitioners in the United Kingdom. These hypotheses behave dynamically and are based on the interactions of risk variables within the overall model illustrated in Figure 9.6. This suggests that variables which cause risks are interrelated within a chain of cause and effect loops. According to

Ghaffarzadegan et al. (2011), each cause influences the latter one in a closed loop of cause and effects making a domino.

System boundaries

After the ANP's pairwise comparison, a model boundary can be formed and a dynamic hypothesis, also known as a CLD developed. The model boundary is used to define the limit within which the economic risk model will operate. It is a representation of "how far in the future should a modeler consider and/ or how far back in the past lie the roots of the problem?" (Sterman 2010). Table 9.9 illustrates the system boundaries for economic risks system model under discussion in this chapter.

Table 9.9 System boundary for economic risks system

Risk Code	Risk Type
	Type I: Endogenous Variables
P_{R3}	Economic risks
E_{V1}	Change in government funding policy
E_{V2}	Taxation
E_{V4}	Wage inflation
E_{V5}	Local inflation
E_{V6}	Foreign exchange
E_{V7}	Material price
E_{V9}	Energy price
E_{V12}	Project delays of all forms
E_{V13}	Cost of delays
E_{V14}	Cost of resolution
E_{V15}	Disputes
E_{V16}	Economic certainties
E_{V17}	Economic uncertainties
E_{V18}	Error generation
E_{V19}	Escalation to project cost overrun
E_{V20}	Escalation to project time overrun
E_{V21}	Material price hike
E_{V22}	Project quality deficiency
E_{V23}	Risks of project cost overrun
E_{V24}	Risks of project time overrun
E_{V25}	Worksite coordination problems
	Type II: Exogenous Variables
E_{V3}	Government discontinuity (change)
E_{V8}	Economic recession
E_{V10}	Catastrophic environmental effects
E_{V11}	Project technical difficulties

To specify a model boundary, the modeler must separate the initial components list into two important groups. In this chapter there are two "risk types" in each of the economic risks system boundary.

- **Risk "type I" represents** Endogenous variables – dynamic variables involved in the feedback loops of the system. The behaviors of these variables are generated within model
- **Risk "type I" represents** Exogenous variables – components whose values are not directly affected by the system. These types of variables can be considered as essential parameters with values coming from the AHP Risk Priority Values (R_{PV}).

System verification

Construction and engineering projects worldwide are unique and have unprecedented historical data on cost and time overruns. Nevertheless, it is necessary to elicit further information from project experts who have knowledge and expertise in risks assessment in transportation megaprojects during construction to support this historical data of cost and time overruns. Although, some researchers argue that experts' judgement tends to have cognitive limitations because it has potential biases associated with the individual subjective views (Fiske & Taylor 2013 and Meyer & Booker 2001). It is still necessary to incorporate these experts' views into this chapter of the book to aid the application of the system dynamics and statistical techniques for assessing subjective data in a systematic way, thus reducing subjective bias.

For consistent decisions, the initial model was sent to seven participants (a group of risk experts and consultants) to criticize. The principal role of the verification in the first (qualitative) stage of modelling is to identify risks of inconsistency within the causal loop diagram (CLD) and to obtain sufficient information for the final model construction and for model simulation. Participants in the verification process for this model include a project manager, a project engineer, a departmental stakeholder, a technical consultant and an operational manager involved in a transportation megaproject in Edinburgh, UK. Others include an insurer, and a financial and legal adviser who are experienced in many facets of large-scale transportation infrastructure projects.

The breadth of the model verification was to build confidence in the model. Although there is no single verification method in literature, verification by the experts mentioned in this chapter improved confidence gradually as the model was constructively criticized by these experts as new points of correspondence between the models and the identified empirical reality. From the standpoint of verification, a number of errors were identified in the initial model development. Application of verification ideas were then used to refine the initial models. It suffices to say that the process helped to establish confidence in the soundness and usefulness of the final models with respect to their purposes.

Final model development and simulation

Based on a verified CLD, a stock and flow diagram for the economic risk system was developed. A stock is the term for any entity that accumulates or depletes over time whereas a flow is the rate of accumulation of the stock. Figure 9.10 illustrates an example of a typical stock and flow model (SFM).

Figure 9.11 contains the stock and flow diagram representation of the economic risks system obtained for the Edinburgh Tram Network (ETN) project during construction. This diagram is simply a representation of the relationships between the system components that create the dynamic behaviors of project cost and time overruns for the ETN project at the construction phase.

The SFM is developed with risk of project time overrun, risk of project cost overrun and quality deficiency in focus. We chose to diagram the basic mechanisms of the economic risks impacts on project performance in terms of stocks and flows by putting the causal-loop relationships in the back of our minds. Normally, a stock and flow diagram has the propensity to be more detailed than a causal-loop diagram representation. A stock and flow

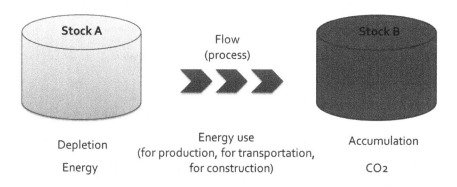

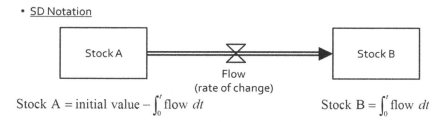

$$\text{Stock A} = \text{initial value} - \int_0^t \text{flow } dt \qquad \text{Stock B} = \int_0^t \text{flow } dt$$

Figure 9.10 A typical stock and flow model

Source: (Boateng 2014)

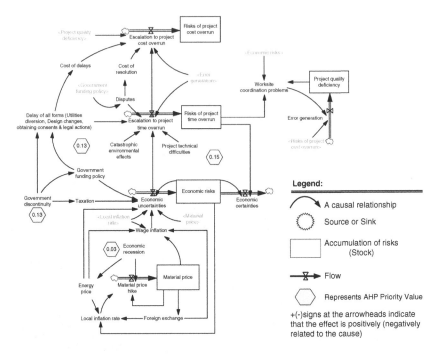

Figure 9.11 Stock and flow diagram for the economic risks system
Source: (Boateng 2014)

diagram always serve as basis for every modeler to think more specifically about a system structure. Through the stock and flow diagramming, many simple mistakes can be circumvented by diagramming the basic risk effects with stocks and flows rather than causal loops. The reason is that relationships between individual risk components of a stock and flow diagram can be more defined than those in a causal loop diagram. Although, stock and flow diagrams are more complex and involve more time to create, they are more explanatory than causal loop diagrams.

In the above SFM, exogenous variables have direct influences on the entire controlling system. For example, in the economic risk model (refer to Figure 9.11) variables such as the economic recession, local inflation rate and material price have a direct influence on the controlling system variable material price hike which stocks material price. The stock 'economic risks' is in turn influenced by several other variables through the economic uncertainties as indicated in Figure 9.11. It can further be noted that economic risks variable is affected by the economic certainties. Similarly, risks of project time and cost overruns are influenced by escalation in project time and escalation to project cost to stock risks of project time and cost overruns respectively. Further, consideration on the SFMs show that a number of variables influence risk of project time and cost overrun through escalation of project time and cost overruns.

The notations in the model follow the methodology of Forrester (1961). There are four different entities on the model: levels, rates, auxiliaries and constants. Levels are indicated by the rectangular boxes, rates are noted by the signal flow meters and auxiliaries are shown without boxes. The exogenous parameters (constants) are from the ANP Risk Priority Values (R_{PV}) and are represented in hexagonal boxes. These constants influence the level of risks in the system. The sources and sinks of risks are shown as "pools". The straight and curved solid lines indicate information flow among the various components of the system models.

Model equation formulation

After the stock and flow diagram is developed for the economic risks system model, mathematical model (equations) for individual system variables (endogenous variables) are also formulated. It is important to note that quantifying risks on an absolute scale is a difficult proposition compared to measurement of vectors as forces in newton. So therefore, we chose to measure risk as dimensionless. Absolute units obtained for each variable in the systems are later expressed as a percentage to demonstrate the level of impact of individual risks on the project performance.

Model Evaluation Tests

After the model equation formulation, two assessment tests are conducted to check the structural conditions of the model. First, structure verification is performed to compare the structure of the model directly with structure of the real system that the model represents. To pass the structure-verification test, the model structure must not contradict knowledge about the structure of the real system. As a result, a review of model assumptions was carried out with the help of two Project Managers who are highly knowledgeable about corresponding parts of the real system.

Second, a dimensional consistency test is conducted on the SFM. This test is conducted to satisfy the dimensional consistency of the models. By inclusion of parameters with little or no meaning as independent structural components, often reveals faulty model structure when this test is performed. Messages from Vensim's built-in function must indicated that the dimensional consistency checks and the structural assessment tests conducted on the risks models for level (stocks), auxiliaries, constants, units and their speed are consistent.

Table 9.10 Mathematical equation for the economic risks system variables

Code	System Variables	Equations	Measurement
P_{R3}	Economic risks	INTEG (Economic uncertainties-Economic certainties, 0.25)	Dimensionless
E_{V1}	Government funding policy	Government discontinuity*Initial AHP value (R_{PV}) for E_{V1} (0.17)	Dimensionless
E_{V2}	Taxation	Government discontinuity* Initial AHP value (R_{PV}) for E_{V2} per Unit time	Dimensionless
E_{V3}	Government discontinuity	Initial AHP value (RPI) for E_{V3} (Constant)	
E_{V4}	Wage inflation;	0.03*((Energy price +Foreign exchange +Local inflation rate)+(Material price per unit time))	Dimensionless/Year
E_{V5}	Local inflation	(Economic risks/Unit time)+Energy price +Foreign exchange	Dimensionless /Year
E_{V6}	Foreign exchange	(Material price*0.03) per Unit time	Dimensionless /Year
E_{V7}	Material price	INTEG (Material price change hike, 0.08)	Dimensionless
E_{V8}	Economic recession	Initial AHP value (R_{PV}) for E_{V8} (Constant)	Dimensionless
E_{V9}	Energy price	Economic recession*0.05	Dimensionless
E_{V10}	Catastrophic environ. effects	Initial AHP value (R_{PV}) for E_{V10} (Constant)	Dimensionless
E_{V11}	Project technical difficulties	Initial AHP value (R_{PV}) for E_{V11} (Constant)	Dimensionless
E_{V12}	Project delays of all forms	Government funding policy * Government discontinuity*0.19	Dimensionless
E_{V13}	Cost of delays	Delay of all forms	Dimensionless
E_{V14}	Cost of dispute resolution	Disputes	Dimensionless

E_{V15}	Disputes	Government funding policy	Dimensionless
E_{V16}	Economic certainties	(Risks of project time overrun * Economic risks) per unit time	Dimensionless/Year
E_{V17}	Economic uncertainties	(Economic risks +Government funding policy)*((Local inflation rate +Taxation + Wage inflation)−(Energy price + Material price/unit time))	Dimensionless/Year
E_{V18}	Error generation	Error generation= (Risks of project cost overrun*Worksite coordination problems) per unit time	Dimensionless/Year
E_{V19}	Escalation to project cost overrun	((Escalation to project time overrun*Error generation)*unit time)+ (Project quality deficiency/ unit time) + ((Cost of delays +Cost of resolution)/ unit time))	Dimensionless /Year
E_{V20}	Escalation to project time overrun	((Error generation +Economic uncertainties) + (Disputes/unit time)) + ((Catastrophic environmental effects* Project technical difficulties)*"Delay of all forms (Utilities diversion, Design changes, obtaining consents & legal actions))/unit time	Dimensionless /Year
E_{V21}	Material price hike	((Material price)*(Local inflation rate + Economic recession)) per unit time	Dimensionless /Year
E_{V22}	Project quality deficiency	INTEG (Error generation, 0)	Dimensionless
E_{V23}	Risks of project cost overrun	INTEG (Escalation to project cost overrun, 0)	Dimensionless
E_{V24}	Risks of project time overrun	INTEG (Escalation to project time overrun, 0)	Dimensionless
E_{V25}	Worksite coordination problems	(Risks of project time overrun*Project quality deficiency +Economic risks)	Dimensionless

Dynamic simulation results and discussion

In system dynamics simulation, trend analysis is given priority and numbers do not have much significance, however, the numbers should be, as far as possible, close to the real life situations. In the context of the economic risks modelling, the AHP input to the system to conduct simulation is represented in Table 9.11.

Following the ANP's input into the economic risks system, dynamic simulations are performed under the following time bounds and units of measurements for all system variables:

I   The initial time for the simulation = 2008, Units: Year
II   The final time for the simulation= 2015, Units: Year
III The time step for the simulation = 0.125, Units: Year
IV   Unit of measurement for system variables = Dimensionless

The dynamic simulations for the economic risk system are demonstrated below. The aim of these simulations is to reveal the patterns of risk influences as well as to generate numerical values of the level of risk impacts on project time, cost and quality.

Results from the system dynamic simulation patterns are further generated to investigate the influence of exogenous parameters such as economic recession, government discontinuity, catastrophic environmental effects and project technical difficulties on the project time and cost performances in two ways. First the Risk Priority Indexes (RPIs) or Risk Priority Values (R_{pv}) used as inputs into the SD were fixed to *no influence* impact level (0%) for the base-runs simulation and secondly to *high influence* impact level (100%) for the current or actual simulation run based on the actual risk priority index obtained from the ANP pairwise calculations. As seen in Figure 9.12, the initial dynamic pattern based on

Table 9.11 ANP inputs to the economic risk system modelling

System Type	Code	System Variables	ANP's R_{pv} (%)	Description
Economic risks	P_R	Economic risks	25	*Base value
	E_{V3}	Government discontinuity	13	Constant
	E_{V8}	Economic recession	03	Constant
	E_{V10}	Catastrophic environmental effects	13	Constant
	E_{V11}	Project technical difficulties	15	Constant

Note:
1   P_R refers to potential economic risks
2   * refers to assumed base value (%) of all risks for the economic risks impact on project objectives
3   E_V refers to economic risk variable

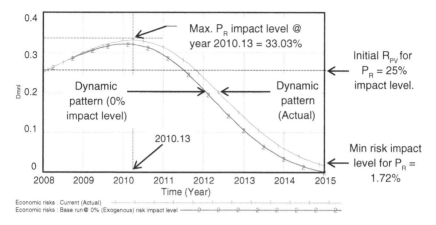

Figure 9.12 Base run and actual scenario simulation patterns for economic risks

the actual simulation run turns to increase steadily even with no risk influence from the exogenous system variables within the first two years of the project before declining after 2010.

However, when the values inputted in the system were replaced with the RPIs, it can be observed that after approximately two years and 48 days (2008–2010.13), the dynamic impact pattern of the economic risks (P_{R3}) increased steadily to reach a maximum point of 33.03% before declining to a minimum value of 1.72% in year 2015. It is important to note that even with no influence from the exogenous system variables; the level of economic risks impact was as close to 32% in year 2010 and 48 days into year 2011 but declined steadily to 0% in year 2015. From a holistic point of view, whether values of the exogenous variables are changed or not, the overall dynamic patterns for the risks of project time and cost overruns would have increased to a considerable level and would then become a critical point for the megaproject developer to assess what the cause might be. This is where the experience of a project manager's ability to plan for effective risks assessment will come into play. Therefore, SD/ANP based simulation should only substantiate or aid managerial decisions.

Further to the dynamic patterns of the economic risks, Table 9.12 illustrates the numerical simulation results for the variables within the economic risks system model. It can further be observed on Table 9.2 and Figure 9.13 that project time and cost are all impacted by economic risks. The mean impact level of economic risks (P_R) on ETN project is revealed to be 21.50%. Time was the most sensitive to the impact of economic risks. The mean scores of project time and cost overruns are 30.74% and 22.36% respectively on the project.

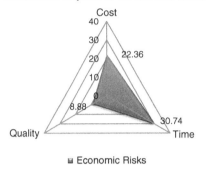

Measured Impact of Economic Risks

■ Economic Risks

Figure 9.13 Measured impact of economic risks

Table 9.12 Dynamic simulation results for the economic risks system model

Risk code	Risk Type	Expected Level of Risk in the project (%)					
		Min	Max	Mean	Median	StDev	Norm
P_{R3}	Economic risks	1.72	33.0	21.51	26.07	10.73	49.86
E_{V1}	Change in government funding policy	2.21	2.21	2.21	2.21	0.00	0.00
E_{V2}	Taxation	0.39	0.39	0.39	0.39	0.00	0.00
E_{V4}	Wage inflation;	1.01	1.67	1.50	1.57	0.19	12.41
E_{V5}	Local inflation	3.29	33.7	22.46	26.67	10.37	46.17
E_{V6}	Foreign exchange	0.24	1.41	0.80	0.77	0.39	49.42
E_{V7}	Material price	8.00	47.1	26.58	25.65	13.14	49.42
E_{V9}	Energy price	0.15	0.15	0.15	0.15	0.00	0.00
E_{V12}	Project delays of all forms	0.05	0.05	0.05	0.05	0.00	0.00
E_{V13}	Cost of delays	0.05	0.05	0.05	0.05	0.00	0.00
E_{V14}	Cost of dispute resolution	2.21	2.21	2.21	2.21	0.00	0.00
E_{V15}	Disputes	2.21	2.21	2.21	2.21	0.00	0.00
E_{V16}	Economic certainties	0.00	9.56	5.13	5.13	3.06	59.58
E_{V17}	Economic uncertainties	−3.4	7.22	1.82	1.61	4.19	230.7
E_{V18}	Error generation	0.00	15.4	4.56	4.18	3.61	79.23
E_{V19}	Escalation to project cost overrun	0.023	0.35	0.115	0.083	9.29	80.49
E_{V20}	Escalation to project time overrun	5.01	16.0	8.59	8.67	0.0244	28.42
E_{V21}	Material price hike	2.27	8.33	5.54	5.55	1.94	35.03
E_{V22}	Project quality deficiency	0.00	30.6	8.88	5.71	8.90	100.3

E_{V23}	Risks of project cost overrun	0.00	77.8	22.36	13.91	21.85	97.72
E_{V24}	Risks of project time overrun	0.00	59.2	30.74	33.99	17.02	55.35
E_{V25}	Worksite coordination problems	15.9	33.5	25.62	26.90	6.29	24.56

Note: Current simulation runs Time (Year) for the economic risks model = 2008 to 2015.

Conclusion

This chapter of the book presents a combined ANP and SD model that can be used to prioritize and simulate economic risk impact on construction and engineering projects. The impact of economic risks such as increase in foreign exchange and inflation, change in government, disputes, change in tax regime and energy price increases on project schedule and cost were investigated on the Edinburgh Tram Network project.

It is worth mentioning however that the most accurate industry-specific parameter values were not available for all types of risk. However, our scenarios for the dynamic risk patterns covered the most possible parameter ranges obtained on the case study and the results follow the general patterns expected. Nevertheless, the model helps to prioritize identified risks and conduct comparative simulation of different scenarios to investigate the effect of changes in different variables of interest on project performance. As an innovative way of combining ANP and SD to assess risks, the approach will provide a complete framework for understanding the criteria used for evaluating and assigning ratings to system elements and the dynamic inter-relationship among those elements. The proposed model could be used by project managers, sponsors and policy makers involved with the procurement of large construction and engineering projects to enable a systems thinking approach to project delivery. A later version of this approach in an upcoming risk management book by the authors will attempt to integrate the impact of other project risk clusters like socio-technical risk as well as environmental and political risk on transportation megaprojects using SD-ANP approach.

References

Boateng, P. (2014). *A Dynamic Systems Approach to Risk Assessment in Megaprojects*. PhD dissertation. Heriot-Watt University, Edinburgh, UK.

Boateng, P., Chen, Z. and Ogunlana, S. O. (2015). An analytical network process model for risks prioritisation in megaprojects. *International Journal of Project Management*, 33(8), 1795–1811.

Chen, Z., Li, H., Ren, H., Xu, Q. and Hong, J. (2011). A total environmental risk assessment model for international hub airports. *International Journal of Project Management*, 29(7), 856–866.

Fiske, S. T. and Taylor, S. E. (2013). *Social cognition: from brains to culture.* London: Sage.

Forrester, J.W. (1961). *Industrial dynamics.* Cambridge, Massachusetts: The MIT Press.

Ghaffarzadegan, N., Lyneis, J. and Richardson, G. P. (2011). How small system dynamics models can help the public policy process. *System Dynamics Review,* 27(1), 22–44.

Madachy, R. J. (2007). Software process dynamics. London: John Wiley & Sons.

Meyer, M. A., and Booker, J. M. (2001). *Eliciting and analyzing expert judgment: a practical guide.* Society for Industrial and Applied Mathematics. London: Acadamic Press Limited.

Ranganath, B. J. and Rodrigues, L.L.R. (2013). System dynamics: theory and case studies. New Delhi: IK International Publishing House Pvt Ltd.

Saaty, T. L. (1996). Decision making with dependence and feedback: The analytic network process. Pittsburgh: RWS Publicati ons.

Saaty, T. L. (2000). *Fundamentals of decision making and priority theory with the analytic hierarchy process* (Vol.6). Pennsylvania: RWS Publications.

Saaty, T. L. (2005). *Theory and applications of the analytic network process: decision making with benefits, opportunities, costs, and risks.* Pittsburgh: RWS Publications.

Sterman, J. D. (1992). *System dynamics modeling for project management.* Unpublished manuscript, Cambridge, MA, 246.

Sterman, J. D. (2010). *Business dynamics: systems thinking and modelling for a complex world.* West Patal, New Delhi: Irwin, McGraw-Hill.

10 Delphi technique in risk assessment

Djoen San Santoso

Introduction

Assessment of risks can be done through qualitative and quantitative approaches. For each approach, many methods are already available for researchers to use. The quantitative approach typically requires more respondents than the qualitative approach for the analysis. However, the qualitative approach demands well qualified persons who have expertise in the subject to produce reliable results. One of the methods that can be used to assess risks is the Delphi method, which in the process can involve both qualitative and quantitative approaches.

The Delphi technique was initially developed by Dalkey and Helmer (1963) from Rand Corporation as an experimental application using experts conducted for a military purpose in the 1950s during the Cold War between the United States and the Soviet Union. In the application of the method, a convergence in the opinion of experts is expected from the iteration process of the topic under study. In the development, the method has been applied in many fields with modified processes and requirements to suit the needs of the research.

Delphi technique

The Delphi method is designed as a structured and systematic group communication process among experts in the field to analyze and discuss a specific issue within the domain of the field. The communication process in Delphi is not a face-to-face approach such as in a focus group discussion. Respondents do not need to allocate a certain specific time to come to a designated location to meet each other. Especially with the advancement of technology nowadays, the Delphi process can be done easily without any hurdles. The method applies a systematic and interactive technique to obtain assessment and judgement of experts on a specific topic under their expertise (Hallowell and Gambatese 2010).

The result of Delphi significantly depends on the judgement of respondents. Therefore, respondents in Delphi technique should have high expertise and solid knowledge in the subject under investigation, which makes them experts (in this chapter, respondents and experts will be used interchangeably to indicate persons with certain knowledge and expertise that participate in a Delphi study). By collecting the respondents' opinions, evaluation and assessment, comments are compiled and summarized, and statistical analysis is applied to produce results on the assessment. These are communicated to all respondents as feedback and as consideration for respondents to review their previous assessment. This iteration highlights the necessity for the respondents to have informed judgement. It facilitates a process of providing more information and consideration to respondents that allows them to reconsider their opinion or assessment based on new information towards building consensus. This whole process is the communication and feedback system of the Delphi technique.

In the process, the Delphi may initiate the procedure using a check list to screen out variables or requesting ideas or alternatives on the topic under study. The technique can be an alternative solution to the possible difficulty of conducting focus group discussions where members have to be present at a certain location and time for the group discussion. The Delphi provides a methodology that emphasizes group communication to achieve a general consensus without having to have a face to face meeting. While preserving the communication aspect among respondents, the Delphi technique can be expected to reduce possible biases or issues that may occur due to face-to-face group discussion.

Essential characteristics

The growth of researchers using Delphi indicates that the technique is well accepted as a scientific methodology in finding solutions for real world issues. How the Delphi technique is different from other techniques in acquiring opinions, evaluation and assessment defines the characteristics of Delphi compared to other techniques.

Anonymity of respondents

As there is no face-to-face meeting among respondents in the Delphi technique, respondents are not expected to know each other and there is no direct interaction among respondents in the process. In this controlled interaction, all communications are provided through and by the research facilitator. In certain critical research studies, respondents may be required to maintain the secrecy of the research by not discussing the research with others. This anonymity facilitates higher freedom for respondents in expressing their professional opinion, evaluation and assessment on the topic being studied without restriction and pressure.

The controlled interaction, where there is no direct confrontation among respondents, is expected to avoid rushed formulation of pre-conceived notions, facilitate openness of respondents' minds on novel ideas, avoid the tendency to defend someone's idea in front of others, and avoid the tendency to easily follow someone's persuasive opinions (Dalkey and Helmer 1963). The effect of dominant individuals (Dalkey 1972) and manipulation to adopt a certain opinion (Adams 2001), which are usually the concerns in group-based approaches, can be reduced. This can be considered as the major benefit of applying the Delphi technique.

An expert is expected to be someone whose opinion or assessment can be relied on. This kind of connotation, which can be an unnecessary burden, may attach to the expert. Consequently, with face-to-face meetings, there is a high tendency a researcher may need to defend her/his uttered opinion or assessment in public even when new or additional information is available. With the privacy facilitated in Delphi, the burden to defend opinion or assessment already expressed is removed. Fear of losing face because of changing already expressed opinion or assessment does not exist. Therefore, each expert is expected to have more freedom to review her/his opinion or assessment based on new or additional information that probably may have been unintentionally neglected. The relatively flexible time in Delphi also stimulates more opportunity for respondents to digest the new or additional information relative to their earlier opinion or assessment.

Besides the benefits that have been mentioned above, a concern is raised on the possibility in manipulating the result to produce feedback beneficial to a certain entity or party involved in the topic under investigation. With no direct contact among respondents and the anonymity of respondents, it is undoubtedly easier to manipulate the Delphi than other qualitative techniques that required direct contact among respondents. However, this does not mean that manipulation cannot be done in other techniques. As asserted by Linstone and Turoff (2002), it is all dependent on the intention of the researchers. The problem is not in the technique or methodology. If the intention is already not good, all methods can be directed to produce biased or pre-designed results or outputs.

Assessment of group opinions

In achieving the purpose of having consensus building, the opinions collected from respondents have to be analyzed to understand how far the opinions vary among respondents. For qualitative opinions, descriptive statistics can be applied if similar answers are expected from respondents. In the case that the qualitative method is used for retrieving alternatives or ideas from respondents, which typically applied at the initial round, listing all ideas or alternatives can be the result of the first round. This result is usually used for the second, which is most likely to be quantitative to sort out the ideas or alternatives.

When quantitative data is collected from respondents, the typical statistics applied are simple measurements of central tendency (mean, mode and median) and level of dispersion (standard deviation and quartile range). It should be kept in mind that when the scale of measurement used to collect the data is neither interval nor ratio, mean should not be used to measure the central tendency. This is commonly the case when the Likert scale is used; therefore, median and mode are the two preferable average measurements. However, in some cases, which is also widely practised, the Likert scale applied to explain ordinal measurement (such as very low to very high, never to always) is considered as ratio rather than ordinal. Therefore, the decision solely belongs to the research as long as it can be justified and is consistent.

Feedback process with opportunity to revise opinion

The assessment of the group opinions should be communicated to all respondents as the feedback. Comments or argumentations from the respondents need to be summarized and integrated in the feedback report. Additional information may be provided as feedback to all respondents when it is requested by any respondents for their consideration and it is available and suitable for the research purposes. Alternatively, information may also be provided when the information is deemed to help respondents to provide better judgement for their assessment. In any cases, the information should be fair and should not lead respondents towards a certain result or direction on their assessment.

This feedback process is the unique characteristic of the Delphi that is supposed to be in the process. Without this feedback process, the opportunity for respondents to revise their opinion does not exist or cannot be justified and the process of having systematic communication to build consensus will not materialize. Feedback that consists of group opinion and possible additional new information acts as a mechanism to update each respondent how his/her opinion stands relative to the opinions of other anonymous respondents. This gives the opportunity to a respondent to reevaluate his/her original opinion, which works as an iteration process to converge the opinion.

A few studies claimed applying Delphi as the methodology of research without providing feedback of the group opinion in the process consequently means respondents do not have opportunity to revise their opinion. It is true that the Delphi technique has evolved over time but when the basic characteristics are removed from the process, the substance of applying Delphi has deviated from the key feature of the technique. Hallowell and Gambatese (2010) stated that without iteration and provision of controlled feedback, the technique cannot be called Delphi.

Important factors in applying Delphi method

Panel of respondents

One factor that defines the reliability of the results of a Delphi research is the respondents on whom the survey is conducted. The respondents in Delphi are the experts in the topic. The qualification of respondents or experts as members of the panel has a crucial role in determining whether the respondents are indeed qualified to provide their evaluation and assessment on the topic under study. In this regard, it is important that the requirements and qualifications of respondents are properly defined and explicitly stated as part of the report. The need to have certain academic requirements or industrial experience or hold a recognized position in the professional community depends mainly on the nature and topic of the research. One study may require both industrial and academic requirements and qualifications. Another study may place more weight on the professional experience than academic or vice versa.

Minimum requirements and qualifications may be imposed when selecting respondents, such as minimum education level, minimum direct professional experience on the topic, professional registration, etc. A certified Professional Engineer, a Chartered Engineer, a doctoral degree in Construction Management, involvement in five Build Operate Transfer (BOT) projects can be examples of the qualifications and requirements. Alternatively, a more flexible approach can be applied where points are assigned to each requirement and qualification and the total points will define whether a person is qualified to be a respondent or not (Hallowell and Gambatese 2010).

Regarding the number of respondents serving on the panel, there is no consensus on the appropriate numbers to apply the Delphi method available in the literature. Delbecq et al. (1975) suggested that for a research that requires homogenous background of respondents, a total of 10–15 respondents is sufficient. However, when the research requires respondents from various backgrounds, more respondents are expected to provide better results. Hallowell and Gambatese (2010) from their study proposed eight respondents as the minimum number.

In determining the number of respondents, it should be kept in mind, the more respondents in executing the Delphi method, the more time is required to collect, compile and analyze the data. Therefore, the time to complete the research may need to be adjusted with the number of respondents or the time may define the numbers of respondents as long as the numbers can be justified. It is interesting to note that Brockhoff (2002) found that there is no significant correlation between the number of respondents and the accuracy and effectiveness of the Delphi method. Nevertheless, a minimum acceptable number should be properly defined that can provide confidence on the results, as indicated in the previous paragraph.

In addition to the proper selection of experts to provide reliable results, commitment from experts is very crucial for the success of Delphi technique. This method does require experts to give their evaluation and assessment on the analyzed topic only one time like in many ordinary surveys. The process typically requires at least three rounds to provide reliable results. Therefore, it is important that from the beginning, the selected experts should be clearly informed how many rounds the survey will require and the time frame, before they give their commitment to participate in the research.

With many rounds of process in applying Delphi technique, strong commitment from respondents to dedicate their time for the research is important. Keeping this in mind, certain research studies can be a better option for using Delphi technique when:

- the research is on a new trend or topic that enhances or improves the field.
- the research contributes to the betterment of the community or environment where the respondents belong to.
- the research is carried out by an agency or organization positioned at a higher level than the current working place of the respondents.
- the respondents consider that the research a good proxy for them to voice or contribute their opinions, ideas or concerns freely without fear.

Number of rounds

The multiple rounds in the Delphi technique relate to the iteration process to build consensus by providing feedback and opportunity to revise opinions. The number of rounds implicitly indicates the communication among respondents on the research topic provided through the research facilitator. There is no strict guidance on the number of rounds necessary to apply Delphi. However, it should be kept in mind that the number of rounds may not be the same as the number of iterations. It is possible to have one or two rounds in the process that has/have nothing to do with iteration. Typically, this kind of round is the initial round and has open-ended questions for soliciting alternatives or identifying factors from respondents, which is more focused towards qualitative exploration. The result of this round will be used for the next round, which is usually quantitative research, where respondents are requested to rate or rank. In this case, the second round is not the iteration process of the first round as the step of allowing respondents to review their opinion or assessment is not there.

Regarding the iteration process, it is reasonable to assume that the more iterations, the more the opinion of a group converges until a certain level (optimal) where more iteration does not warrant or worth the increase of the precision. Technically, the number of iterations should relate to the measurement of convergence agreed for the research. When a high level of convergence is expected, which means only minor deviation is allowed among the assessment of respondents, it is logical to expect that more iterations are needed. With this situation, consistency in the number of iterations used to run the Delphi technique cannot be expected. Logically, researchers will need a minimum of two rounds to apply the Delphi method, where the second round is the feedback and opportunity to revise. If an initial qualitative inquiry is necessary, as explained in the above paragraph, then three rounds are the minimum number of rounds. In planning research, it should be kept in mind that the number of rounds is associated with the expense to carry out the research and the time necessary to complete the project. Therefore, the process and consensus should be designed properly to comply with the time and budget of the research.

One aspect that should be considered in the implementation of the Delphi technique related to the number of rounds is the time gap between rounds. An analysis is necessary for each round to be reported as feedback for the next round. With no direct interaction among respondents as in a meeting or focus group discussion, the process provides flexibility for respondents to respond and for facilitators to analyze the data. The schedule for these should be strictly maintained and the time gap between rounds should be set as small as possible to make sure that the topic is still hot in the mind of respondents. Real-time Delphi (Linstone and Turoff 2002) has been offered as a solution for this, where an online survey or a program may be necessary to be developed to automatically retrieve the opinions of respondents and immediately analyze them to produce feedback in a short time.

Consensus measurement

The method to measure consensus is another important factor that needs to be established as a decision rule in applying Delphi. A consensus generally can be achieved when the diversion of data is small or can be neglected or is less than the maximum value that has been agreed. Standard deviation or variance may be used as a measurement. Unfortunately, there is no strict rule for Delphi on this regard, most likely due to the reason that each research collects unique data, which make guidance or a standard inappropriate (Hallowell and Gambatese 2010).

Defining a certain percentage of respondents that vote on certain ranks or scales can be used as a decision rule for consensus. For example, on a seven-point Likert scale on agreement (0 for strongly disagree and 7 for strongly agree), it can be decided that consensus is reached when 80% of respondents voted for the last two scales. Alternatively, when the data is in ordinal scale, absolute deviation as suggested by Hallowell and Gambatese (2010) can be applied. In their method, the variability is measured relative to the median rather than the mean as in standard deviation.

Another view in deciding when Delphi data has reached convergence is "stability", which is different from consensus. Stability as defined by Dajani et al. (1979) is the consistency of responses between successive rounds of the Delphi process. They claimed that consensus without stability is meaningless. Therefore, stability should be the measurement to decide that a Delphi survey has reached its expected result. Statistical tests to check agreement or association of data between two rounds can be used to analyze stability of the result, to name some of them: paired sample t-test, Wilcoxon signed-rank test, Spearman's rank correlation, etc. For a more detailed discussion on the consensus measurement, a study by von der Gracht (2012) where he rigorously reviewed applied methods in measuring consensus and stability both for descriptive and inferential statistics can be referred to.

An example of applied Delphi method

In analyzing risk for public-private partnership (PPP) in road infrastructure projects, Delphi technique is used for the assessment of risk events. In the methodology, three rounds of Delphi were developed for the purpose. Respondents who served on the panel of experts were selected from practitioners and academicians based on the following criteria:

Non-academician expert:

- Has at least ten years of experience in road construction projects.
- Has direct experience and responsibility in managing risk for at least two years.
- Has a direct involvement in at least one PPP road project in the last three years.
- Has a position at least as a senior engineer or the same level.

Academician expert:

- Has a doctoral degree in the field related to construction or risk management.
- Has at least two journal papers related to risk and PPP, as a first or second author, published in well-recognized international journals.
- Works in a reputable university.

Fifteen candidates were identified to serve as experts. They were contacted and informed about the research and the Delphi method that would be applied for executing the research. Out of the 15 candidates, nine experts were willing to participate, consisting of six persons from non-academic (industry and government) and three persons from academic backgrounds.

The detailed discussion on the three-round Delphi technique designed for the research is provided as follows:

Round 1

A list of risk events related to the implementation of PPP was compiled from literature review. The list was categorized into nine categories, which consist of 123 risk events. In the first round, experts were requested to identify which risk events should be significantly considered for PPP in road projects. The definition of "significant" is not provided to give freedom for experts to interpret this based on their own judgement. If a risk event is considered significant, a cross (x) should be given in the box relevant to the risk event.

In analyzing the result, a rule was established that risk events that were considered as important by at least six experts (66.67% of the total experts) were selected as important risks. Based on this rule, 55 risk events were considered as important.

Round 2

In round 2, the result of round 1 was presented and experts were requested to assess the risk events in two different perspectives: degree of occurrence and degree of impact. A five-point Likert scale was used for both perspectives, where 1 indicates very unlikely and 5 indicates very often for the degree of occurrence, and for the degree of impact, 1 represents very low and 5 represents very high. Additionally, experts were also requested to evaluate which party should best manage the risk events with three options provided: government, investor or shared management.

Round 3

This round was aimed to be an iteration process for experts to review their assessment. As feedback for experts, the mean value of each perspective for each risk event together with the bar chart showing the distribution of the assessment from experts were presented together with the expert's own assessment. Regarding the risk allocation, the bar chart indicating distribution of the experts' opinion to which party risk should be allocated or shared was presented.

In addition to the results from analyzing the data, the allocation of risks suggested by three international journal papers with case studies coming from the same region (in this case: Asia Pacific region) were also provided for consideration. A brief summary of each case study was explained to provide description and understanding of the case. In providing fair and useful information, the case studies should not be selected to match the *a priori* expectation of the risk allocation, if there is any. The information should be purely intended to provide more materials for contemplation.

Even though assessment of risks from the degree of occurrence and impact from other studies in different countries are available, the occurrence and severity of risks are considered to be more country specific. Meanwhile, the allocation of risks are done by considering the party best able and suitable to manage the risk, which are less likely to be affected by locality. With this consideration, only the allocations of risks were provided to experts as additional input and thought.

In the form circulated, experts were requested to review their initial assessment based on the provided feedback. However, it was also clearly stated that they should only revise their initial assessment when they considered it fit and necessary.

The comparison between round 1 and round 2 was presented in terms of the level of agreement. In the risk assessment for both perspectives, an agreement was considered to be achieved when at least 75% of experts voted on either 1 and 2 or 4 and 5. Therefore, the agreement can be on the low probability (1 and 2) or high probability (4 and 5) for the degree of occurrence, and low impact (1 and 2) or high impact (4 and 5) for the degree of severity. In the same way, the level of agreement for risk allocation is reached when 75% of experts agree on the allocation. The comparison shows improvement of consensus among experts in the second round as presented in Table 10.1.

As shown in Table 10.1, improvement on consensus can be observed where the number of disagreements reduced in the second round for both the degrees of occurrence and impact with only one risk event that cannot reach consensus from the perspective of the degree of occurrence. Reduction in disagreement can also be observed for the allocation of risk

Table 10.1 Comparison of consensus between first and second rounds

Level of agreement	First round			Second round		
	Degree of occurrence	Degree of impact	Risk allocation	Degree of occurrence	Degree of impact	Risk allocation
Not agree (< 75%)	7	5	8	1	0	3
Agree (75% and above)	48	50	47	54	55	52

to the appropriate party. However, there are still three risk events where experts have significantly different opinions on the allocation. This may be due to different conceptions and interests that are perceived by each individual expert due to their background. The non-academician experts consist of persons from government, consultants and contractors. In this regard, their attachment may influence their evaluation on how the risk events should be allocated.

Following the result of the second round, the only risk event that could not reach consensus was not included for further consideration. From the 54 risk events, critical risk events were identified by screening only risk events that have high probability of occurrence and high impact (remember, it is possible to have consensus on low occurrence or low impact). Regarding the allocation of risk, the finding was presented as the result, where experts could not find agreement on the allocation of the three risk events, whether they should be best managed by investors or their management shared. This indicates the need to have further discussion, analysis and better formulation of these three risk events.

References

Adams, S. J. (2001). Projecting the next decade in safety management: A Delphi technique study. *Professional Safety*, Vol. 10, 26–29.

Brockhoff, K. (2002). The performance of forecasting groups in computer dialogue and face-to-face discussion. In Linstone, H. A. and Turoff, M. (Eds). *The Delphi method: Techniques and applications* [Electronic version]. Newark, NJ: New Jersey Institute of Technology.

Dajani, J. S., Sincoff, M. Z. and Talley, W. K. (1979). Stability and agreement criteria for the termination of Delphi studies, *Technological Forecasting & Social Change*, Vol. 13, 83–90.

Dalkey, N.C. (1972). The Delphi method: An experimental study of group opinion. In Dalkey, N. C., Rourke, D. L., Lewis, R. and Snyder, D. (Eds.). *Studies in the quality of life: Delphi and decision-making*. Lexington, MA: Lexington Books.

Dalkey, N. C. and Helmer, O. (1963). An experimental application of the Delphi method to the use of experts. *Management Science*, Vol. 9(3), 458–467.

Delbecq, A. L., Van de Ven, A. H. and Gustafson, D. H. (1975). *Group techniques for program planning: A guide to nominal group and Delphi processes*. Glenview, IL: Scott, Foresman, and Co.

Hallowell, M. R., and Gambatese, A. (2010). Qualitative research: Application of the Delphi method to CEM research. *Journal of Construction Engineering and Management, ASCE*, Vol. 136(1), 99–107.

Linstone, H. A., and Turoff, M. (2002). *The Delphi method: Techniques and applications* [Electronic version]. Newark, NJ: New Jersey Institute of Technology.

von der Gracht, H. A. (2012). Consensus measurements in Delphi studies: Review and implications for future quality assurance. *Technological Forecasting & Social Change*, Vol. 79, 1525–1536.

11 Risk management in enterprise resource planning implementation

Prasanta Kumar Dey, Ben Clegg and Walid Cheffi

Introduction

Enterprise Resource Planning (ERP) is designed to provide seamless integration of processes across functional areas with improved workflow, standardization of business practice, and access to real-time, up-to-date data. As a consequence, ERP systems are complex and implementing them can be a challenging, complex, time consuming and expensive project for any company (Yusuf et al. 2004; Koh and Simpson 2007). Although there are a multitude of installed ERP systems, ERP projects continue to be considered risky to implement in business enterprises as they fail for a variety of closely interconnected organizational and technical factors (Hallikainen et al. 2009). In fact, the introduction of any large-scale integrated information system (i.e. ERP) can lead to significant changes in processes, tasks and people related issues (Kraemmerand et al. 2003; Winter et al. 2006). The particular risks associated with ERP projects make it necessary for organizations to deploy risk management approaches throughout the project life-cycle (Aloini et al. 2007). The need for risk management approaches arises due to the lack of effective guidance on the implementation of ERP (Ngai et al. 2008). The reluctance of companies to communicate about implementation failures (Hakim and Hakim 2010) does not make it easy for researchers to propose and test efficient frameworks. So far, the literature on ERP projects (Kutsch and Hall, 2005; Bakker et al. 2010) reveals that (1) knowledge of the risks alone is not sufficient for companies deploying a risk management approach; and (2) the deployment of systematic and scientific approaches by managers is uncommon.

Based on publications from 1997 to 2009, Baker et al. (2010) also state that the current studies are largely based on how risk management is assumed to work instead of how it is actually used in project practice. The main purpose of this chapter is to develop a new Risk Assessment Framework (RAF), which enables better management of those risks. The study distinguishes itself from the existing literature on ERP risks management by adopting a more balanced and integrative framework as it demonstrates a practical and holistic approach to identifying and managing risk in ERP implementation.

It integrates risk identification, analysis and control by classifying risk hierarchically (external engagement, programme, work stream and work package levels) across technical, schedule, operational, business and organizational categories. This not only helps develop responses to mitigate risks but also facilitates risk control through the organizational hierarchy.

The chapter firstly reviews the literature in order to identify various risk factors and challenges of ERP implementation, along with any available risk management frameworks. Then using a case study it demonstrates the application of the new RAF for managing ERP implementation.

Literature review

While there are studies on various aspects of ERP implementation in industry, we still know little on how to practically manage risks and uncertainties of ERP projects as the majority of companies still fail to effectively implement ERP systems.

ERP implementation risks

Davenport was amongst the first to report the challenges of ERP implementation and business process change (Davenport 1998). Several studies state that amongst the main reasons for IT project failure is the misperception of project risks and the inadequateness of risks management by project managers (Aloini et al. 2007; Kwak and Stoddard 2004). In recent years, several researchers have tried to identify the critical success factors for ERP implementation (Mabert et al. 2003; Al-Mashari et al. 2003; Mandal and Gunasekaran 2003; Umble et al. 2003; Woo 2007). Motwani et al. (2002) investigate factors that facilitate or inhibit the success of ERP projects based on a case study methodology comparing a successful ERP implementation with an unsuccessful one. They unveil several ERP implementation risks amongst which are the following: ineffective strategic planning and communication and insufficient project team skills. The authors conclude that careful change management, network relationships, and cultural readiness are key success factors.

Yusuf et al. (2004) focus on the issues behind the process of ERP implementation by means of case study methodology. The paper examines business and technical as well as cultural issues of ERP implementation in Rolls-Royce plc. It highlights the need for adequate communication approaches and business process reengineering (BPR), and for improved project and change management techniques. The authors highlight the necessity for matching the processes to specific software configurations, training senior management and end-users, and educating people to accept change.

Ehie and Madsen (2005) conducted empirical research on the critical issues which impact on the success of ERP implementation. They highlight several ERP risks such as inappropriate consulting services experiences, inadequate BPR, unsuitable ERP selection and low top management commitment.

Aloini et al. (2007) produced the top ten most frequent risk factors based on a literature review. The authors point out that top five risk factors are inadequate ERP selection, ineffective strategic thinking and planning, ineffective project management techniques, bad managerial conduct, and inadequate change management.

A more recent study conducted by Malhotra and Temponi (2009) suggest six key factors which can lead to successful ERP implementation: (1) project team structure, (2) implementation strategy, (3) database conversion strategy, (4) transition technique, (5) risk management strategy and (6) change management strategy. Another study devoted to ERP risks identification is conducted by Hakim and Hakim (2010). The authors categorize the risks involved in ERP implementation from the perspectives of the client-organization and that of the experts. In doing so, they classify risks into six categories related to organization, specialized skills, project management, system, user and technology.

ERP implementation risk could be categorized across the project phases with respect to project management processes, organizational transformation and information technology in order to suggest mitigating measures for each category; see Table 11.1.

Successful implementation of ERP systems can result from effective management of these risks, which are very generic as they have been collated from a wide variety of ERP projects across industries.

Risk management frameworks

Like other project risk management, ERP implementation project risk management needs to be undertaken in three phases – the planning, implementation and post implementation phases. Risk analysis in ERP planning phase is closely associated with ERP systems selection; as prior research recognizes that ERP implementation is a risky endeavor. Teltumbde (2000) adopts a combined nominal group technique and the analytic hierarchy process model to evaluate ERP systems and considers risk as one of the constructs. Lee and Kim (2000) applied the analytic network process and 0–1 goal programming approach for ERP system selection. Badri et al. (2001) use 0–1 goal programming with multiple criteria such as benefits, hardware, software and other costs, risk factors, preferences of decision makers and users, and completion and training time commitments. The analytic hierarchy process-based approach with risk as one of the constructs has been adopted by Wei et al. (2005). Risk analysis in planning phase addresses part of the issues. There are considerable amounts of risk in any ERP implementation phase because of technical complexity and organizational transformation requirements.

Literature reports adoption of generic methods (e.g. PMI 2008 and AS/NZS ISO 31000, 2009) to manage ERP implementation risks (Aloini et al. 2007). Aloini et al. (2007) report a risk diagnosing methodology, which

Table 11.1 Risk factors in accordance to project phase and risk category

Project phases	Risk categories		
	Project management processes	Organizational transformation	Information technology
Planning	Inaccurate business case. Unclear objectives. Weak implementation team.	Lack of management/executive commitments and leadership. Lack of synergy between IT strategy and organizational competitive strategy. Unclear change strategy.	Lack of communication with the end users. Inadequate training plan for the users.
Implementation	Inappropriate management of scope. Lack of communication between ERP implementation team, ERP provider and ERP users. Poor contract management.	Inappropriate change management. Inappropriate management of culture and structure.	Business process reengineering incompetence. ERP installation incompetence Inappropriate selection of ERP software. Inappropriate system integration. Inaccurate performance data. Inappropriate users training.
Hand-over, evaluation and operations	Inappropriate contract closeout.	Inadequate organizational readiness. Resistance to change. Lack of user training.	Inappropriate system testing and commissioning. Multi-site issues. Lack of clarity on inspection and maintenance. Inaccurate performance measurement and management framework.

consists of context analysis, risk identification, risk analysis, risk evaluation, risk treatment, monitoring and review, and communication and consulting. They also suggest that risk management strategy consists of two approaches – the first aims at reducing risky circumstances, while the second deals with risk treatment after a risk appears. Markus et al. (2000) propose a multi-site ERP implementation framework to deal with the associate risk. They observe that multi-site ERP implementations are tricky on at least four different levels – business strategy, software configuration, technical platform, and management execution. Successful multisite ERP implementations address the interactions and trade-off among the four different sites; their approach is more conceptual than practical. Huang et al. (2004) suggest a more comprehensive approach using risk identification and analysis using Delphi techniques and the analytic hierarchy process respectively. Although their model is theoretically sound, it lacks practical implication.

In summary, prior research has developed several methods for managing risk in ERP implementation that are both theoretically sound and practical for specific cases. However, this study extends this work by providing a more holistic approach, that considers risks *hierarchically* (external engagement, programme, work stream, and work package levels) for technical, scheduling, operational, business, and organizational factors.

Methodology

The case study that follows reports an ERP project in the energy service sector in the UK. It illustrates the use of the new RAF for ERP implementation. In this case study, the project was successfully implemented with an appropriate risk identification and management strategy. Interviews with key actors were conducted using a structured interview guide (Appendix 1); the questions were developed from the literature review. The representatives of client, consultant and ERP vendor project group took part in the interview; all of them had more than 15 years' experience in ERP implementation / industry operations. Data was also collected by means of direct observations and from internal documents.

Case study on risk management for ERP implementation

The following case study shows a customized project risk management framework that successfully helped the implementation of an ERP project in a UK-based energy service provider (hereafter referred to as 'The Group').

The Group

The Group was formed following the privatization of the gas energy market in the UK and a subsequent de-merger of part of the business in 1997. It has since developed into an international business with a total turnover of GBP 13.4 billion. The Group employs over 30,000 people and has expanded globally through a strategy of acquisitions and partnerships in both Canada and the United States. More recently the company has focused on entrance into the deregulating European markets.

As The Group had grown by acquisition and mergers, it now possesses an IT landscape consisting of disparate IT systems and disconnected processes. Accordingly, it has embarked on an ERP implementation and re-implementation strategy with SAP as the chosen ERP solution. The Group already had some sub-optimal ERP (SAP) implementations in parts of their business.

The Group adopted a phased approach focusing on the highest priority process areas first and gradually increasing the ERP modular footprint over a timescale of several years. The first two priorities on its roadmap lay in different process areas and had different strategic drivers and business case models. However, there were several areas of commonality; such as a common ERP platform, a common implementation methodology and approach, and a common approach to the project management team structure and management processes.

ERP implementation project

This case study was a ten-month business transformation initiative consisting of an SAP ERP platform implementation for finance, procurement and HR processes with 1,500 system users and 35,000 payroll records involved. To support its vision, The Group undertook this business transformation project to radically overhaul its back office systems and to reduce cost. The objectives were to achieve cross-functional simplification, automation, standardization and integration. To have their three back office functions working in a fully integrated and largely automated way would provide an invaluable platform enabling the group to begin developing wider improvements, based on a common and flexible backbone.

The project involved implementing SAP's mySAP ERP application suite to support their HR and Finance, and the e-Procurement and Business Warehouse modules to support their operational activities. Overall the new solution provided a platform from which the functions could transform their partnership and integrate to the rest of The Group's businesses far more effectively. The new solution was based on the SAP's Netweaver open platform, allowing legacy SAP and non-SAP applications to be fully integrated. A leading multi-national ERP consultant company was also engaged to plan and implement the project. They worked closely with the ERP provider and The Group's project management team from concept through to commissioning of the project in order to ensure effective implementation and operations. The Group's internal project team, external consultant project team and ERP vendor's project team formed a core multi-skilled

ERP implementation team. The Group's project team was formed through careful selection of experienced and capable people from both functional and IT focused areas.

The project resulted in the migration of significant volumes of complex legacy data (250 million transactions worth GBP 1.53 trillion); the solution was successfully implemented and achieved its objectives to provide simplified and standardized processes across the back-office. The SAP ERP suite provided automated and integrated support for these processes.

ERP implementation risk management process

The core team managed risk in the project using the process shown in Figure 11.1; the process has the following high-level stages: identify and classify risk, analyze risk, determine approach to identified risk, track risk, and mitigate risk.

The risk management process involves a variety of stakeholders – each with different roles and levels of authority; each played a pivotal role in the identification analysis and control of risks.

The Programme Management Office (PMO) Risk Manager 'owns' this risk management process. The PMO generated Risk Update Reports (RURs) – requiring fortnightly risk statistic updates from the Work Package Managers about the programme's performance which were accurately recorded, categorized, and actioned. Action took the form of instruction to the Work Package Managers or escalation to a higher authority – the Release / Area Manager.

The Work Package Manager's role was to assess risks at the point-of-work, raise and update fortnightly RURs, and action instructions given by the PMO Risk Manager.

The Release / Area Manager received RURs fortnightly from the PMO Risk Manager, reviewed risks with Work Package Managers, solved work stream level risks, and decided if some risks required escalation to a higher authority – the Programme Manager.

The Programme Manager provided instruction to Release / Area Managers on risk response strategies, managed programme management level risks, raised and updated risks fortnightly using the RURs, raised

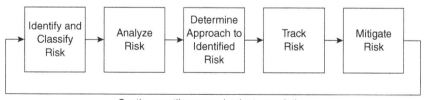

Continue until program/project completion

Figure 11.1 Risk management process (high-level)

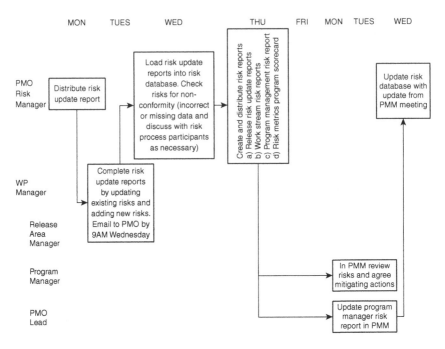

Figure 11.2 Risk control cycle

external risks with the Programme Board. The PMO RURs with critical programme level issues were reviewed in the Programme Management Meetings (PMMs) when the client (The Group) and the vendor (SAP) were present.

The PMOs Risk Manager performed a critical integrative role dedicated to managing the ERP implementation risks. The PMO Risk Manager, Work Package Managers and Programme Manager were all external consultants. The RURs were conducted on a fortnightly cycle as shown in Figure 11.2.

The risk control cycle began with the production of RURs which were distributed to each risk assessment participant. This allowed each individual to raise a new risk – completing each mandatory field in the document – or make updates to existing risks. The RURs were distributed on a Monday morning and returned to PMO with updates on Wednesday morning. The PMO risk manager liaised with each risk participant to ensure the correct information was being reported before the consolidation of RURs into area specific RURs. This normally took place in the form of a phone call or a face to face meeting to verify. RURs were then consolidated which provides relevant risk information to the management team of each workstream; as well as the risk information relevant to other work streams, and the whole project.

However, even though a purposeful risk management process was in place, the core team *did not, until this project took place,* have a risk assessment framework to objectively differentiate the different types of risk in this process. As a result of this work wider knowledge was brought to bear and an ERP risk analysis framework was constructed and used; this became incorporated into the RURs.

The following paragraphs further describe the stages in the risk management process and detail how the new RAF was used.

Risk identification

Risk owners were assigned and their responsibility was to determine the most appropriate treatment for the risk. Possible treatments include: acceptance, mitigation, transference, and reduction. The risk management process was repeated fortnightly and risks were debated in various forums at the appropriate escalation level. This process allowed work package, work stream, and programme risks to be identified and mitigated.

ERP implementation risks were categorized into five key areas – technical, schedule, operational, business, and organizational. Technical risks may arise due to selected technologies (hardware and software) developing performance, quality, reliability, or security problems. Schedule risks may impact the ability to achieve the programme's goals within the proposed schedule. Operational risks evolve because of degrees of uncertainty in estimated implementation costs often due to 'scope-creep' (adding more and more features as the project progresses) which directly impacts on cost. Business risks may occur due to changes to economic or other conditions outside the direct control of the project, which can negatively affect the business case. These can include legal issues, government regulations, marketplace changes, user skills, political considerations, customer stability, and funding. Organizational (internal) risks may prevent completion of the project or realization of ROI. This may include the ability to provide both the physical facilities and appropriate personnel required to support the programme's work efforts.

Additionally, these risks were also given one of four levels – external engagement, programme management, work stream, and work package level, in line with the incumbent reporting structure. External engagement level risks are those that involve client-based concerns and hence require customer actions to help mitigate. These are risks that the client should be made aware of in the interests of the programme as a whole, and have contractual implications. Programme management level risks have potential impact on several work streams, potential impact on significant release milestones, timing, completion, or success as a whole; these are risks that cannot be fully identified or articulated by individual Work-Stream Level Managers. Work-stream level risks impact multiple work packages within the same work stream and require management and risk mitigation by a Work Package

Table 11.2 Generic RAF for ERP implementation

Categories	Levels			
	External Engagement	Program management	Work stream	Work package
Technical (hardware and software)	Legacy system change impact interfaces	Business resources required not available – Business resource may 'overlap'	The project execution deviates from design/principles	Not meeting IT (hardware, software, network, security system) specification
	The project end-users fail to support deployment	Mismanagement of overall IT architecture	Quality at risk due to time/ cost drivers	Data cleansing does not meet the requirements
			Insufficient servers' processing power	
		Fail to transfer knowledge (consultant to the business project resources)	Insufficient data base capacity within SAP for the volume of transactions being migrated across from the legacy systems	
			Telecommunication links with outsourcing partners fails, resulting in a lack of access to SAP by the offshore team	
		SAP profiles do not correspond to organisation roles	IT fails to resolve functional issues	
		Failure to move towards SOX[1] compliance	Delay in hardware procurement	
		Insufficient training facilities available	Decision on system architecture configuration selection was not taken on time	
			Plan is not achievable because of many concurrent activities	
			Inappropriate system testing	

	Late Decisions/Sign-off	Organization fails to adopt change		
Schedule	Legacy systems require changes which would be likely to delay the project			The new system fails to reconcile business information Scope creep
Operational	Communication risk between the project and the business	Failure to deliver benefits as outlined in the business case Inadequate Training The new system fails to provide appropriate financial information The information generated by the new system fails to comply with Data Protection Act	No disaster recovery arrangements System malfunction in post-go-live phase	
Business	Risk that sponsor cancels the project The business suffers 'change fatigue' Business inadequately prepared to take on new solution	Lack of resources available from within the business to fill specific roles		
Organizational	Other projects that are happening in parallel within the business impact the ERP project	Project resources required not available e.g. for training	Project team 'burns out'	Lack of resources in new technology areas being implemented due to their specialist nature Project team turn-over

[1] The Sarbanes-Oxley Act (SOX) not only affects the financial side of corporations, but also affects the IT departments whose job it is to store a corporation's electronic records. The Sarbanes-Oxley Act states that all business records, including electronic records and electronic messages, must be saved for 'not less than five years'. The consequences for non-compliance are fines, imprisonment, or both. IT departments are increasingly faced with the challenge of creating and maintaining a corporate records archive in a cost-effective fashion that satisfies the requirements put forth by the legislation.

Manager. Work package level risks have an impact within a work package and could impact on the completion date, quality levels or costs of the work package; these risks can be managed and mitigated by the Work Package Manager, and do not require escalation to Programme Manager.

Risk factors, as identified in The Group were placed on the RAF as shown in Table 11.2.

Risk analysis

Risk analysis involves analysing the potential impact and probability of identified risks in order to guide risk responses (see Figure 11.3). This stage of the process occurred on an ad-hoc basis within the teams, although the collection and formal logging of risks happened periodically (every fortnight). In order

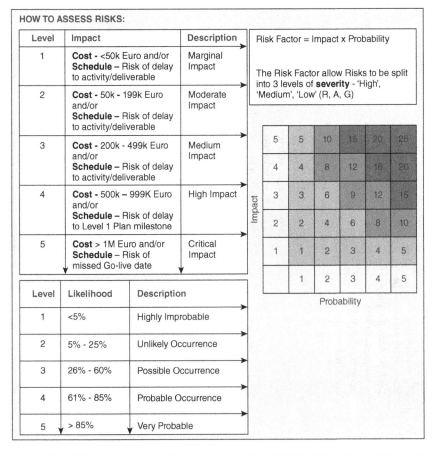

Figure 11.3 Risk assessment scoring: impact and probability (R, A, G stand for red, amber, and green respectively)

to evaluate risks more objectively, standardised scores were used to evaluate each risk, as shown in Figure 11.3; each risk factor is scored in the same way.

Risks were prioritized according to their potential impact on the programme and likelihood of them occurring. Each was rated as a 'High' ('H'), 'Moderate' ('M') or 'Low' ('L') risk severity to The Group's overall ERP implementation programme. These are shown in Table 11.3 using 'H', 'M', and 'L' respectively to represent each 'Impact' and 'Likelihood' (I, L).

By attributing costs to each risk, based on its likelihood of occurring and its level of impact, it was possible to demonstrate the potential risk exposure to the overall programme. The RAF scores and associated cost (not shown here for reasons of confidentiality) were used as an important management decision making tool for making key decisions about the direction of The Group's ERP implementation. The inclusion of costs encouraged the team to consider the full implications of each risk. One key point to note is that the Risk Management Team accepted that the initial estimates of likelihood and impact of risk can be inaccurate. This meant that risk assessors didn't feel initially pressured to spend time on estimating impact unnecessarily; as the Risk Management Team understood that estimates could change over time, along with their associated costs.

The remaining three stages in the risk management process (cf. Figure 11.1) (determine approach to identified risk, track risk, and mitigate risk) are not discussed in this chapter as they were company-specific solutions of less interest to generic practice, and were company confidential. The first two stages have been presented because they provide generic risk analysis principles, by using the new RAF for ERP implementation.

By using the above reporting structure, new RAF, and Risk Control Cycle the Group's ERP project was successfully commissioned; it has subsequently been reported that the application of the proposed risk management framework had brought many benefits. The team believed the RAF helped to successfully achieve the programme's objectives, provided a systematic approach to determining cost-effective risk reduction actions, provided a systematic approach to monitoring and reporting progress in reducing risk, helped identify timeframes for the evaluation of actions and results, encouraged the on-going, systematic evaluation and analysis of risks – whilst focusing upon a continual reduction in risk exposure.

Discussion

ERP implementation projects are inherently risky. Appropriate ERP system selection can considerably reduce subsequent implementation and operational risks. Although there are some frameworks available to help manage risk in implementation, they are typically too theoretical; and their usages are limited mainly due to lack of knowledge by the users. Therefore, a practical risk assessment framework to help manage ERP implementation risk

Table 11.3 RAF applied to The Group's ERP implementation

Categories	Risk Level (Impact, likelihood)			
	External engagement	Program management	Work stream	Work package
Technical (hardware and software)	Legacy system change impact interfaces [H, M]*	Business resources required not available – Business resource may 'overlap' [H, H]	The project execution deviates from design/principles [M, M]	Not meeting IT (hardware, software, network, security system) specification [H, L]
	The project end-users fail to support deployment [L, M]	Mismanagement of overall IT architecture [H, H]	"Quality" at risk due to time/cost drivers [M, H]	Data cleansing does not meet the requirements [M, H]
			Insufficient servers' processing power [H, L]	
		Fail to transfer knowledge (consultant to the business project resources) [H, M]	Insufficient data base capacity within SAP for the volume of transactions being migrated across from the legacy systems [H, M]	
		SAP Profiles do not correspond to organisation roles [H, L]	Telecommunication links with outsourcing partners fails, resulting in a lack of access to SAP by the offshore team [H, L]	
			IT fails to resolve functional issues [H, M]	
		Failure to move towards SOX[1] compliance [H, L]	Delay in hardware procurement [H, M]	
		Insufficient training facilities available [H, L]	Decision on system architecture configuration selection was not taken on time [M, L]	
			Plan is not achievable because of many concurrent activities [H, L]	
			Inappropriate system testing [L, L]	

Category	Risks [Impact, Likelihood]
Schedule	Late Decisions/Sign-off [H, H]; Legacy systems require changes which would be likely to delay the project [H, H]; The new system fails to reconcile business information [M, M]; Scope creep [H, L]
Operational	Organization fails to adopt change [L, L]; Failure to deliver benefits as outlined in the business case [M, L]; Communication risk between the project and the business [M, L]; Inadequate Training [H, M]; The new system fails to provide appropriate financial information [H, M]; The information generated by the new system fails to comply with Data Protection Act [H, L]; No disaster recovery arrangements [H, M]; System malfunction in post-go-live phase [L, L]
Business	Risk that sponsor cancels the project [H, L]; The business suffers 'Change fatigue' [L, H]; Business inadequately prepared to take on new solution [M, L]; Lack of resources available from within the business to fill specific roles [M, H]
Organizational	Other projects that are happening in parallel within the business impact the ERP project [H, H]; Project resources required not available e.g. for training [H, M]; Project team 'burns out' [M, M]; Lack of resources in new technology areas being implemented due to their specialist nature [H, L]; Project team turn-over [M, M]

was needed. This study proposed such a framework which contributed to the successful implementation of an ERP project in a UK-based energy service provider. In the study, using the new Risk Analysis Framework (RAF), the risks were classified into external engagement, programme management, work stream and work package levels as well as technical, schedule, operational, business, and organizational categories. The risks were analyzed using the Risk Analysis Framework, which enabled The Group to quantify risk by likelihood (L) and impact (I) on a high (H) medium (M) and low (L) severity rating. These results helped to develop responses against each risk and assign associated costs. The regular Risk Control Cycles helped manage the changing risk up the organizational hierarchy and over time.

The newly proposed RAF integrates risk identification, analysis, and control by classifying risk hierarchically (external engagement, programme, work stream, and work package), which helps to allocate risks to specific stakeholders for effective mitigation and management. The RAF also categorizes risk as technical, schedule, operational, business, or organizational which helps one to analyze the impacts of risk factors and adopt effective control of all the risks. Understanding the specific nature of a risk helps one to quantify impact and likelihood and prioritize resource deployment for mitigating the risks. Relating risk control mechanism with organizational hierarchy helps appropriate management of risk from initial identification until closure of the specific risk. The attribution of cost to a risk further highlights its potential severity, and the regular cyclical assessments of risk ensures that up to date information is being used.

In risk management, the risk identification phase can have more significance in comparison to the risk analysis and response development phases, because if the risks are not identified correctly any subsequent sophisticated analysis techniques or management responses will be unlikely to produce the desired effects. On the other hand, appropriate risk identification can facilitate both appropriate subsequent analysis and management action. Our newly proposed RAF not only helps the stakeholders (client, consultant, or ERP provider) identify the risks correctly, but also facilitates objective analysis and allows appropriate management of those risks.

This chapter contributes to rethinking the dominant classical models in ERP project management, as research scholars and practitioners need to integrate the multidimensionality of ERP project risks with the dynamics of risk management practices better. In order to address these ERP project issues, and to cope with the inherent risks, we recommend (1) the adoption of an integrative and systematic approach for managing risks, and (2) considering the ERP project as a mix of complex social processes (Winter et al. 2006).

This research has enabled us to propose a pluralist and multi-faceted vision of risk management activities; depending on the risk ratings allocated to various project hierarchy and risk categories. Our work has provided concrete guidance about the introduction of ERP systems in organizations, as well as the management of associated risks throughout the project life-cycle. Inspired by this case study, companies can consider actions that may help bring troubled ERP projects under control.

So far, this research emphasizes the relationship between risk management and the success of ERP implementation projects. It questions the assumptions underlying several studies that claim that good risk management is merely good information system (IS) project management (Kutsch and Hall 2005; Bakker et al. 2009) and fail to recognize that risk management itself should be a generic practice that should be practiced in any large organizational change project. Because of the size and complexity of ERP implementations they are unlikely to be achieved successfully if the whole enterprise into which it is being implemented does not consider the risk as part of a broader initiative; as provided by this new RAF.

Summary and Conclusion

This chapter addresses the implementation issues and challenges of ERP projects. We have adopted an integrative and balanced approach in order to classify risks into four levels and five categories. Firstly, by reviewing the literature, we identified generic risk factors of ERP projects. Secondly, using a case study, a five-stepped Risk Management Process (Figure 11.1) and Risk Control Cycle (Figure 11.2) were introduced. Thirdly, by applying the generic Risk Assessment Framework (RAF) (Table 11.2) and risk assessment scoring (Figure 11.3), risk factors were identified, their likelihood of occurrences and impact were derived and ascertained for The Group (as per Table 11.3). We believe that such a framework is comprehensive, redundancy-free and easily transferable into a wider field. Not only does this chapter propose a Risk Assessment Framework, but also it specifies and tests a systematic method of how to deploy such a framework. In summary, our chapter (1) adds to the disparate literature on the ERP implementation risks and (2) innovates by proposing and testing an integrative and comprehensive approach.

The literature review (Yusuf et al. 2004; Aloini et al. 2007; Malhotra and Temponi 2009) reveals that the key success factors for ERP implementation are: commitment from top management, selecting the appropriate systems and proper management of its integration with existing business information systems – including the reengineering of the business processes. Additionally, this study shows that managing ERP project processes, along with managing information technology and managing organizational transformation effectively makes implementation of ERP projects more successful.

In a proactive approach to risk management, all concerned stakeholders participate in risk identification and analysis for each phase of the project before making decisions on project variables (e.g. resource deployment and allocations, implementation methodology selection, contractors and supplier selection etc.). The success of ERP implementation is partly related to the fact that the stakeholders understand and effectively carry out their ongoing responsibility in the project (Malhotra and Temponi 2009).

ERP projects are technically complex, multidisciplinary, of long duration and are capital intensive; therefore, they can be characterized as highly risky projects. It is sometimes difficult to develop a firm project plan at the beginning of a project due to lack of information at the initial stage; and so, dynamic risk analysis can help to improve knowledge about the project and provide better plans as the project progresses. Although risk management practices increase the project cost in terms of deploying extra human resources and overheads, and additional resources for risk mitigation etc., the benefits (proactive approaches to prevent failure) will ultimately outweigh the costs. Consistent with Peng and Nunes (2009), we call for extending the risk management practices to the post-implementation period. This will help ensure the sustainability of enterprise information systems. A further research avenue would be to determine the specific conditions under which risk management can be effective (Loch et al. 2006) as in The Group observed in our study.

References

Al-Mashari, M., Al-Mudimigh, A. and Zairi, M. 2003. Enterprise resource planning: A taxonomy of critical factors. *European Journal of Operational Research*, 146 (2), 352–364.

Aloini, D., Dulmin, R. and Mininno, V. 2007. Risk management in ERP project introduction: Review of the literature, *Information and management*, 44, 547–557.

AS/NZS ISO 31000, 2009. Risk management – Principles and guidelines, Sydney, NSW.

Badri, M. A., Davis, D. and Davis, D. 2001. A comprehensive 0 – 1 Goal Programming model for project selection, *International Journal of Project Management*, 19, 243–252.

Bakker, K., Boonstra, A. and Wortmann, H. 2009. Does risk management contribute to IT project success? A meta-analysis of empirical evidence. *International Journal of Project Management*, doi:10.1016/j.ijproman.2009.07.002.

Davenport, T.-H. 1998. Putting the enterprise into the enterprise system. *Harvard Business Review*, 16 (4), 121–131.

Ehie, C. and Madsen, M. 2005. Identifying critical issues in enterprise resource planning (ERP) implementation. *Computers in Industry*, 56 (August 6), 545–557.

Hakim, A. and Hakim, H. 2010. A practical model on controlling the ERP implementation risks. *Information Systems*, 35, 204–214,

Hallikainen, P., Kivijärvi H. and Tuominen, M. 2009. Supporting the module sequencing decision in the ERP implementation process – an application of the ANP method. *International Journal of Production Economics*, 119 (2), 259–270.

Huang, S., Chang, I., Li, S. and Lin, M. 2004. Assessing risk in ERP projects: identify and prioritize the factors. *Industrial Management and Data System*, Vol. 104 (8), 681–688.

Koh, L.S.C. and Simpson, M. 2007. Could enterprise resource planning create a competitive advantage for small businesses? *Benchmarking: An International Journal*, 14 (1), 59–76.

Kraemmerand, P., Møller, C. and Boer, H. 2003. ERP implementation: an integrated process of radical change and continuous learning. *Production, Planning & Control*, 14 (4), 338–348.

Kutsch, E. and Hall, M. 2005. Intervening conditions on the management of project risk: dealing with uncertainty in information technology projects. *International Journal of Project Management*, 23, 591–599.

Kwak, Y.-H. and Stoddard, J. 2004. Project risk management: lessons learned from software development environment. *Technovation*, 24, 915–920.

Loch, C.-H., DeMeyer, A. and Pich, M.-T. 2006. *Managing the Unknown.* Wiley, New York.

Lee, J.W. and Kim, S.H. 2000. Using Analytic Network Process and Goal Programming for independent information system project selection, *Computer and Operation research*, 27, 367–382.

Mabert, V.-A., Soni, A. and Venkataraman, M.-A. 2003. The impact of organisation size on ERP implementations in the US manufacturing sector. *OMEGA*, 31 (3), 235–246.

Malhotra, R. and Temponi, C., 2009. Critical decisions for ERP integration: small business issues, *International Journal of Information Management*, 30 (1), 28–37.

Mandal, P. and Gunasekaran, A., 2003. Issues in implementing ERP: a case study. *European Journal of Operational Research*, 146 (2), 274–283.

Markus, M. L., Tanis, C. and Fenema, P. C. V. 2000. Multisite enterprise resource planning implementation, *Communication of the ACM*, 43 (4), 42–46.

Motwani, J., Mirchandani, D., Madan, M. and Gunasekaran, A. 2002. Successful implementation of ERP projects: evidence from two case studies. *International Journal of Production Economics*, 75 (1/2), 83–96.

Ngai, E.W.T., Law., C.C.H. and Wat., F.K.T. 2008. Examining the critical success factors in the adoption of enterprise resource planning, *Computers in Industry*, 59 (6), August, 548–564.

Peng, G.C. and Nunes, M.B. 2009. Identification and assessment of risks associated with ERP post-implementation in China. *Journal of Enterprise Information Management*, 22 (5), 587–614.

PMI (Project Management Institute). 2008. A guide to project management body of knowledge, Guide. 4th ed. Upper Darby, PA: PMI.

Teltumbde, A. 2000. A framework for evaluating ERP projects, *International Journal of Production Research*, 38 (17), 4507–4520.

Umble, E.-J., Haft, R.-R. and Umble, M. 2003. Enterprise resource planning: Implementation procedures and critical success factors. *European Journal of Operational Research*, 146 (2), 241–257.

Wei, C., Chien, C. and Wang, M. J. 2005. An AHP-based approach to ERP system selection. *International Journal of Production Economics*, 96, 47–62.

Winter, M., Smith, C., Morris, P. and Cicmil, S. 2006. Directions for future research in project management: the main findings of a UK government-funded research network. *International Journal of Project Management*, 24, 638–649.

Woo, H.S. 2007. Critical success factors for implementing ERP: the case of a Chinese electronics manufacturer. *Journal of Manufacturing Technology Management*, 18 (4), 431–442.

Yusuf, Y., Gunasekaran, A. and Abthorpe, M. 2004. Enterprise information systems project implementation: a case study of ERP in Rolls-Royce. *International Journal of Production Economics*, 87 (3), 251–266.

Appendix 1

Questionnaire used to develop the ERP Case Study at The Group.

Background

Brief company profile

Brief details of ERP project implemented

Risk management in ERP project implementation

Risk identification:

What risks are likely to occur in ERP implementation?

How was risk identified in the project? Was there a formal approach to identifying risk?

What risks were actually identified?

Who was involved in identifying the risk?

What are the issues and challenges in identifying risk?

Risk analysis:

How was the risk analyzed?

Was there a formal approach to risk analysis?

Who was involved in analyzing the risk?

What are the issues and challenges in identifying risk?

Risk control:

What measures can be taken to control risk when implementing ERP projects?

Was there a formal approach to controlling risk during implementation?

Who was involved in controlling risk during implementation?

How effective were the risk control measures in terms of accomplishing time, cost and quality targets of the ERP project?

Risk in ERP implementation

Based on actual ERP implementation experience what are the various risk events / factors across risk categories (technical, schedule, operational, business and organizational) and levels (engagement external, programme, work stream and work package).

12 Risk management in oil and gas construction projects in Vietnam

Nguyen Van Thuyet, Stephen O. Ogunlana and Prasanta Kumar Dey

Introduction

Projects are exposed to both internal risks (financial, design, contractual, construction, personal, involved parties and operational risks) and external risks (economical, social, political, legal, public, logistical and environmental risks). All the risks may influence cost, schedule or quality of the project in negative ways (Charoenngam and Yeh 1999). Therefore, risk management should be well recognized and handled as an integrated function of project management.

Vietnam is an emerging economy with increasing gross domestic product (GDP) (World Bank 2001). The construction industry is the main contributor to the growth with real expansion of 7.2% in 2001 and 14% in the first quarter of the year 2002. However, Vietnam's construction industry has recently encountered many issues that cause negative impacts on construction projects, one of which is the lack of systematic and effective risk management system.

Risks in construction often cause time and cost overruns. Many projects have been delayed or exceeded their planned budgets, as project managers could not manage risk effectively. These problems seem to happen more frequently these days, because of the emerging nature of the economy. Projects today are exposed to considerably more risks and uncertainties because of factors such as planning and design complexity, presence of various interest groups (project owner, consultants, contractors, vendors, etc.), resource availability (material, equipment, funds, etc.), climatic environment, social concerns as well as economical and political statutory regulations.

The oil and gas industry in Vietnam contributed 10% of the GDP in 2001 (Statistical Publisher 2001). Today, the industry continues to grow strongly implying demands for construction of new oil and gas facilities. Oil and gas construction projects are often capital intensive. Hence, their successful implementation is strategically important. However, oil and gas construction projects are exposed to risks because of large capital investment, involvement of many stakeholders, use of complex technology, high environmental and social impact.

Oil and gas projects in Vietnam are implemented through joint ventures partnerships involving multinational companies like British Petroleum,

Petronas, Total, Chevron, Conoco, etc. Such partners supply capital and high technologies needed for oil and gas projects which Vietnamese partners are still lacking. The participation of foreign partners makes the projects suffer from risks such as differences in practices between domestic and foreign partners, policy and political risks, financial risks, legal and political risks. The inflation rate in Vietnam is quite high; while the national currency is relatively weak. Vietnam is located in South East Asia, a region considered the most dynamic and challenging in the world. The quality of management in Vietnam is still below world standard as the country is emerging from a planned economy. In view of the above, oil and gas construction projects in Vietnam pose lots of risks that can cause adverse impacts on project implementation. Therefore, there is an urgent need for good risk management in oil and gas project management. Accordingly, the objectives of the study are to determine the major risks affecting oil and gas construction projects in Vietnam and propose appropriate strategies to effectively mitigate the major risks.

Methodology

The questionnaires used to gather data were designed in accordance with the study objectives, the literature and the research hypotheses. Two kinds of questions (open and closed-ended questions) were adopted for top management, project managers, functional managers and project team members, and were multiple choice type of questions. Additionally, interviews were conducted for further analysis. The questionnaires were pilot tested to improve them before the field survey.

The target population for this study was people involved in oil and gas construction projects in Vietnam in the middle stream and downstream project groups of Vietnam Oil and Gas Corporation (Petro Vietnam). Since Petro Vietnam is the major player in the oil and gas industry in Vietnam, it represents the oil and gas industry in Vietnam. Questionnaire respondents were top managers, project managers, functional managers, risk experts, project team members and other relevant staff of Petro Vietnam.

Primary data were collected by questionnaires and in-depth interviews, and secondary data were also collected from various sources. Secondary data were collected for the primary purpose of triangulation, i.e. the use of multiple sources of data for comparison. Descriptive statistics and hypothesis tests were utilized for data analysis. The Statistical Package for Social Science version 10.1 (SPSS) was used for data analysis. Three groups of variables were categorized to test differences in the opinions of respondents. The categories were working experience of the respondents, the respondents' job position and the type of respondents' company (state-owned and joint venture):

(1) Work experience variable group:

- Respondents $> = 10$ years of working experience; and
- Respondents $< = 10$ years of working experience.

(2) Job position variable group:

- management – top managers, project managers, managers and vice-managers of functional departments; and
- employee – project team members, related staffs.

(3) Company type variable group:

- state-owned companies; and
- joint venture companies.

The Chi-square test test was chosen for testing hypotheses about differences in respondents' responses. The null hypothesis in this study is that the variables in the same group do not differ among themselves. Cross-tabulation in SPSS is used for the Chi-square test. The statistical significance level of 10% was used for the statistical tests. Since the study dealt with people's opinions, 10% significance level is acceptable. In order to avoid the influence of analytic biases from various sources, research findings were verified and tested as follows: (1) Checking for researcher effect: biases stemming from the researcher effects on the site; and biases stemming from the effects of the sites on the researchers. To cope with them, careful examination of the study and checking feedback were needed: (2) Checking the means of outliers: it was reviewed by colleagues to avoid self-selecting biases.

Data collection and analysis

The questionnaires were distributed to six of Petro Vietnam's subsidiaries specializing in oil and gas projects in the middle to downstream activities. Five were state-owned companies and one was a joint venture company. A total of 70 questionnaires were delivered and 42 questionnaires were returned, the return rate was 60%. Table 12.1 shows details of the questionnaire responses and Table 12.2 shows the distribution of respondents' age, working time and highest education level. A deliberate attempt was made to include more managers in the study because managers have more comprehensive knowledge on various aspects of projects.

The ages of respondents in this investigation varies from 25 to 50 years. The modal age group is 31–35 (35.7%), the group from 25 to 30 is the second largest in size with 28.6%. All the respondents in the 45–50 age group were managers.

Of the respondents, 21 (50%) had more than ten years' work experience. The remaining 50% of respondents had less than ten years of experience. All

Table 12.1 Distribution of administered and returned questionnaires

Position	Distribution	Return	Percent
Management	30	16	53.3
Employee	40	26	65
Total	70	42	60

Table 12.2 Distribution of respondents' age, working time and highest education

	Frequency	*Percent (%)*
Age		
Group 25–30	12	28.6
Group 31–35	15	35.7
Group 36–40	2	4.8
Group 41–45	10	23.8
Group 45–50	3	7.1
Over 50	0	0
Total	42	100
Working time		
<=10 years	21	50
>=10 years	21	50
Total	42	100
Highest education		
Bachelor	38	90.5
Master	3	7.1
Doctor	1	2.4
Total	42	100

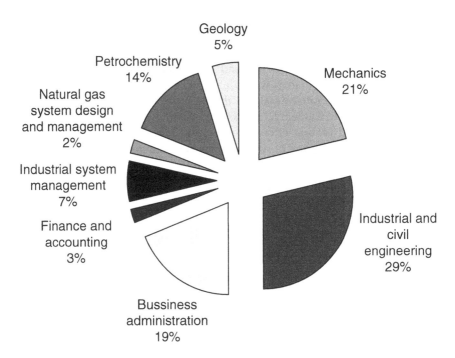

Figure 12.1 Distribution of respondents' major field study

the respondents had at least a Bachelor's degree. Figure 12.1 shows the distribution of respondents' major field of studies.

Identification of major risks in oil and gas construction projects

In order to determine the major risks in middle to downstream oil and gas construction projects, a checklist of risks was prepared from extant literature and given to the respondents. They were requested to judge two attributes of each risk: the frequency of occurrence, denoted by Fr, and the degree of impact, denoted by the Im. The risk, denoted R, is the function of these two attributes (Shen et al. 2001):

$$R = (Fr; Im)$$

This function can be quantified by a model, which Raftery (1994) and many researchers agreed on. It defines that the risk is equal to the multiplication of frequency of occurrence and degree of impact:

$$R = (Fr * Im)$$

where R, Fr, Im are all measured numerically. The respondents judged the frequency of occurrence using the five-level judgment scale of very high, high, medium, low and very low. The same scale was also applied to the degree of impact of the risk.

To apply the model, the opinion judgment scale needs to be converted into numerical scales. PMBOK (2000) suggested that the "very high" take a value of 0.9, and that "high", "medium", "low" and "very low" take values of 0.7, 0.5, 0.3, 0.1, respectively. A risk assessed by a respondent is called risk score and is calculated from the general model:

$$R_{ij} = Fr_{ij} * Im_{ij}$$

where, Fr_{ij} = frequency of occurrence assessed by respondent j for risk I; Im_{ij} = degree of impact if risk i assessed by respondent j.

By averaging scores from all 42 responses, it is easy to have an average score for each risk, and this average score is called the risk-index score and is used to rank risks. The calculation of risk-index score can be written as:

$$RI_i = \frac{\Sigma R_{ij}}{42}$$

where, RI = risk-index score for risk i; Rij = risk score of risk i assessed by respondent j. Risk ranking was done based on the risk-index score as shown in Table 12.3. The ten risk groups are used in order to easily interpret the risk ranking.

Table 12.3 Risk ranking for oil and gas construction projects

Rank (1)	Risk Code (2)	Risk Factors (3)	RI (4)	Std. Deviation (5)
Top Ten Risks				
1	R2.4	Bureaucratic government system and long project approval procedure	0.4857	0.2213
2	R7.1	Poor design	0.4767	0.179
3	R8.5	Incompetence of project team	0.4419	0.1789
4	R6.1	Inadequate tendering	0.4324	0.2017
5	R8.9	Late internal approval process from the owner	0.4276	0.2029
6	R8.4	Inadequate project organization structure	0.4252	0.1924
7	R8.1	Improper project feasibility study	0.4195	0.2135
8	R11.6	Inefficient and poor performance of constructors	0.4167	0.1823
9	R8.2	Improper project planning and budgeting	0.3938	0.1791
10	R7.4	Design changes	0.3938	0.1756
Second Ten Risks				
11	R11.1	Inadequate tendering price	0.389	0.1742
12	R11.7	Inefficient and poor performance of constructors	0.3795	0.1735
13	R6.2	Delay in signing contract	0.361	0.1729
14	R8.6	Inadequate coordination among contractors	0.3457	0.161
15	R6.4	Ambiguous conditions of contract	0.3367	0.1948
16	R9.1	Lack of knowledge and experience on construction	0.3348	0.1967
17	R8.3	Improper selection of project location	0.3343	0.2176
18	R3.2	Insufficient laws on projects	0.3324	0.1588
19	R3.3	Inefficiency of legal process	0.331	0.1694
20	R1.8	Increase of resettlement cost	0.3252	0.1884
Third Ten Risks				
21	R1.1	Exchange rate changes	0.3224	0.1977
22	R7.3	Constructability	0.3181	0.1987
23	R2.1	Changes of policies	0.3171	0.2085
24	R1.6	Increase of material cost	0.3157	0.1208
25	R2.3	Corruption and bribery	0.311	0.1959
26	R1.9	Economic and financial crisis	0.2986	0.2105
27	R1.7	Increase of equipment cost	0.2986	0.1489
28	R7.2	Difference of standards and codes in JV	0.2957	0.1844

29	R11.10	Deference practice among foreign contractors and domestic contractors	0.2924	0.18
30	R3.1	Change in laws and regulations	0.2857	0.1443
Fourth Ten Risks				
31	R2.2	Government interference	0.2833	0.1887
32	R11.9	Lack of coordination among contractors, sub-contractors	0.2814	0.1444

Rank (1)	Risk Code (2)	Risk Factors (3)	RI (4)	Std. Deviation (5)
33	R2.5	Lack of cooperation from government	0.279	0.1776
34	R6.5	Work conditions deferring from contract	0.2733	0.1828
35	R1.3	Inflation rate fluctuation	0.2657	0.1643
36	R11.4	Poor quality of procured materials	0.2643	0.1864
37	R9.3	Ineffectiveness and lack of supervision of consultants	0.2552	0.1499
38	R8.7	Poor relation and dispute with partners	0.2548	0.1546
39	R4.1	Unusual weather (flood, earthquake . . .)	0.2543	0.1538
40	R6.3	Inadequate type of contract	0.2538	0.1904
Other Risks				
41	R1.2	Interest rate fluctuation	0.251	0.1156
42	R5.1	Dispute with residents around site	0.2443	0.2063
43	R11.2	Equipment failure	0.2386	0.167
44	R1.5	Increase of labor cost	0.2276	0.1176
45	R11.5	Late in material delivery	0.2238	0.1625
46	R1.10	Low credibility of lenders	0.2221	0.1726
47	R11.3	Material shortage	0.2205	0.1399
48	R11.11	Shortage of new technologies	0.22	0.1235
49	R10.1	Late approval by lender	0.2167	0.1681
50	R9.2	Financial problems with consultants	0.2081	0.1419
51	R1.4	Increase of tax rate	0.2014	0.1461
52	R11.8	Increase of resettlement cost	0.1881	0.1203
53	R8.8	Poor relation with government department	0.1852	0.1433
54	R10.2	Lender interference	0.1795	0.1685
55	R10.3	Inflexible and complex rules and regulations of lender	0.1662	0.1059
56	R10.4	Lack of lender co-operation	0.1629	0.1162
57	R5.2	Lack of coordination among public agencies concerned	0.149	0.1224
58	R5.4	Environmental protection pressure of other groups	0.14	0.1102
59	R5.3	Damage to work by third party	0.1033	0.1011

Table 12.3 shows the list of ranked risks including 59 risks sorted in ascending order according to their overall impact on projects in Vietnam. The major focus of this paper are as mentioned in the research objective are the top ten risks. They will be thoroughly analyzed to find out their causes and characteristics, whereby appropriate measures to mitigate them can be proposed.

Analysis of the top ten risks

The "bureaucratic government system and long project approval procedure" risk occupies the first place with the highest risk-score index (0.4857).

It has the highest mean of occurrence frequency (0.625) and the fifth highest mean of impact degree (0.705) (Table 12.4). These indicate that the oil and gas projects in Vietnam are prone to this risk. Projects are required to be approved by several administration levels. Bureaucratic administration systems, poor law implementation and the incompetence of government staff were the main causes for this risk.

The next four risks are "poor design" (RI = 0.4767), "incompetence of project team" (RI = 0.4419), "inadequate tendering" (RI = 0.4324) and the "late internal approval process from the owner" (RI = 0.4276). The other nine risks in the top ten risks group are internal. Among them are five risks related to management of the owner, and they are quite highly placed in the risk ranking: third (incompetence of project team), fifth (late internal approval process from the owner), sixth (inadequate project organization structure),

Table 12.4 Statistics of top ten risks

N°	Risk Code	Risk	Mean of Occurrence Frequency	Mean of Impact Degree	Risk-Index Score (RI)
1	R2.4	Bureaucratic government system and long project approval procedure	**.652**	.705	0.4857
2	R7.1	Poor design	.629	**.743**	0.4767
3	R8.5	Incompetence of project team	.590	.729	0.4419
4	R6.1	Inadequate tendering	.567	.724	0.4324
5	R8.9	Late internal approval process from the owner	.595	.686	0.4276
6	R8.4	Inadequate project organization structure	.581	.705	0.4252
7	R8.1	Improper project feasibility study	.576	.700	0.4195
8	R11.6	Inefficient and poor performance of constructors	**.562**	.724	0.4167
9	R8.2	Improper project planning and budgeting	**.562**	.686	0.3938
10	R7.4	Design changes	.586	**.652**	0.3938

seventh (improper project feasibility study) and ninth (improper project planning and budgeting). The presence of management-related risks with a high ranking shows the incapability of owners to plan, organize, motivate, direct and control projects. A few in-depth interviews were conducted to find the reasons. The interviews revealed that incompetence of the manpower and lack of systematic project structure were the core causes. Therefore, in order to manage risks one must focus on the improvement of the managerial knowledge, skills and abilities of the owners and its manpower.

There were two risks related to the design in the top ten risk category – the "poor design" risk placed second (RI = 0.4767) and "design change" placed tenth (RI = 0.3938). Although placed second overall, "poor design" risk has the highest mean for degree of impact (0.743), more than that of any risk ranked. It is logical because design work is done at the early stage in the project life cycle, and the quality of earlier stage work often has strong impact on the total project (Hendrickson and Au 1989). Only a small omission in the design may cause big changes in using the execution method, resource allocation, etc. causing time and cost overruns as well as poor quality.

The "design change" risk may occur anytime after the final designs are approved for project execution. The earlier it happens the lesser the impact. In Vietnam, it often occurs in the construction phase. The lowest mean of impact degree for this risk was derived to be 0.652, which shows that the design changes in oil and gas projects in Vietnam has medium to small effect. The root causes of design changes were given as poor design, changing scopes of work, changing specifications or not predicting underground conditions.

The risk "inadequate tendering" with high risk-index score (0.4323), mean of occurrence frequency (0.567) and mean of impact degree (0.724) occupied the fourth place. This implies strong impact of inadequate tendering on project outcome. In-depth interviews revealed that the poor selection of contractors and contractors' poor ethics in tendering were the major concerns for tendering-related risk. Inadequate selection criteria and poor evaluation process were the main causes for selecting unreliable and incompetent contractors. Contractors' ethics in tendering is a sensitive and complex problem, it relates to the "right or wrong" moral of contractors during the tendering process (Ray et al. 1999). This is becoming a major problem in Vietnam as the transition to a free market economy is being made.

"Inefficient and poor performance of contractors" was identified as another important risk to address in the oil and gas industry of Vietnam. The competence of contractors is always a source of concern for owners. In Vietnam projects often suffer from time delays, cost overrun, low productivity and low quality due to contractors failure to performance due to lack of financial ability, relevant experience, trained manpower, technology and equipment, good management skills, good construction methods and inadequate resource availability. Moreover, oil and gas construction projects are often large and complex. They require high technology, new and complex construction methods. These have attracted leading international contractors to Vietnam's oil and gas industry. The differences in level of knowledge between national and international contractors cause conflicts impeding project success. It has risk-index score of 0.4167, the lowest mean of occurrence frequency (0.562), and considerable mean of impact degree (0.724).

Hypothesis testing

The x^2 hypothesis tests were carried out to check the differences in the judgments of the respondents on the risks. The tests focus on three groupings: job position, working experience and company type. The null hypotheses are:

> *H01.* There is no difference in the judgment about risks among respondents having different job positions.

> *H02.* There is no difference in the judgment about risks among respondents having different working experience.

> *H03.* There is no difference in the judgment about risks among respondents working for different company types.

The significance level for the hypothesis test is 10%. Using SPSS software, x^2 test was done and the results are shown in Table 12.5. It is found that there is not much difference in the judgment of respondents in all the three groupings. In the working experience grouping, there were only three risks having significant differences (inefficient and poor performance of constructors; delay in signing contract; and lack of knowledge and experience in construction). In the job position grouping, there were six risks having significant differences (incompetence of project team; government interference; work conditions deferring from contract; unusual weather, war; shortage of new technology; and financial problems with consultants). In the company type, there were 11 significant differences (improper project planning and budgeting; ambiguous conditions of contract; insufficient laws on projects; inefficiency of legal process; economic and financial crisis; increase of equipment cost; difference of standards and codes in joint venture (JV); change in laws and regulations; poor relation with government departments; inflexible and complex rules and regulations of lender; and environmental protection pressure of other groups).

The findings of less difference in the judgment of risk having different job positions (three out of 59) and working experience (six out of 59) imply the relative agreement from the respondents irrespective of experience and job position. However, there are significant differences in the company type

Table 12.5 x^2 test result for hypothesis testing

Rank	risk code	risk factor	RI according to working experience		p	Significance @10%</b
			≥10 years >	<10 years		
8	RII6	Inefficient and poor performance of constructors	0.43	0.4033	0.056	significant
13	R6.2	Delay in signing contract	0.4513	0.4362	0.065	significant
16	R9.1	Lack of knowledge and experience on construction	0.4035	0.322	0.048	significant

Rank	risk code	risk factor	RI according to job position		P	Significance @10%</b
			Management	Employee		
3	R8.5	Incompetence of project team	0.4513	0.4362	0.032	significant
31	R2.2	Government interference	0.3001	0.2729	0.063	significant
34	R6.5	Work conditions deferring from contract	0.2531	0.2857	0.033	significant
39	R4.1	Unusual weather (floods, earthquake . . . .)	0.2412	0.2623	0.082	significant
48	R11.11	Shortage of new technologies	0.2045	0.229	0.011	significant
50	R9.2	Financial problems with consultants	0.1859	0.2217	0.062	significant

Rank	risk code	risk factor	RI according to company type		P	Significance @10%</b
			State-owned	Joint venture		
9	R8.2	Improper project planning and budgeting	0.4035	0.322	0.025	significant
15	R6.4	Ambiguous conditions of contract	0.3512	0.2294	0.054	significant
18	R3.2	Insufficient laws on projects	0.321	0.41676	0.064	significant
19	R3.3	Inefficiency of legal process	0.3225	0.3939	0.029	significant
26	R1.9	Economic and financial crisis	0.2845	0.40294	0.085	significant
27	R1.7	Increase of equipment cost	0.3087	0.22386	0.001	significant
28	R7.2	Difference of standards and codes in JV	0.32	0.11588	0.096	significant
30	R3.1	Change in laws and regulations	0.2714	0.39152	0.052	significant
53	R8.8	Poor relation with government department	0.1752	0.2592	0.056	significant
55	R10.3	Inflexible & complex rules & regulations of lender	0.1698	0.13956	0.072	significant
58	R5.4	Environmental protection pressure of other others	0.145	0.103	0.072	significant

Note: P – probability value calculated from x^2 test

grouping (11 out of 59 risks). The different characteristics of the two company types may create these differences. For example, in joint ventures both international and local partners need to use the same construction standard, meaning that one partner may not use his familiar standard but have to adapt to a new standard. This need for adaptation can cause difficulties in practice. The state-owned companies, by contrast, still prefer to use the old construction standard they are familiar with. That is responsible for the significant difference in perception on the different standards and codes risk between state-owned companies and joint ventures. Another example is relation of joint ventures with local government departments. For joint ventures, the good relationship with the government is a critical factor of project success. As such, they emphasize this issue while the state-owned companies put less emphasis on it because they themselves belong to the government, thus the government may give them priorities in running projects.

Among the top ten risks, inefficient and poor performance of contactors has significant difference in each grouping. In the working experience grouping, there exists a disagreement on the judgment about "inefficient and poor performance of constructors" risk between respondents having much experience and those having less than ten years experience. The former has the mean of risk-index score of 0.43 while the latter has 0.403 (Table 12.5). This implies that the more experienced respondents perceive the risk more than the group with less experience. In the job position variable group, the significant difference occurs in the "incompetence of project team" risk. The managers gave higher risk-index score than employees (0.4513 against 0.4362). Thus, the managers are more concerned by the incompetence of project team than employees. However, it is useful for the project teams' competence improvement because managers are in the position to give support and more consideration for improvement (through training) when they feel their project teams are not competent. There is much difference in risk-index score between the state-owned companies and joint venture companies (0.4035 against 0.322) implying significant difference in perception of the "improper project planning and budgeting" risk. It would seem that the competence of project team, the high technology support and other potentials of joint venture companies give them confidence that they do not perceive improper project planning and budgeting as being risky as the state-owned companies.

Strategies for managing major risks

This survey had identified the risks as well as their overall impact on oil and gas construction projects in Vietnam. Every risk should be analyzed and given appropriate strategies for better project execution. This study focuses only on the top five major risks due to the severity of their impacts on projects. Specific strategies are proposed to effectively mitigate each of the top five risks. The strategies are to design the means, plans of actions to cope with risk-related problems in the effective manner (Cleland and Ireland 2002). In order to formulate appropriate and practical strategies, each major risk would be thoroughly analyzed to find the causes, characteristics in accordance with the project environment in Vietnam.

The following paragraphs demonstrate the risk responses of the oil and industry under study.

Bureaucratic government system and long project approval procedure (R2.4)

"Bureaucratic government system and long project approval procedure" risk is common to projects in Vietnam because of incompetent staff of government regulatory agencies, unclear responsibility and power, relatively poor law implementation processes and complex approval procedures (Qui Hao 2002). This causes long delay to receive project approval. This also reduces Vietnam's image in the eyes of foreign investors. Of companies interviewed in a survey, 20% said that they did not intend to extend their business in the next three years (Qui Hao 2002) because of long approval time. Moreover, the total foreign investment capital into Vietnam has seriously decreased: $2,345 million in 1997 vis-a-vis $307 million in 2002. It clearly indicates the need for improving project approval procedures.

Mitigating measures for this risk is complex as it is external to the organization. The executives who were interviewed suggested the strategies like requesting government for administrative reform; good relationship with government; good relationship with environment authority, NGOs; familiarity with approval procedures and understanding local laws and regulations. Since the oil and gas industry is the main contributor to the country's economy, a collective request from the organizations in the oil and gas industry would sensitise the government to reform various ministries. The request must be given to the central levels so as to initiate reform in every level of government services. Vietnam is currently in the process of administrative and economic reforms. Although it has achieved considerable positive results, the reforms should be deeper and wider to encourage foreign investors to invest in Vietnam and form joint ventures with organizations in the country.

Creating and maintaining good relationships with both central and local governments also helps reduce project approval delays. The owner should adapt well to the local government and try to understand them as well as their requirements (Bing et al. 1999). The owner should show the benefits that project offers in short- and long-term socioeconomic developments of the local community and the region. Benefits such as job creation, improvement of living standard, tax income, etc. are important government objectives. The organizations need to fulfil those in order to maintain a good relationship with the government. It is clear that when the relationship with both local people and government is good, the project approval would be quicker and the owners will save time and money.

Besides maintaining a good relationship with government, good relationships with environmental authority and NGOs is also important. This is so because oftentimes there are conflicts between project owners and these agencies that may negatively influence projects. Environmental impact assessment and social impact assessment are required to be carried out for infrastructure projects not only to satisfy regulatory requirements, but also to remain productive and competitive throughout the project's life.

The owner organizations should be familiar with the government project approval procedures as well as being conversant with local laws and regulations. This would reduce the approval time considerably. Building a database for past project approvals and forming templates of approval documentation would considerably reduce time and cost of project approval process.

Poor design (R7.1)

The risk of "poor design" causes time, cost and quality non-achievement on project. Contractors have to absorb a large portion of the extra cost unless they are able point out the flaws in specification in the early stages of the project. This causes conflict between owner and contractors. Additionally, ambiguous design and specification cause delays in execution along with both cost overrun and quality non-achievement (Tilley 2003). Designer's incompetence and lack of experience, unclear specifications and scopes are the main sources of design related risks. In Vietnam, although both domestic and foreign designers design oil and gas facilities, there is dominance of foreign designers because of complexity of design in the oil and gas industry. Despite having experience, these foreign designers still encounter design problems due to the owner's ambiguous specifications and scopes; and use of different standard design systems. Domestic designers, in contrast, do not have much knowledge and experience in oil and gas facility design because the oil and gas industry in Vietnam is in its infancy.

In view of the above, several strategies proposed by interview respondents include partnering, usage of experienced and familiar design companies, concurrent engineering, good translation of the owner's ideas to design parties and hiring of competent consultants to evaluate design works. It was recommended to deploy all the possible strategies simultaneously.

In the last decade, the partnering philosophy has been a major vehicle in the worldwide effort of creating significant improvement in the construction industry (Brown et al. 2001). Partnering is defined as a long-term commitment between two or more organizations for the purpose of achieving specific business objectives by maximizing the effectiveness of each participant's resources. This requires changing traditional relationships to a shared culture without regard to organizational boundaries. The relationship is based upon trust, dedication to common goals, and an understanding of each other's individual expectations and values. Expected benefits include improved efficiency and cost effectiveness, increased opportunity for innovation, and the continuous improvement of quality products and services (CII 1991). The three key elements of any successful partnering relation are trust, long-term commitment and shared vision (CII 1991). In the relationship of partnering, the owner carries his commitment by giving his best resources to support the design party such as clear specifications and scopes, on-time payment and rewards. The designers in return also use their best possible resources such as competent and experienced designers, high technology, etc.

Another strategy to mitigate the "poor design" risk is the involvement of experienced design organizations. Therefore, designer selection should be based on experience and past performance.

Concurrent engineering approach would considerably improve constructability, because of simultaneous design and construction and good communication between design and construction teams.

The other strategies suggested were good translation of the owner's ideas to designers and hiring of competent consultants to evaluate design works. The owner should have a competent architect/engineer who acts as his representative to translate his ideas into technical terms such as specifications and scopes for designers. The designers, thus, can easily understand and transform them into drawings. Consultants are also to be employed in order to evaluate the quality of the designs. Once again, the owner should select competent, experienced and reliable consultants to carry out the work.

Incompetence of project team (R8.5)

The owners' project team consists of project managers, project executives and functional members (Kerzner 1997). In order to complete a project successfully, it is critical that every project team member has a good understanding of the fundamental project requirements, which include project planning, organizing, motivating, directing and controlling (Cleland and Ireland 2002), and has a positive attitude. The problems of the incompetence of project teams in Vietnam's oil and gas industry have been divided into two categories: individual incompetence and ineffective teamwork.

The first issue concerns the ability of each individual in the project team in performing his/her job. Only a few people can perform their duties well with high productivity, on time and with high efficiency. The reasons for this poor performance often come from the lack of knowledge, skills and abilities of project team members. Although most of the project team members have Bachelor's degrees or higher education, their knowledge, skills and abilities seem not to be sufficient for managing oil and gas projects. In addition, oil and gas projects require project team members to work with foreign partners to an international business standard. The second issue relates to the effectiveness of work teams. Robbins (2001) stated that a work team generates positive synergy through coordinated effort. Their individual efforts result in a level of performance that is greater than the sum of those individual inputs. Today work team effectiveness is vital for the success of a project (Kerzner 1997). The project involves many areas of work, which no single individual can fully handle and that is why project team members have to work together as a team. To build an effective work team, the staffing must be good. However, in oil and gas projects in Vietnam, staffing is below international standard. The team members are not appropriately selected for specific projects. Moreover, the work team is not well organized, coordinated, directed, motivated and controlled, because of the complexity and dynamism of work team. Teamwork demands more time and more resources than individual work, more communication is needed, and there could be many conflicts to manage and meetings to run. Therefore, the productivity and effectiveness of both the individual and the work team are still low.

In order to cope with the problem of the project team's incompetence, several critical strategies were proposed namely: training and education, good staffing and effective teamwork. Training and education are now considered a competitive way for organizations to achieve improvement. Periodically, executives should provide necessary professional courses and training to enhance the knowledge, skills and to update new knowledge, changes, and new technologies for all staff. To develop an effective and practical training and education program, a simplified training model was proposed by Goldstein (1993). The model contains three phases, namely: needs assessment; training and development; and evaluation. Bajracharya et al. (2000) also suggested that training should be aligned to organizational goals with training outcomes properly integrated to organizational processes.

Besides training and education, another important strategy that could be employed to mitigate the risk of incompetence of the project team is good staffing. That means the human resource must be appropriately allocated to projects to enable them to function smoothly. Obviously, each project needs a good manager with enough knowledge and skills to successfully plan, organize, motivate, direct and control the project to achieve the defined objectives. Project team members also need to be well matched to particular projects. Moreover, the functional managers must commit to provide adequate functional employees when the project manager requests. Good relations between the project manager and functional managers is an imperative.

The last strategy proposed was building an effective project team. Creating an effective team covers a wide range of areas but it can be grouped into four keys components: work design, composition, context, and process (Robbins 2001). The work design category includes variables such as freedom and autonomy, the opportunity to utilize different skills and talents, the ability to complete whole and identifiable tasks and working on a task or project that has a substantial impact on others. Composition includes variables that relate to how teams should be staffed. It addresses the ability and personality of team members, allocation of roles and diversity, size of team, member flexibility and members' preference for teamwork. The context is very important to create the effectiveness. Three contextual factors that appear to be the most significantly related to team performance are the presence of adequate resources, effective leadership and a performance evaluation and reward system that reflects team contribution. The last component (process) is related to member commitment to a common purpose, establishment of a specific team goal, team efficacy. In brief, building an effective work team requires the simultaneous efforts of many people involved.

Inadequate tendering (R6.1)

Tendering plays a very important role in project procurement since it helps select the most appropriate contractors to execute projects. The tendering system proves effective in shortening completion time, improving quality and lowering costs of construction works (Wang et al. 1998). However, poor tendering practice causes many problems in achieving projects' objectives. Oil and gas projects in Vietnam often adopt one of the three main tendering methods: competitive tender (selective or open), negotiated tender and a combination of both. In principle, open competitive tender is the most effective one, when an unlimited number of bidders can compete fairly in the tendering process. Thus, it is easy to choose the contractor who is best able to handle the project. However, this type of tendering is time consuming, complex and expensive. Therefore, its use is limited only to some very large and complex projects requiring international contractors. The selective competitive tender and negotiated tender are more often used because they are still effective while saving money and time as well as overcoming the lack of tendering expertise of the owner. These two kinds of tender are mainly based on the good relationship between clients and contractors.

This study revealed that "poor evaluation criteria setting" and "unethical behaviour of bidders" were the two main reasons for tendering risk in the oil and gas industry in Vietnam. The criteria vary across projects. They are normally divided into four groups: finance, time, quality and the credibility of the bidders. In-depth interviews revealed that the contracts are awarded primarily on the basis of financial and credibility of the bidders. Often, the lowest bidder is chosen in order to save on project cost. In addition, since tendering is a very sensitive issue, accepting the lowest-price tender may help the public owners to defend themselves from criticisms and to show accountability (Wong et al. 2000). However, in some cases, bidders submit the lowest price in order to win the bid and at the later stage they negotiate with the owner to jack-up their offer. Sometimes, the evaluation weights are not well set. For example, the evaluation weight for the credibility in the tender of the first refinery in Dung Quat was too high; therefore, a familiar contractor that had worked with the owner in the upstream had been chosen to form a joint venture. However, this contractor had experience in upstream activities only. As a result of the incapability of the contractor, the joint venture later dissolved, and the project failed.

Another problem in tendering in Vietnam is unethical behaviour through collusion of bidders. Collusive tendering occurs when a number of firms that have been invited to tender agree between themselves either not to bid, or to bid in such a manner as not to be too competitive with the each other (Ray et al. 1999). Collusion corrodes the basis and attacks the rationale of the competitive tendering system by restricting free competition. When free competition is circumvented by collusive agreement of ostensible competitors, the client's choice for contractor is restricted. Collusion also causes a reduction in the number of available bidders, an artificial increase in the average tender price and a reduction in tender variance irrespective of the particulars of the collusive tendering arrangement (Fraser and Skitmore 2000). Collusion can appear in many kinds such as communication with other bidders, bribery, withdrawal, artificial inflation of tender prices and covering price. Although collusive tendering is illegal in Vietnam, it is difficult to detect whether there exists collusion in tendering exercises. Consequently, the client (or owner) often suffers from this problem.

The interviewees believed that contractor evaluation using multiple criteria decision-making (MCDM) technique and detection enhancement for collusive tendering and legal enforcement enhancement in tender practice would benefit the tendering process considerably. Use of MCDM brought mutual benefits to both client and contractor (Jennings and Holt 1998) in the UK when 83% of respondents (clients) agreed on that statement.

To maximize the value to the owner, tender evaluation should not focus only on financial aspects but also all other aspects need consideration, thus the tender selection philosophy that only "lowest-price wins" needs changing. The most responsive contractor based on preset criteria should be selected. In practice, low price will still carry the highest evaluation weight in tender evaluation. That prevents the owner from choosing a successful bidder having lowest price but being less competent compared to the other bidders with higher prices. Maximum "value" can be measured only from contractors' attributes (i.e. selection criteria) during both prequalification and final tender evaluation stages. Criteria can be established according to the specific project, they are so-called "project-specific criteria". The weight of each evaluation criterion is also appropriately determined depending on each project and expectation of the owner (Palaneeswaran et al. 2003).

The second strategy is to deal with the unethical issues caused by the collusion of bidders. Unethical behaviour of bidders is a very difficult problem for the client to control. To cope with this issue, first the owner has to improve detection of bidders at the prequalification and final evaluation stages. Second, the liabilities of bidders in tender participation must be more highly enforced. The detection can be done in several ways such as thorough examination of contractors' bids, searching information about contractors and the relationship between them and prediction of the potential collusion. The owner may also employ selective competitive bidding practice in which he invites only the professional and credible contractors. In bidding practice, bidders should be warned of punishment if they participate in any collusion. To this end, appropriate laws may need to be enacted with strong enforcement.

Late internal approval process from the owner (R8.9)

Risk of "late approval process from the owner" ranked fifth. Being an internal risk, it originates from problems within the organization to which the project belongs. There are two main explanations for this development – low commitment of the people involved, and lack of authority of the manager in problem solving.

Normally, during project execution, many problems such as shortage of resources, complex conflicts with external parties, legal problems, etc. crop up. The project manager alone cannot solve them using his authority and responsibility. He has to report to his functional and top management for support. Prompt decisions from the top management helps the project manager to progress as scheduled. In addition, functional managers can also cause several difficulties in allocation of resources to projects. The relationship between the project manager and functional managers is very critical in this respect. The better the relationship, the faster is the resource allocation. In Vietnam, project authorisation is not appropriate because most projects belong to the government, and the project manager's authority is quite limited. He does not have enough power to make quick decisions in problem solving and needs advice and approval from higher authorities.

Two main strategies are proposed to mitigate this risk: enhanced involvement of all people in the organization and empowerment. Involvement of every organization member in sharing the common vision of the company is an important principle in total quality management. This philosophy is now gaining increased importance. According to this philosophy, every individual within the organization must be involved in every operational process of the organization, contributing his/her part, and facilitating others' work in a common effort to achieve the goals of the organization. That means every person must share the responsibility and vision of the organization. However, the involvement of top management must be secured first and most vigorously. The support of top management is the one of the most critical factors of project success. Thus, the support of top management reduces waiting time for approvals, helping project to progress as planned.

Empowerment is the authority to make decisions within one's area of operations without having to get approval from anyone else (Luthans 1998). The enhancement of empowerment for the project ensures management's confidence in the project manager. It helps to make on-time decisions. However, this does not dilute accountability. Enhanced delegation of power improves decision-making authority of the project manager, which enhances innovation in project management process to achieve project success.

Table 12.6 Strategies for mitigating the top five risks

Major risks	Strategies
1 Bureaucratic government systems and long Project approval procedure	request for administrative reform from the government
	Good relationship with government
	Good relationship with environmental authority, NGOs
	Familiarity with approval procedures, local laws and regulations
2 Poor design	Partnering
	Employment of experienced design companies
	Concurrent engineering
	Good translation of the owner's ideas to designers
	Hiring of competent consultants to evaluate design works
3 Incompetence of project team	Training and education
	Good staffing
	Effective teamwork management
4 Inadequate tendering	Enhancement of MCS
	Detection enhancement for collusive tendering and legal enforcement enhancement
5 Late internal approval process from the owner	High commitment of all people involved
	Empowerment

Table 12.6 summarizes the strategies for addressing the major project risks in the oil and gas industry in Vietnam.

Summary and conclusions

For decades, risk management has been considered a fundamental and indiscrete activity in project management in developed countries. However, this practice is quite strange and limited in developing countries including Vietnam. This research is therefore timely and very important. Its findings are more critical considering that the oil and gas industry is the major contributor to the overall economy of Vietnam but highly exposed to risks due to its own characteristics and fast changing environment.

This research focused on two main objectives, namely finding out the major risks affecting oil and gas construction projects and deriving appropriate strategies to effectively mitigate the major risks. Using a questionnaire survey, the study has systematically examined the risks affecting oil and gas construction projects. The risk-index score developed, supported by in-depth interviews, have helped in identifying major risks and their mitigation measures. The top ten risks identified were:

(1) bureaucratic government system and long project approval procedure;
(2) poor design;
(3) incompetence of project team;
(4) inadequate tendering;
(5) late internal approval process from the owner;
(6) inadequate project organization structure;
(7) improper project feasibility study;
(8) inefficient and poor performance of constructors;
(9) improper project planning and budgeting; and
(10) design changes.

For projects to be successful, it is critical that the major risks affecting projects be thoroughly examined. Their causes and characteristics must be carefully analyzed, in order to support the owner to propose the most appropriate and practical strategies to mitigate them. In this research, the top five major risks have been put through this process. This research proposes requesting for administrative reform from the government, maintaining good relationship with government, environment authority, and NGOs, familiarizing with approval procedures, local laws and regulations, partnering with major project stakeholders, engaging experienced design organizations, applying concurrent engineering, translating owner's ideas clearly to designers, hiring of competent consultants to evaluate design works, training project people, appropriate staffing, emphasizing on teamwork, enhancing of MCDM practices in contract evaluation, fostering environment for non-collusive tendering and enhancing legal enforcement, enhancing commitment of all people involved, and empowering in order to effectively mitigate the major risks.

References

Bajracharya, A., Ogunlana, S.O. and Bach, N.L. (2000), "Effective organizational infrastructure for training activities: a case study of the Nepalese construction sector", *System Dynamics Review*, Vol. 16, pp. 91–112.

Bing, L., Tiong, K.L., Fan, W.W. and Chew, D.A. (1999), "Risk management in international Construction joint ventures", *Journal of Construction Engineering and Management*, Vol. 5 No. 4, pp. 277–284.

Brown, D.C., Ashleigh, M.J., Riley, M.J. and Shaw, R.D. (2001), "New procurement process", *Journal of Management in Engineering*, Vol. 17 No. 4, pp. 192–201.

Charoenngam, C. and Yeh, C.Y. (1999), "Contractual risk and liability sharing in hydropower construction", *International Journal of Project Management*, Vol. 17 No. 1, pp. 29–37.

CII (1991), In Search of Partnering Excellence, Special Publication 17-4, pp. 118–134.

Cleland, D.I. and Ireland, L.R. (2002), Project Management: Strategic Design and Implementation, McGraw-Hill, Boston, MA.

Fraser, A.Z. and Skitmore, M. (2000), "Decision with moral content: collusion", *Construction Management and Economics*, Vol. 18, pp. 101–111.

Goldstein, I.L. (1993), Training in Organization: Needs Assessment, Development, and Evaluation, 3rd ed., Brooks/Cole, Pacific Grove, CA.

Hendrickson, C. and Au, T. (1989), Project Management for Construction: Fundamental Concepts for Owners, Engineers, Architects and Builders, Prentice-Hall, Englewood Cliffs, NJ.

Jennings, P. and Holt, G.D. (1998), "Prequalification and multi-criteria selection: a measure of contractors' opinion", *Construction Management and Economics*, Vol. 16, pp. 651–660.

Kerzner, H. (1997), Project Management: A System Approach to Planning, Scheduling and Controlling, 5th ed., Van Nostrand Reinhold, New York, NY.

Luthans, F. (1998), Organization Behavior, International Edition, McGraw-Hill, Boston, MA.

Palaneeswaran, E., Kumaraswamy, M.M. and Ng, T.S.T. (2003), "Targeting optimum value in public sector through 'Best value' focused contractor selection", *Engineering, Construction and Architectural Management Journal*, Vol. 10 No. 6, pp. 418–431.

PMBOK (2000), A Guide to Project Management Body of Knowledge, Project Management Institute, Newtown Square, PA.

Qui Hao (2002), available at: www.vneconomy.com.vn/index.php?action = thongtin & chuyenmuc = 01&id = 030127082539 (accessed January 20, 2019).

Raftery, J. (1994), Risk Analysis in Project Management, E&FN Spon, London.

Ray, R.S., Hornibrook, J., Skitmore, M. and Zarkada-Fraser, A. (1999), "Ethics in tendering: a survey of Australian opinion and practice", *Construction Management and Economics*, Vol. 17, pp. 139–153.

Robbins, S.P. (2001), *Organization Behaviour*, 9th ed., Prentice-Hall, Englewood Cliffs, NJ.

Shen, L.Y., Wu, G.E.C. and Ng, C.S.K. (2001), "Risk assessement for construction joint ventures in China", *Journal of Construction Engineering and Management*, Vol. 127 No. 1, pp. 76–81.

Statistical Publisher (2001), *National Annual Economic Report*, Hanoi, Vietnam.

Tilley, P. (2003), available at: www.cmit.csiro.au/innovation/2000-10/csiro_survey.htm (accessed February 4, 2019).

Wang, S.Q., Tiong, R.L.K., Ting, S.K., Chew, D. and Ashley, D. (1998), "Evaluation and competitive tendering or BOT power plant project in China", *Journal of Construction Engineering and Management*, Vol. 1998, pp. 333–341.

Wong, C.H., Holt, G.H. and Cooper, P.A. (2000), "Lowest price or value? Investigation of UK construction clients' tender selection process", *Construction Management and Economics*, Vol. 18, pp. 767–774.

World Bank (2001), *Annual Report*, pp. 258–283.

Further reading

Edwards, P.J. and Bowen, P.A. (1998), "Risk and risk management in construction: a review and future directions for research", *Engineering Construction & Architectural Management*, Vol. 5 No. 4, pp. 339–349.

PetroVietnam (2001), *Annual Report*, pp. 156–157.

PetroVietnam (2002), available at: www.petrovietnam.com.n (accessed August 28, 2018).

Project Management Institute (2000), A Guide to the Project Management Body of Knowledge, 2000 Edition, Project Management Institute, Newtown Square, PA.

Tuyen (2003), "Tuyen, pre-project planning process in capital development projects: a case study of urban development project in Hanoi city", master thesis, No. ST-00-31 AIT, Bangkok, Thailand.

Vietnam Economy (2003), available at: www.vneconomy.com.vn/index.php?action = thongtin&chuyenmuc ¼ 03&id ¼ 030120164353 (accessed February 5, 2019).

13 Risk allocation assessment through good project governance concept

A case study of PPP project in Thailand

Martinus Abednego and Stephen O. Ogunlana

Introduction

The economic growth in developing countries is mostly influenced by infrastructure development that provides basic services towards the industry and households (Martin & Lee 1996) as well as providing a positive effect on economic output (Kessides 1993). Due to the size and its level of complexity, infrastructure projects require massive capital investment (Grimsey & Lewis 2002), which unfortunately is highly limited in developing countries. In order to be able to keep up with the economic growth that is very much dependent on infrastructure development as well as fulfilling the increasing demand from the continuously growing population, it is very much necessary for the government to acquire the required funds through private sector participation.

The Asian Development Bank (ADB), World Bank and other international financial agencies had been supporting the infrastructure development around the South-East Asia region. Private sector participation had started in the region around the mid-1980s and grew rapidly in the 1990s. The annual investment commitments for infrastructure projects with private participation experienced a significant increase from US$2.6 billion to US$41 billion from 1990 to 1997 (Figure 13.1 and Table 13.1), but then it suffered quite severely after the first financial crisis hit most of the countries in this region.

Nevertheless, the infrastructure sector had begun to redeem itself after its downfall caused by the economic crisis, and investment in this sector is still very much desired by the private sector. Private sector participation had once again played a significant role to reduce poverty through infrastructure development and private investment in competitive markets will continue to contribute to the economic growth (World Bank 2002). Therefore, it is very much necessary to have a healthy yet competitive environment to achieve successful infrastructure projects with private participation.

Public-Private Partnership, widely known as PPP, is a concept of private sector participation that is considered as the appropriate approach that allows better allocation of the already limited government resources and is also capable of ensuring quality improvement on projects that are carried

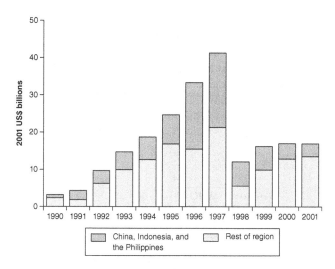

Figure 13.1 Annual Investment in Infrastructure Projects with Private
Participation, East Asia and Pacific, 1990–2001

Table 13.1 Annual Investment in Infrastructure Projects with Private Participation,
East Asia and Pacific, 1990–2001

(2001 US$ billions]

Year	Electricity	Natural gas transmission and distribution	Telecommunications	Transport	Water and sewerage	Total
1990	0.1	..	1.3	1.2	..	2.6
1991	0.5	..	1.2	2.5	..	4.3
1992	5.0	0.2	1.3	1.1	2.0	9.6
1993	6.2	0.7	2.2	2.6	2.9	14.6
1994	7.8	..	5.6	5.0	0.2	18.6
1995	7.9	1.9	6.5	8.2	0.3	24.7
1996	12.0	0.7	9.8	10.0	0.7	33.2
1997	15.1	..	11.8	7.3	7.0	41.3
1998	5.6	0.0	3.6	2.3	0.7	12.2
1999	1.6	0.9	9.9	2.4	1.1	16.0
2000	3.9	0.9	6.7	5.4	0.1	16.9
2001	2.9	0.7	5.0	7.5	0.4	16.6
Total	68.6	6.0	65.0	55.7	15.3	210.6

Source: *World Bank, PPI Project Database*

out through this procurement system. However, several problems still occur
in this procurement system even though it has been implemented in quite a
number of projects. The majority of users of this system do not realize that

PPP projects also possess governance concerns that are considered as long-term issues requiring a more strategic approach in addition to the general management concerns that are considered as short-term issues requiring day-to-day supervision. Most of them had confusingly intertwined these two characteristics and such condition had resulted in improper allocation of the risks that may occur in the project. Therefore, it is clearly important to allocate the risk properly since it has a substantial impact on project performance (Ward et al. 1991) and that assigning the risks to the most suitable

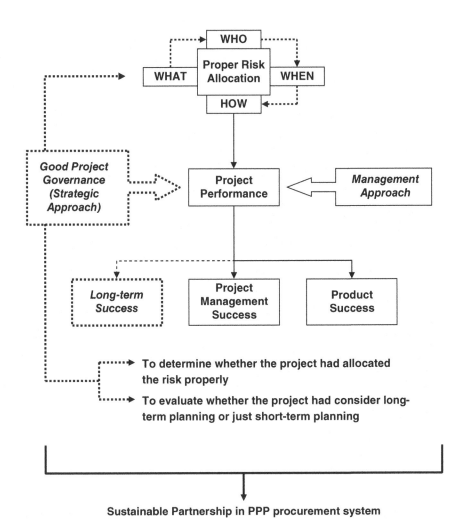

Figure 13.2 Inclusion of good project governance (strategic approach) to enhance project performance

party, which is the one with the best ability to control the risks, is necessary to increase the chance of success in a project, including infrastructure projects with private participation (Kumaraswamy & Zhang 2001).

Acknowledging the need to consider both short-term issues through management approach as well as long-term issues through strategic/governance approach, it is necessary to combine these two approaches in order to enhance the project's performance and ultimately achieving overall project success, thus the development of Good Project Governance (GPG) concept. Figure 13.2 illustrates the relationship between management approach and strategic approach.

Methodology

The general concept of Good Project Governance was developed through the combination of good corporate governance characteristics identified by UN-ESCAP and the Critical Success Factors for PPP that was identified by Kumaraswamy & Zhang (2001), and it has the following characteristics (Abednego & Ogunlana 2006):

1 Right decision at the right time (active participation)
2 Contract fairness
3 Information transparency
4 Responsiveness, in terms of making concrete action within a reasonable time-frame
5 Continuous project control and monitoring to achieve common goals and satisfy all interests
6 Equality between all stakeholders
7 Effectiveness and efficiency
8 Accountability.

These characteristics are further classified into five core principles, *fairness, transparency, accountability, sustainability* and *effectiveness/efficiency* (Abednego & Ogunlana 2006). Each core principle is further broken down into sub-components where each sub-component consists of several key analytical issues in order to enable a more detailed analysis of the relationship between proper risk allocation and project success. Moreover, the core principles and their respective sub-components will then be assessed based on its key analytical issues in order to know whether or not the identified risks have been properly allocated.

The detail breakdown of the above mentioned core principles are as follows in Table 13.2.

A questionnaire survey was then carried out to verify the initially identified core principles and their respective sub-components as well as the key analytical issues. The survey was carried out in three countries, Thailand, Indonesia and Vietnam. However, experts, either with related academic

Table 13.2 Project governance principles and their subconstructs along with key analytical issues

Core Principles	Sub-Components	Key Analytical Issues
Fairness	Project design & planning	Identification of experts responsible for design & planning
		Project design and planning development process
	Contract document	Contract document development process
	Government regulations, laws & policies	Existing government regulations, legal system and policies related to PPP
	Project procurement	Project proposal assessment criteria
		Project procurement method
Transparency	Information management	Information management system (procedure, distribution method)
	Financial management	Project financing/investment strategy
		Incentive/compensation program

Core Principles	Sub-Components	Key Analytical Issues
Accountability	User & community participation	Public participation process
		Project demand & impact analysis
	Quality assurance	Quality management
		Human resources competence enhancement
Sustainability	Stakeholder management	Coordination between stakeholders
		Conflict resolution approach
		Infrastructure development plan
		Decision-making approach
Effectiveness & Efficiency	Project management	Project administration
		Project monitoring & control

background or significant practical experience, in order to ensure its validity and identify any potential shortcomings in it prior to its distribution, inspected the questionnaire.

The paired-comparison method, in which the core principles were paired with each other and the respondents have to choose which core principle is more significant in that particular pairing was used to determine its ranking. The survey result shows that *fairness* is considered to have the most influence for achieving proper risk allocation, followed by *transparency, accountability, sustainability* and *effectiveness/efficiency,* respectively.

Next, a reliability test was carried out using the Cronbach's Alpha Coefficient as a method to measure the internal consistency of the sub-components as well as the key analytical issues for each of the core principles. The sub-components and key analytical issues must have a

Table 13.3 Project governance principles and their subcomponents as revealed

Core Principles	Sub-Components
Fairness	Project design & planning
	Government regulations/laws/policies
	Project procurement
Transparency	Information management
	Financial management
Accountability	User & community participation
	Quality assurance
Sustainability	Stakeholder management

minimum alpha value of 0.5 in order to be considered having an internal consistency. Although the test resulted in the omission of one core principle and some of the sub-components, its essence is still taken into account from different perspectives. Based on the result, the initial content of GPG concept was adjusted and updated as follows, as shown in Table 13.3.

After this adjustment, the Mean Analysis approach was used to compare the opinions of the survey respondents with regards to the key analytical issues relevant to each sub-component. A 4-point Likert scale is used in this method to prevent the natural tendency of the respondents to be neutral and force them to either agree or disagree with the given statements provided in the questionnaire survey. By going through such rigorous analysis process, the case studies were assessed using the same standards, thus the final result and comparisons will be more valid.

Data collection and analysis

A tollway project in Thailand was selected as one of the case studiess because it implements private participation policy, especially in terms of project financing, design, construction, operation and maintenance. A concession between government agency and private company was established to manage the tollway, while an international engineering company was awarded the contract by the concession to carry out the construction process.

Seeing that the study is focused on PPP issues, therefore, the targeted respondents were selected individuals from the government agency and private company that were involved in the project with vast experience levels and who had direct involvement in the project development process. Direct meetings with these individuals were carried out whenever possible to ensure that the survey was delivered to the right person and proper explanations regarding the questionnaire can be given to respondents. Moreover, a native speaker with the relevant knowledge background translated the questionnaire to prevent potential misinterpretation and misunderstanding on the questions, thus increasing the validity and reliability of the results.

With regards to the respondents, approximately 47% of them have more than 11 years of experience, showing that most of them have significant experience in tollway projects, with 40% of the respondents being involved in all of the project stages. In terms of respondent groups, around 60% of respondents were from the private sector while 40% were government officials, with approximately 50% of all respondents having had significant involvement in the project. The respondents were then requested to assess the tollway project in terms of its risk allocation arrangements and the level of success it had achieved, using the verified GPG core principles as the evaluation guideline. The following section discusses the analysis and assessment result based on the GPG concept.

Project design and planning

The survey result shows that experts were always involved in the design and planning stage, which means that these experts were technically and professionally responsible for all the potential risks related to design and planning work. Moreover, the result also shows that the government had a more inferior role during these two stages compared to the private sector. Such conditions were not really unexpected due to the fact that government tends to position itself at a more superior level, especially in a hierarchical structure, and also possesses the authority in determining the direction of project development. Nevertheless, the government was willing to be held responsible for any consequences caused by their inability to provide the necessary support, thus minimizing potential problems during these stages. Additionally, the availability of detail design as well as a detailed project planning program prior to its initiation also prevented potential problems during the construction stage. Since the design and planning stage were completed in a very limited time due to the experience of some of the top management staff in similar projects, the private stakeholder felt that the allocated time for carrying out the work at these stages was sufficient. On the contrary, the government stakeholder had an opposite perception due to the fact that they did not have sufficient time to check and verify the design/planning work proposed by the private stakeholder.

Government regulations, laws and policies

With regards to this matter, it was discovered that the existing regulations and policies related to private participation are still very much insufficient in Thailand though its government had issued the Private Participation Act in 1992. In addition to that, the currently implemented legal system is also still inadequate and incapable to appropriately solve conflicts/disputes occurring between the government and private sector. To put it simply, these statutes are unable to provide the basically required assurance to any private investments, thus showing the lack of government support and commitment. Such condition is worsened by the fact that the existing legal/regulatory frameworks are rarely updated with lessons learned acquired from past projects.

Project procurement

The survey shows that there is a different opinion with regards to the project proposal evaluation process, where the private sector felt that they were not given clear specifications while the government thought otherwise, although both agreed that the government had provided clear explanations whenever a project proposal is rejected. Though both parties acknowledged the need to have a specific regulation for standardizing the procurement procedure in addition to improving the currently applied procedure, the government still insists on the existence of regulations controlling the procurement procedure for PPP-type projects. On the other hand, the private sector still believes that the existing regulations/laws are unsuitable and unable to accommodate such type of project.

Information management

It was discovered through the survey that each stakeholder tends to keep information within their organization and more often than not they are also quite reluctant in sharing the information with other stakeholders. As a result, inaccurate information dissemination occurred within the project, thus leading to misinformation and causing disputes. Moreover, each stakeholder has their own information management system instead of developing a standard system that is to be used during the project life cycle, causing different opinions regarding standard communication process.

Financial management

Since not all of the project investors were involved from its initiation stage, they experience difficulties in properly analyzing and recognizing the potential financial problems that may occur. Nevertheless, the private stakeholder was still able to initiate an appropriate financial strategy that produced a reasonable and balanced revenue sharing scheme for all stakeholders. On the contrary, the government believed that it were the one who initiated the financial strategy while also positioning itself on top of the hierarchical structure instead of being at the same level in the partnership.

Unfortunately, the government was unwilling to be responsible for any occurring financial problems caused by the changes in the national economic, though ironically they agreed that the national economic changes had made a significant impact towards the project.

User and community participation

Both the government and private stakeholder confirmed that public hearing was performed to provide the community with relevant information regarding the tollway development plan and also providing the opportunity to give comments as well as suggestions. In addition to that, project demand, economic and impact analysis were also carried out to ensure its feasibility. However, there was a dissenting opinion with regards to the party performing the analysis, in which the private stakeholder claimed that it was the government who had carried out the analysis while on the other hand the government did not accept the notion as a party who performed the impact analysis. As a result, conflicts and disputes materialize when problems occurred caused by some inaccurate analysis.

Quality assurance

In order to ensure the quality during the construction stage, a private engineering consultant was appointed by the concessionaire to represent the government as the project supervisor. Based on the survey result, though the private stakeholder felt that the consultants had performed effectively, the government did not share the same opinion. According to the government, the consultant had not performed quite as expected and had not acted on the behalf of the government. Moreover, since the government also had no interference whatsoever towards the selection as well as approval of sub-contractors, the government felt that the private stakeholder was liable for the works performed by the sub-contractors.

With regards to professional resources, the private stakeholder was more dominant compared to their government counterpart. The reason for this is because some of the top management level staff in the concession company were previously involved in similar projects, thus enabling them to be more prepared to foresee potential problems in the project.

Stakeholder management

As discovered from the survey, coordination between government agencies had caused some problems in the project, such as overlapping of projects or development program thus creating inefficiencies. In terms of stakeholder relationship, the relation between government stakeholder and its private counterparts was more of a hierarchy rather than a partnership. Unfortunately, the relationship between private stakeholders, especially between the local private stakeholder and foreign private stakeholder, was also more of a competition rather than an alliance, which interfered with the synergy and created disadvantages when dealing with external stakeholders. As a result, the private stakeholder's unity was disrupted and weakened their bargaining position during the dispute resolution process with the government stakeholder.

Through the survey, opinions from all stakeholders that were involved in the project were represented and clearly show the diversity of opinions. The analysis acquired from the survey was then utilized further to discover the willingness of both the government as well as private stakeholder to accept the consequences of the risks that occurred in the project while also trying to understand the stakeholder's capability to manage the risks.

Findings

The verified GPG key analytical issues were utilized to discover whether or not the risks in the tollway project had been properly allocated, while also determining the requirements for achieving the target that would eventually enhance the project's performance. Therefore, it is clear that the approach would help determine whether or not the project had achieved *Good Project Governance* that is based on its core principles.

Through the analysis, it was discovered that the tollway project suffered long-term consequences and it is mainly due to the fact that government agency involved in the project lacked the experience in dealing with a project under PPP procurement system, thus preventing them from acknowledging any potential long-term issues that required a more strategic approach. Other major obstructions that prevented the project from achieving overall success were government commitments and stakeholder relationship. With regards to the commitment issue, it was shown by the lack of government support in the form of regulations, laws and policies relevant to private participation. Ironically, even though the private stakeholders were fully committed to help achieve project success, the efforts become pointless when the government was not willing to share the responsibility. Such condition would then inevitably create conflicts and disputes between stakeholders, thus causing further damage towards the project in the long run.

The assessment result showed that the risks had not been properly allocated, thus creating disputes between the government and private stakeholder that required an international arbitration resolution. Because of this, the project suffered long-term consequences and therefore cannot be considered to have achieved good project governance.

References

Abednego, M. and Ogunlana, S.O. (2006). 'Good project governance for proper risk allocation in Public-Private Partnerships: A case study on the 2nd stage of Cipularang Tollway Project', *International Journal of Project Management*, Vol.24, 622–634.

Asian Development Bank (1996). *Current issues in economic development: An Asian perspective*, edited by M.G. Quibria and J.M. Dowling, Oxford: Oxford University Press.

Grimsey, D. and Lewis M.K. (2002). 'Evaluating the risks of public private partnerships for infrastructure projects', *International Journal of Project Management*, Vol.20, 107–118.

Kessides, C. (1993). *The contributions of infrastructure to economic development: A review of experience and policy implementations*, Washington DC: The World Bank.

Kumaraswamy, M.M. and Zhang, X.Q. (2001). 'Governmental role in BOT-led infrastructure development', *International Journal of Project Management*, Vol.19, 195–205

Martini, C.A. and Lee, D.Q. (1996). 'Difficulties in infrastructure financing', *Journal of Applied Finance and Investment*, Vol.1(1), 24–27.

Walker, C., Smith, A.J., Mulcahy, J., Lam, P.T.I. and Cochrane, R. (1995). *Privatized infrastructure: The BOT approach*, London: Thomas Telford Publications.

Ward, S.C. and Chapman, C.B. (1991). 'On the allocation of risk in construction projects', *International Journal of Project Management*, Vol.9, No.3, 140–147.

World Bank (2002). *Private participation in infrastructure: Trends in developing countries in 1990–2001*, Washington DC: The World Bank.

Further reading list

Batley, R. (1996). 'Public-Private relationships and performance in service provision', *Urban Studies*, Vol.33(4–5), 723–751.

Bennet, R. and Krebs, G. (1991). *Local Economic Development Public-Private Partnership Initiatives in Britain and Germany*, London:Belhaven Press.

Bennett, E. (1998). *Public-Private Cooperation in the Delivery of Urban Infrastructure Services (Water and Waste)*, PPPUE Background Paper, UNDP/Yale Collaborative Program, www.undp.org/pppue.

Brodie, M. (1995). 'Public/private joint ventures: the government as partner – bane or benefit?', *Real Estate Issues* Vol.20(2), 33–39.

Charoenpornpattana, S. and Minato, T. (1999). *Privatization-induced risks: State-owned transportation enterprises in Thailand*, Proceedings of the CIB-W92 & CIB-TG23 Joint Symposium: Profitable partnering in construction procurement, edited by Prof. Stephen O. Ogunlana, E&FN SPON.

Collin, S. (1998). 'In the twilight zone: a survey of public-private partnerships in Sweden', *Public Productivity and Management Review*, Vol.21(3), 272–283.

Cooke-Davies, T. (2002). 'The "real" success factors on projects', *International Journal of Project Management*, Vol.20, 185–190.

Grant, T. (1996). 'Keys to successful public-private partnerships', *Canadian Business Review*, Vol.23(3), 27–28.

Hardcastle, C. & Boothroyd, K. (2003). 'Risks overview in public-private partnership', *Public-Private Partnerships: Managing risks and opportunities*, edited by Akintola Akintoye, Matthias Beck and Cliff Hardcastle, School of the Built and Natural Environment, Glasgow Caledonian University, Blackwell Science.

Jaafari, A. (2001). 'Management of risks, uncertainties and opportunities on projects: time for a fundamental shift', *International Journal of Project Management*, Vol.19, 89–101.

Keene, W. (1998). 'Reengineering public-private partnerships through shared-interest ventures', *The Financier*, Vol.5(2–3), 55–59.

Li, B. & Akintoye, A. (2003). 'An overview of Public-Private Partnership', *Public-Private Partnerships: Managing risks and opportunities,* edited by Akintola Akintoye, Matthias Beck and Cliff Hardcastle, School of the Built and Natural Environment, Glasgow Caledonian University, Blackwell Science.

Nikomborirak, Deunden (2004). *Private sector participation in Infrastructure: The case of Thailand,* Asian Development Bank Institute Discussion Paper No.19, December 2004.

14 Risk-based inspection and maintenance

A case of oil pipeline operations in India

Prasanta Kumar Dey

Introduction

Achieving throughput is one of the challenges for today's businesses. This satisfies customers as well as keeps inventory to an optimum level. This is possible not only through usage of efficient facilities, but also having failure-free operations. An effective project and efficient operations make this possible. A framework is therefore required for zero shutdown of operations throughout the products / services life. The success of any logistic system is achieved through flexibility of operation, low incidence of failures and promptness in meeting delivery schedule.

Cross-country oil pipelines are the most energy-efficient, safe, environmentally friendly, and economic way to ship hydrocarbons (gas, crude oil, and finished products) over long distances, either within the geographical boundary of a country or beyond it. A significant portion of many nations' energy requirements is now transported through pipelines. The economies of many countries depend on the smooth and uninterrupted operation of these lines, so it is increasingly importance to ensure the safe and failure-free operation of pipelines.

While pipelines are one of the safest modes of transporting bulk energy, and have failure rates much lower than the railroads or highway transportation, failures do occur, and sometimes with catastrophic consequences. A number of pipelines have failed in the recent past, with tragic consequences (Hopkins 1994). While pipeline failure rarely causes fatalities, disruptions in operation lead to large business losses. Failures can be very expensive and cause considerable damage to the environment.

The use of the term *failure* in the context of pipelines varies from country to country and across organizations. For example, in Western Europe, any loss of gas/oil is considered a failure, while, in the US, an incident is considered to be a failure only when it is associated with the loss of commodity and also involves a fatality or injury, or damage over $50,000 (Payne 1993). In India, generally any loss of commodity, however small, is considered to be a failure which is defined as any unintended loss of commodity from a pipeline that is engaged in the transportation of that commodity.

Traditionally, most pipeline operators ensure that during the design stage, safety provisions are created to provide a theoretical minimum failure rate for the life of the pipeline. Safety provisions are considered when selecting pipes and other fittings. To prevent corrosion, a pipeline is electrically isolated by providing high resistance external coating materials. As a secondary protective measure, a low-voltage direct current is impressed in the pipe at pre-calculated distance to transfer any corrosion that occurs due to breaks in the coating caused by a heap of buried iron junk, rails, etc. This is called impressed current cathodic protection. The quality of the commodity that is being transported through the line is also ensured, and sometimes corrosion-preventing chemicals (corrosion inhibitors) are mixed with the commodity. To avoid deliberate damage of the pipeline in isolated locations, regular patrolling of the right-of-way from the air as well as on foot is carried out, and all third party activities near the route are monitored.

Various techniques are routinely used to monitor the status of a pipeline. Any deterioration in the line may cause a leak or rupture. Modern methodologies can ensure the structural integrity of an operating pipeline without taking it out of service (Jamieson 1986).

The existing inspection and maintenance practices commonly followed by most pipeline operators are formulated mainly on the basis of experience. However, operators are developing an organized maintenance policy based on data analysis and other in-house studies to replace rule-of-thumb based policies. The primary reasons for this are stringent environmental protection laws, scarce resources, and excessive inspection costs. Existing policies are not sharply focused from the point of view of the greatest damage/defect risk to a pipeline. The basis for selecting health monitoring and inspection techniques is not very clear to many operators. In many cases, a survey is conducted over an entire pipeline or on a particular segment, when another segment needs it more. Avoidable expenditures are thus incurred.

A strong reason exists, therefore, to derive a technique that will help pipeline operators select the right type of inspection/monitoring technique for segments that need it. A more clearly focused inspection and maintenance policy that has a low investment-to-benefit ratio should be formulated. Accordingly, the objective of the study is to model a framework for managing risks of oil pipeline operations for achieving throughput.

Methodology

This study adopts interactions with the stakeholders and literature reviews to establish the model for managing risk of operations and a case study to demonstrate its effectiveness.

Literature

Risk-based inspection and maintenance has long been researched. Tomic (1993) proposed the use of risk-focused maintenance in improving system reliability or availability through systematically identifying the applicable and effective course of action for each failure mode of a system. The major advantage of employing a risk management approach is to provide a thorough assessment of risk factors of equipment failures. Vaughan (1997) defined the fundamental part of risk management function as the design and implementation of procedures to minimize the occurrence of loss. Cooper and Chapman (1987) described risk management as identification, evaluation, control, and management of risks from the perspective of social hazard management. Rowe's (1993) approach does not consider the phase of controlling and monitoring. Various models have been suggested for various industry applications. Tummala and Mak (2001) applied the risk management process (RMP) in developing a risk management model for improving electricity supply reliability and transmission operations and maintenance, respectively. Ng et al. (2003) used a risk-based maintenance management model for tunnel operations. Dey and Gupta (2001), Dey (2001a) and Dey (2003) have developed a model for risk-based maintenance of oil pipelines. Dey et al. (2004) have used a risk-based approach for maintaining offshore gas pipelines. This study develops an integrated risk-based inspection and maintenance model for an entire oil pipeline system (stations, pipelines, and tank farm).

Model

Figure 14.1 depicts the suggested model of holistic risk management of oil pipelines system for zero shutdown. The model can be demonstrated using the following steps:

- Defining the physical boundaries of the system for analysis and classifying the system into subsystems
- Listing likely risk events for each subsystem
- Identifying subsystem for detailed analysis
- Identifying risk events of risky subsystem
- Deriving likelihood of risk events using analytical framework
- Developing responses to mitigate likelihood and severity of risk using effective inspection and maintenance strategy
- Analyzing cost-benefit of proposed responses.

If the benefit is more than the cost, the responses will be implemented with suggestions for facility improvement measures for future projects. If the cost of responses are more than the benefit, the subsystem as well as the entire system is unfit for operations. The operators must find alternate design, implementation and operation of the facilities. A knowledge base is required to be developed out of the risk analysis exercises for future design, implementation, and operations of facilities.

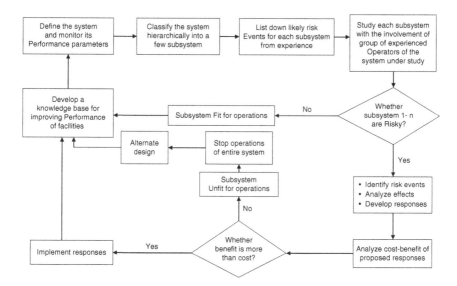

Figure 14.1 Model for facility risk management

Applications

The entire model has been illustrated through a case study of a crude oil pipeline (length 1300 km) in the western part of India. The throughput of the pipeline is nine million metric ton per annum (MMTPA) with augmentation capability of 12 MMTPA, having three intermediate booster stations and an offshore terminal. This pipeline is 19 years old and has a history of corrosion failure. The poor condition of the coating, as revealed during various surveys and an unreliable power supply to cathodic protection stations are the reasons for this. The line passes through long stretches of socio-economically backward areas and is vulnerable to pilferage and sabotage. In some regions, the right-of-way is shared with other agencies, so the chance of external interference is high. Failure data revealed numerous pre-commissioning failures, raising doubts about the quality of construction. Figure 14.2 shows a schematic of the pipeline system. Table 14.1 shows a brief database of the pipelines under study. Detailed information of the pipeline is available in Dey et al. (1996) and Dey (1997).

A group had been formed to study the pipelines system. The group members had more than 15 years of experience in working in oil pipelines operations.

The following paragraph describes the proposed risk management model.

Step 1: Defining the physical boundaries of the system for analysis and classifying the system into subsystems

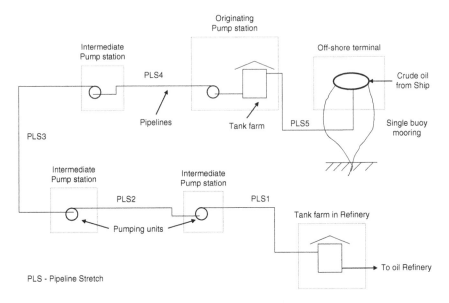

Figure 14.2 Schematic of oil pipelines under study

Figure 14.3 depicts the hierarchical classification of the oil pipeline subsystem. When every subsystem performs properly, only then does the entire system achieve its throughput.

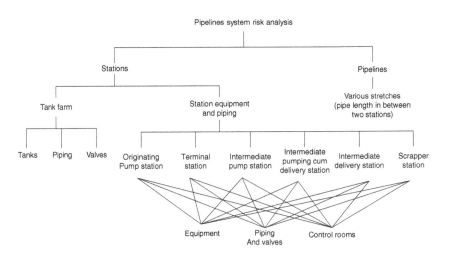

Figure 14.3 Oil pipelines system and subsystem

Table 14.1 Database of pipeline stretches (figures in km)

Descriptions	Pipeline stretches 1	Pipeline stretches 2	Pipeline stretches 3	Pipeline stretches 4	Pipeline stretches 5
Length	360	310	280	330	20
Terrain detail:					
Normal	270	195	135	269	3
Slushy	36				
Rocky (hilly)			115	20	
River & canal crossings	4	2		6	
Populated	50	88		35	
Offshore					17
Coal belt		25			
Forest			30		
Desert					
Soil condition	corrosive	corrosive	Less corrosive	Less corrosive	Less corrosive
Third party activities		More due to coal belt			
Chances of pilferage	Higher due to populated area	Higher due to populated area		Higher due to populated area	
Construction complexity			More due to rocky and forest	More due to river crossing	More due to offshore piping
Operational complexity					More due to offshore terminal

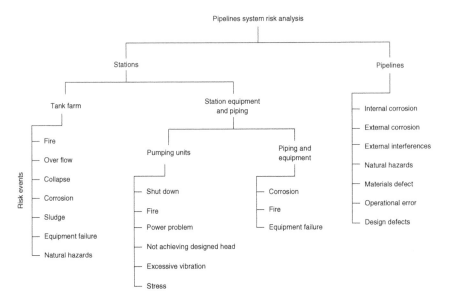

Figure 14.4 Risk events for operating oil pipelines

Step 2: Listing likely risk events (check list) for each subsystem

Figure 14.4 depicts the various likely risk events for pipelines operations associated with each subsystem. Experienced pipelines operators developed these from their past experience. These events / factors are very general, which may be present in any oil pipeline system. However, this enables the group to analyze the condition of the pipeline system under study along with other information and their own experience in operating oil pipelines.

Step 3: Identifying subsystem for detailed analysis

The pipelines operators, who are entrusted to study the pipelines, brain-stormed to determine the specific subsystem for detailed study on the basis of their vulnerability of failure. The operators decided to study the oil pipeline stretches, because according to them this subsystem was most vulnerable for the entire pipelines system under study.

Step 4: Identifying risk events of risky subsystem

The pipeline operators identified the following risk factors for the pipelines under study. They are:

- corrosion;
- external interference;

- construction and materials defects;
- natural hazards; or
- human and operational error.

One of the major causes of pipeline failure is corrosion (Annual Report of CONCAWE 1994; Pipes and Pipelines International 1993), an electrochemical process that changes metal back to ore. Corrosion generally takes place when there is a difference of potential between two areas having a path for the flow of current. Due to this flow, one of the areas loses metal.

External interference is another leading cause of pipeline failure (Annual Report of CONCAWE 1994; Pipes and Pipelines International 1993). It can be malicious (sabotage or pilferage) or caused by other agencies sharing the same utility corridor. The latter is known as third-party activity. In both cases, a pipeline can be damaged severely. External interference with malicious intent is more common in socio-economically backward areas, while in regions with more industrial activity, third-party damage is common.

All activities, industrial or otherwise, are prone to natural calamities, but pipelines are especially vulnerable. A pipeline passes through all types of terrain, including geologically sensitive areas. Earthquakes, landslides, floods, and other natural disasters are common reasons for pipeline failures.

Poor construction, combined with inadequate inspections and low-quality materials, also contribute to pipeline failures. Other reasons include human and operational error and equipment malfunctions (US Department of Transportation 1995). Computerized control systems considerably reduce the chance of failure from these factors.

Human and operational errors are another source of pipeline failure. Inadequate instrumentation, foolproof operating system, lack of standardized operating procedures, untrained operators etc. are the common causes of pipeline failure due to human and operational errors.

Step 5: Deriving likelihood of risk events using analytical framework

The next step of this methodology is the formation of a risk structure model in Analytic Hierarchy Process (AHP) framework. Based on the identified risk factors, a hierarchical risk structure is formed (see Figure 14.5). In the context of the study, the goal is to determine the relative the likelihood of pipeline failures. Level II is criteria (risk factors), level III is subfactors, and level IV is alternatives (the pipeline stretches). Figure 14.5 shows the AHP model for analyzing risk from a failure perspective.

The entire pipeline was classified into five stretches. The risk structure and pairwise comparison were established through a workshop of the executives who operate various oil pipelines. About 30 executives participated. They have more than 15 years of experience in pipelines operations. Before formulating the model, they were given full knowledge of pipeline

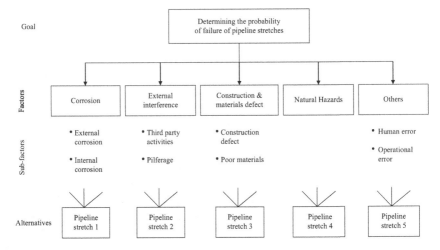

Figure 14.5 Risk structure for pipelines subsystem in AHP framework

conditions through a database of various pipelines stretches (Table 14.1) and pipeline record sheet (Figure 14.6).

The risk factors and subfactors are compared pairwise to determine the likelihood of pipeline failure due to each factor and subfactor. Then the alternative pipeline stretches are compared with respect to each risk subfactor, to determine the likelihood of failure for each pipeline stretch. Then likelihood of failure of various pipeline stretches was determined through synthesizing the results of pairwise comparison across the hierarchy.

Table 14.2 shows the comparison matrix in factor level. It determines the relative importance of each factor.

Table 14.2 Pairwise comparison matrix in factor level

Factors	Corrosion	External Interference	Construction & materials defect	Acts of God	Others	Likelihood
Corrosion	1	2	3	7	3	**0.40**
External Interference	1/2	1	3	5	3	**0.29**
Construction & materials defect	1/3	1/3	1	3	2	**0.14**
Acts of God	1/7	1/5	1/3	1	1/4	**0.05**
Others	1/3	1/3	1/2	4	1	**0.12**

TYPICAL PIPELINE DATASHEET

1. Name of pipeline
2. Nominal diameter (inches)
3. Wall thickness (inches)
4. Pipe grade
5. Specific minimum yield stress
6. Operating pressure (kg/cm^2)
7. Age (years)
8. Age of oldest section (years)
9. Age of coating (years)
10. Length of pipeline (km)
11. Number of pipeline sections
12. Length of longest pipeline section (km)
13. Product type
14. Coating type
15. Type of soil
16. Discharge temperature
17. Population density
18. Number of crossings
19. Surveillance level
20. Inhibitor efficiency
21. Corrosion rate
22. History of leaks due to internal corrosion
23. History of burst due to internal corrosion
24. Number of cathodic protection (CP) station
25. CP availability
26. Efficiency
27. CP interface
28. Coating condition
29. Soil aggression
30. Instrumented pig surveying (IPS) conducted, if any
31. Major findings of IPS (in brief)
32. History of leaks due to external corrosion
33. History of burst due to external corrosion
34. Average metal loss due to external corrosion
35. Number of hydro-test failures
36. Number of years since last hydro-test
37. Number of pressure cycles per month
38. History of bursts due to fatigue
39. History of leaks due to fatigue
40. Evidence of stress corrosion cracking
41. History of bursts due to stress corrosion cracking
42. History of leaks due to stress corrosion cracking

43. History of bursts due to 3rd party damage
44. History of leaks due to 3rd party damage
45. History of sabotage/pilferage
46. History of bursts due to sabotage/pilferage
47. History of leaks due to sabotage/pilferage
48. Mining activities
49. Soil stability
50. Earthquake/fault zone
51. History of floods
52. History of failure due to natural calamity
53. History of failure due to equipment failure
54. History of failure due to human error
55. Failure cost (details against each item)
56. Number of employees
57. SCADA systems installed
58. Leak-detection mechanism available (software)
59. Training level of employees (from training history card)
60. Availability of equipment (maintenance history record)
61. Existing CP and coating survey schedule
62. Other health-monitoring survey schedule.

Figure 14.6 Areas covered in the data forms

The scale of measurement is reflected below (Table 14.3):

Table 14.3 Scale of measurement for AHP

Numerical Values	Definition
1	Equally important or preferred
3	Slightly more important or preferred
5	Strongly more important or preferred
7	Very strongly more important or preferred
9	Extremely more important or preferred
2,4,6,8	Intermediate values to reflect compromise
Reciprocals	Used to reflect dominance of the second alternative as compared with the first

Source: Saaty 1980

Table 14.4 shows the entire risk analysis for the pipeline subsystem using AHP framework.

PLS – Pipeline stretch

The final outcomes of each of the pipeline stretch against the risk factors are summarized in Table 14.4. Both local probability and global probability for each of the five stretches are summed up to derive the probability of a pipeline stretch failure and its position with respect to other stretches. The results of the analysis (Table 14.4) reveal that the chances of pipeline failure due to corrosion and external interference are greater than other factors.

The following additional observations were made from the risk analysis study:

Table 14.4 Likelihood of failure of various pipeline stretches

Factors	Likelihood	Subfactors	Likelihood	PLS_1	PLS_2	PLS_3	PLS_4	PLS_5
Corrosion	0.40	External	0.221	0.108	0.064	0.007	0.011	0.031
		Internal	0.181	0.038	0.022	0.020	0.042	0.060
External Interference	0.29	Third party activities	0.186	0.030	0.078	0.011	0.061	0.006
		Malicious	0.100	0.033	0.039	0.005	0.018	0.005
Construction & mat. defect	0.14	Construction defects	0.072	0.012	0.007	0.028	0.007	0.018
		Poor mats.	0.072	0.006	0.007	0.027	0.016	0.017
Acts of God	0.05		0.05	0.006	0.001	0.014	0.006	0.020
Others	0.12	Human error	0.048	0.001	0.005	0.003	0.008	0.030
		Operational error	0.072	0.001	0.003	0.009	0.003	0.056
Likelihood of failure of various pipeline stretches				**0.236**	**0.227**	**0.123**	**0.172**	**0.242**
Ranking				**2**	**3**	**5**	**4**	**1**

The pipeline stretches 1 and 2 are vulnerable from external corrosion due to slushy terrain, whereas pipeline stretches 4 and 5 are vulnerable to internal corrosion due to long submerged pipe sections.

External interference due to third party activities are a major problem in pipeline stretch 2 because of coal mining activities, whereas in stretch 4, it is due to major river crossings and canal crossing.

External interference due to malicious reasons is prevailing in stretches 1 and 2 because it passes through a long and highly populated industrial areas.

The pipeline stretch 3 is passing through mostly rocky terrain, exposing the pipe to various types of failure due to construction and poor materials. As this stretch is vulnerable to subsidence problems, the likelihood of pipeline failure from acts of God is quite high along with high chance of failure due to construction defect and poor materials.

The stretch 5, i.e. the offshore pipeline, is very sensitive to operational and human errors as well as failure due to various natural calamities.

All pipeline stretches are ranked with respect to their failure chances, the pipeline stretch 5 comes first, pipeline stretch 1 comes second and pipeline stretch 2, 4, and 3 come third, forth and fifth respectively.

> Step 6: Developing responses to mitigate likelihood and severity of risk using effective inspection and maintenance strategy

In this step, specific inspection/maintenance requirements are determined for specific segments of pipelines from the likelihood of failure data, to mitigate risk.

The output of the above analysis helps in deciding specific inspection and maintenance programs for each pipeline stretch. An instrument pig survey has been suggested for pipeline stretches 4 and 5 to detect internal corrosion. A survey technique is chosen to reveal areas affected by external corrosion. One technique is a current attenuation survey or pearson survey (these surveys detect breaks in pipeline coating, i.e. areas where the pipeline is exposed to soil). Survey techniques that can identify both internal and external corrosion are not needed and are not cost effective.

Pipeline stretches 1, 2, and 3 are prone to external interference (pilferage and sabotage), so they require frequent patrolling. Stretches 2 and 4 are susceptible to third party damage. Therefore, more publicity about the route among agencies working near it could be a solution. Cooperation with these agencies needs to be improved. However, a few contingency plans for handling the situations of failure incidents are to be kept ready for the above two stretches.

Pipeline stretches 3 and 5 are vulnerable from normal and abnormal natural calamities. Although various measures were taken in designing and constructing the pipelines in both the stretches for minimizing failure, a few contingency plans are also to be formulated in line with the anticipated incidents.

Table 14.5 indicates the inspection and maintenance programs for pipeline under study vis-à-vis cost for each program. Table 14.6 indicates

Table 14.5 Inspection and maintenance cost* (figures are in Indian Rupees (INR) in million) with the application of risk-based inspection model

US$ 1 = INR 47

Inspection and maintenance strategy	Problems	PLS_1	PLS_2	PLS_3	PLS_4	PLS_5
Instrument pig survey	Internal corrosion				25	5
Cathodic protection survey	External corrosion	4	4			
Contingency plans	Third party activities		1		1	
More patrolling	Malicious	2	2		2	
Contingency plans	Acts of God			1		1
Improved instrumentation						5
Pipe coating	External corrosion	3	2			
Pipe replacement	Construction defect and poor pipe materials			3		
Total cost (INR 61 million for five years)		**9**	**9**	**4**	**28**	**11**

*The cost figures are estimated from the budgetary offers of the vendors.

Table 14.6 Inspection and maintenance cost* (figures are in INR in million) without use of risk-based inspection model

US$ 1 = INR 47

Inspection and maintenance strategy	Problems	PLS_1	PLS_2	PLS_3	PLS_4	PLS_5
Instrument pig survey	Internal corrosion	25	25	25	25	5
Cathodic protection survey	External corrosion	4	4	4	4	1
Contingency plans	Third party activities	1	1	1	1	1
More patrolling	Malicious	2	2	2	2	
Contingency plans	Acts of God	1	1	1	1	1
Improved instrumentation						5
Pipe coating	External corrosion	3	2			
Pipe replacement	Construction defect and poor pipe materials			3		
Total cost (INR 153 million for 5 years)		**36**	**35**	**36**	**33**	**13**

*The cost figures are estimated from the budgetary offers of the vendors.

the conventional inspection and maintenance programs vis-à-vis cost in absence of the proposed risk-based model. This establishes the advantage of using the risk-based model in designing inspection and maintenance of cross-country petroleum pipeline.

The inspection and maintenance cost has two components – fixed cost and variable cost. The variable cost depends on the length of pipeline. However, the fixed cost depends on design and consulting charge and apportionment of the overhead cost for the inspection tools. The fixed cost for specific inspection is very high compared to the variable cost. Therefore, the inspection cost for almost all pipeline section is approximated as the same. The following calculations show the computation for inspection and maintenance of pipelines:

Instrument pig survey

Consulting charge (fixed cost)	= 5
Design (fixed cost)	= 7
Overhead charge for tools (fixed cost)	= 10
Survey (variable cost)	= 3
Total	= 25 (INR in million)

Cathodic protection survey:

Consulting charge (fixed cost)	= 0.5
Design (fixed cost)	= 1.0
Overhead charge for tools (fixed cost)	= 2.0
Survey (variable cost)	= 0.5
Total	= 4 (INR in million)

Pipe coating for stretch 1

Coating materials and application	= 2.5
Overhead	= 0.5
Total	= 3.0 (INR in million)

Pipe replacement in stretch 3

Pipe materials	= 2.0
Pipe laying	= 0.4
Overhead	= 0.6
Total	= 3.0 (INR in million)

Selection of a particular inspection technique depends on the owner's experience. However, this approach will give a rational basis to the owner when selecting the most appropriate survey technique as well as the pipeline stretch where the survey is most needed.

Table 14.7 The cost of a pipeline Failure

Result	Cost (in million INR)
Loss of production	10
Loss of commodity	5
Loss of life and property	10
Loss of image	30
Environmental damage	50
Total	**105**

* *The costs are estimated by simulating various situations of pipeline failure in Indian context*

US$ 1 = INR 47

Step 7: Analyzing cost – benefit of proposed responses

Generally, a pipeline failure involves various costs that are difficult to compute. Each cost component is unique to specific failure and depends upon factors such as the magnitude, area, and time of the failure, where it happens, and others. A broad classification involving the factors shown in Table 14.7 is possible. The amounts shown in each of these categories are the estimated maximum failure costs.

These factors depend on various subfactors and parameters. For the purpose of this article, a typical pipeline was considered. The cost encountered in this case (maximum) was estimated for India. An analysis of 20 years of failure expenditure data for the pipeline was conducted, and suitable escalation was applied wherever necessary, on the basis of published literature and increases in the cost of various commodities. The failure are classified (on the basis of cost incurred) into four categories:

small failures: up to 25 million INR;

medium failures: between 25 and 40 million INR;

large failures: between 40 and 70 million INR; and

very large failures: up to 105 million INR.

The probability of failure in each of these four categories is taken into consideration, along with the cost of failure. The severity of failure of various pipeline stretches was estimated in brainstorming session by the executives. The outcomes are tabulated along with the likelihood of occurrences of various risk factors (previously determined) as shown in Table 14.8. A Monte Carlo simulation was performed using the PC-based software Micro-Manager. The expected cost of failure of each stretch has been shown in Table 14.8. Skilled personnel are needed to compute costs against each of the factors

Table 14.8 Severity of failure of various pipeline stretches

Risk factors	PLS₁		PLS₂		PLS₃		PLS₄		PLS₅	
	Likelihood	Severity*	Likelihood	Severity*	Likelihood	Severity*	Likelihood	Severity*	Likelihood	Severity*
External	0.108	40	0.064	25	0.007	40	0.011	40	0.031	105
Internal	0.038	25	0.022	25	0.020	40	0.042	40	0.060	105
Third party activities	0.030	105	0.078	105	0.011	105	0.061	105	0.006	105
Malicious	0.033	25	0.039	25	0.005	40	0.018	40	0.005	105
Constr. defects	0.012	25	0.007	25	0.028	40	0.007	40	0.018	105
Poor mats.	0.006	25	0.007	25	0.027	40	0.016	40	0.017	105
	0.006	105	0.001	105	0.014	105	0.006	105	0.020	105
Human error	0.001	25	0.005	25	0.003	40	0.008	40	0.030	105
Operational error	0.001	25	0.003	25	0.009	40	0.003	40	0.056	105
Likelihood of no failure	0.764	0	0.773	0	0.877	0	0.828	0	0.758	0
Expected failure cost	10.35		12.04		6.56		11.28		25.44	

Total expected cost of pipeline failure = INR 66 million per year

*Severity figures are in INR in million;

PLS – Pipeline stretch

US$ 1 = INR 47

shown in Table 14.7. The cost of environmental damage varies from place to place, so readers are cautioned to use their own experience and expertise when estimating the cost of a pipeline failure. Table 14.6 shows the conventional inspection and maintenance cost without using risk-based model. Our proposed method has the potential to reduce costs and is thus preferred over conventional method. Additionally, Table 14.9 shows expected cost of failure in the event of using proposed risk-based inspection and maintenance.

Assumptions:

- The pipeline failure in each stretch is an independent event.
- With one failure in specific pipeline stretch and subsequent maintenance, the pipeline stretch will be vulnerable for failure in the subsequent years with equal likelihood.
- Each risk factor causes failure of pipeline system upon occurrence, the degree of which is measured by small, medium, large and very large failures. Accordingly, cost is incurred for its rectification. As for example, external corrosion will cause medium to large failure. In order to rectify the failure, 40 million INR is required to be spent.

Assumptions:

- With the implementation of proposed inspection and maintenance program the probability of failure of each pipeline stretch would reduce to half of the previously computed figure.
- Severity of the failure would be the minimum computed failure cost i.e. INR 25 million.

Discussions

Risk management process consists of identification of the specific risks, analyzing their effects, and developing responses. In a pipeline system, it is used to address the following questions:

1   What are the adverse situations that the pipeline subsystem can encounter?
2   How likely are these events to occur?
3   How severe would the consequence be if these events do occur?
4   What responses are feasible to mitigate the occurrences and effect?

In a pipeline system we can rarely hope to get all necessary data for analyzing the risk elements in the system. Not only detailed data will be difficult to get, it may be a very costly proposition as money spent in data acquisition will involve huge capital outlay that a company can ill afford and alternative transportation modes may turn out to be more feasible.

It may be noted that the elements that cause adverse situation do not have equal probability of occurrence. It varies across various pipeline

Table 14.9 Computation of expected failure cost of pipeline in the event of proposed inspection and maintenance program

Risk factors	PLS_1		PLS_2		PLS_3		PLS_4		PLS_5	
	Likelihood	Severity*	Likelihood	Severity*	Likelihood	Severity*	Likelihood	Severity*	Likelihood	Severity*
Likelihood of failure	0.118	25	0.114	25	0.062	25	0.086	25	0.121	25
Likelihood of no failure	0.882	0	0.887	0	0.939	0	0.914	0	0.879	0
Expected failure cost	2.95		2.85		1.55		2.2		3.1	
Total expected cost of pipeline failure = INR 13 million per year										

*Severity figures are in INR in million;

PLS – Pipeline stretch

US$ 1 = INR 47

systems in accordance with design conditions, operational procedures, soil condition and host of other factors. For analyzing risk, it is imperative that the word risk be defined. Risk can be succinctly described in mathematical form as

risk= probability of occurrence × consequences

To minimize risk, either probability of occurrence has to be reduced or the consequences be controlled or both. Controlling of consequences is extremely unlikely as most failures leave very little scope to act instantly for mitigation measures. The best alternative is therefore minimizing the risk factors for developing efficient facilities.

A decision support system (DSS) assists management decision-making by combining data, sophisticated analytical models and tools, and user-friendly software into a single powerful system that can support semi structured or unstructured decision-making. A DSS provides users with a flexible set of tools and capabilities for analyzing important blocks of data. In this study AHP is used to develop a DSS for inspection and maintenance strategy selection.

The analytic hierarchy process (AHP) developed by Saaty (1980) provides a flexible and easily understood way of analyzing complicated problems. It is a multiple criteria decision-making technique that allows subjective as well as objective factors to be considered in the decision-making process. The AHP allows the active participation of decision-makers in reaching agreement, and gives managers a rational basis on which to make decisions. AHP is based on the following three principles: decomposition, comparative judgment, and synthesis of priorities.

The AHP is a theory of measurement for dealing with quantifiable and intangible criteria that has been applied to numerous areas, such as decision theory and conflict resolution (Vargas 1990). AHP is a problem-solving framework and a systematic procedure for representing the elements of any problem (Saaty 1982).

Formulating the decision problem in the form of a hierarchical structure is the first step of AHP. In a typical hierarchy, the top level reflects the overall objective (focus) of the decision problem. The elements affecting the decision are represented in intermediate levels. The lowest level comprises the decision options. Once a hierarchy is constructed, the decision-maker begins a prioritization procedure to determine the relative importance of the elements in each level of the hierarchy. The elements in each level are compared as pairs with respect to their importance in making the decision under consideration. A verbal scale is used in AHP that enables the decision-maker to incorporate subjectivity, experience, and knowledge in an intuitive and natural way. After comparison matrices are created, relative weights are derived for the various elements. The relative weights of the elements of each level with respect to an element in the adjacent upper level are computed as the components of the normalized eigenvector associated with the largest eigenvalue of their comparison matrix. Composite weights are then determined by aggregating the weights through the hierarchy. This is done by following a path from the top of the hierarchy to each alternative at the lowest level, and multiplying the weights along each segment of the path. The outcome of this aggregation is a normalized vector of the overall weights of the options. The mathematical basis for determining the weights was established by Saaty (1980).

Risk analysis is usually a team effort, and the AHP is one available method for forming a systematic framework for group interaction and group decision-making (Saaty 1983). Dyer and Forman (1992) describe the advantages of AHP in a group setting as follows: 1) both tangibles and intangibles, individual values and shared values can be included in an AHP-based group decision process; 2) the discussion in a group can be focused on objectives rather than alternatives; 3) the discussion can be structured so that every factor relevant to the discussion is considered in turn; and 4) in a structured analysis, the discussion continues until all relevant information from each individual member in a group has been considered and a consensus choice of the decision alternative is achieved. A detailed discussion on conducting AHP-based group decision-making sessions including suggestions for assembling the group, constructing the hierarchy, getting the group to agree, inequalities of power, concealed or distorted preferences, and implementing the results can be found in Saaty (1982) and Golden et al. (1989). For problems with using AHP in group decision-making , see Islei et al. (1991).

AHP was used for risk management because:

- The risk factors are both objective and subjective;
- The factors are conflicting, achieving of one factor may sacrifice others;
- Some objectivity should be reflected in assessing subjective factors;
- AHP can consider each factor in a manner that is flexible and easily understood, and allows consideration of both subjective and objective factors; and
- AHP requires the active participation of decision-makers in reaching agreement, and gives decision-makers a rational basis upon which to make their decision.

Researchers use AHP in various industrial applications. Partovi et al. (1990) used it for operations management decision-making. Dey et al. (1994) used it in managing the risk of projects. Korpela and Tuominen (1996), and Dey (2002) used AHP for benchmarking logistic operations and project management respectively. Mian and Christine (1999) used AHP for evaluation and selection of a private sector project. Dey (2001b) described AHP as an effective tool for project selection. Dey et al. (2001) used AHP for cross-country petroleum pipeline route selection. Mustafa and Ryan (1990) used AHP for bid evaluation.

A decision-maker can express a preference between each pair as equal, moderate, strong, very strong, and extremely preferable (important). These judgements can be translated into numerical values on a scale of 1 to 9 (Table 14.2). Elements at each level of hierarchy are compared with each other in pairs, with their respective "parents" at the next higher level. With the hierarchy used here, matrices of judgements are formed. A brainstorming session was held to compare the risk factors. The pipeline executives established a common consensus for the AHP hierarchy, pairwise comparison in factors, sub-factors and alternative levels through group decision-making. Disagreements were resolved by reasoning and collecting more information. Their hierarchy contained the detail necessary for risk analysis.

Petroleum pipelines are the nervous system of the oil industry, as this transports crude oil from sources to refineries and petroleum products from refineries to demand points. Therefore, the efficient operations of these pipelines determine the effectiveness of the entire business.

As pipelines pass through varied terrain, the condition of pipelines varies widely across their entire length and throughout their life cycle. However, inspecting the entire pipelines through specific inspection methodology/ tool cannot detect pipeline problems for the entire length as inspection tools are designed to detect specific problems only. On the other hand, inspecting the entire pipeline by various tools to detect the entire associate problems is not cost effective.

This study presents a DSS model in AHP framework, which determines the likely problems associated with each stretch with the involvement of the experienced pipeline operators. This leads to developing a cost effective inspection and maintenance strategy for the pipelines.

This methodology has been applied in an Indian cross-country petroleum pipelines case. This study shows that the cost of inspection and maintenance after using the proposed risk-based DSS is INR 61 million (US$ 1 = INR 47) for five years as compared to INR 153 million for five years using conventional methods. The expected failure cost also would be reduced to INR 13 million per year with the proposed inspection and maintenance strategy using risk-based DSS as compared to INR 66 million per year without any inspection and maintenance. These show the rationale for using risk-based DSS for risk management.

Additionally, this study establishes a cost effective insurance plan for the pipeline under study. The basis of the insurance premium depends on likelihood of its failure, expected failure cost in a given period, risk perception of the management/organization and inspection/maintenance programs undertaken.

In this case study, the maximum amount of insurance premium for the pipeline under study would be the expected failure cost per year i.e. INR 66 million without any inspection and maintenance as indicated in Table 14.8. If the pipeline operators undertake the inspection and maintenance program in line with as indicated in Table 14.5, the likelihood and severity both decrease considerably. The expected cost of failure would reduce to INR 13 million as shown in Table 14.9. Hence, the annual insurance premium would lie between INR 66 and 13 million in line with management risk perception.

The same methodology can be used for any operating unit to develop strategic DSS for inspection and maintenance.

Advantages of this method of analysis described here include the following:

- reducing subjectivity in the decision-making process when selecting an inspection technique;
- identifying the right pipeline or segment for inspection and maintenance;
- formulating an inspection and maintenance policy;
- deriving the budget allocation for inspection and maintenance;
- providing guidance to deploy the right mix of labor in inspection and maintenance;
- enhancing emergency preparations;
- assessing risk and fixing an insurance premium; and
- forming a basis for demonstrating the risk level to governments and other regulatory agencies.

If a productive system is designed, constructed and operated ideally, many inspection and maintenance problems will not crop up. The overall performance of pipeline operations and maintenance would be improved through the following actions:

- Pipeline routes are to be decided on the basis of life cycle costing approach, not on the basis of shortest route. Dey and Gupta (1999) have shown one of these approaches.
- The maintenance characteristics of the pipeline are to be considered along with pressure and temperature parameters while designing pipe thickness for various stretches of pipeline.
- Pipeline coating shall be selected on the basis of terrain condition, environmental policy of the organization, cost of coating materials, construction methodology, inspection and maintenance philosophy.
- Construction methodology of pipeline in the critical section is to be formulated during the feasibility stage of the project and this shall be commensurate with the design and operational philosophy of the pipeline as a whole. The factors like availability of technology, availability of consultants, contractors and vendors, experience of owner project group, government regulations, and environmental requirements throughout the life of pipeline should be rationally considered during selection of the best construction methodology.
- Networking in pipeline operations demands a foolproof mechanism in the system for minimizing operational and human errors. Improved instrumentation shall be designed which is commensurate with the design philosophy of entire pipeline system.

- All pipeline operators are to be suitably trained in pipeline operation before taking charge of working in specific pipelines. Pipeline simulation training may be one of these kinds of training. Criticality of pipelines and expertise of personnel is to be considered for manning pipeline operations.

The technique does have limitations, because subjectivity is not totally eliminated. For instance, the weightage against each of the failure factors is based upon experience, available data and perception of the pipeline executives and decision-makers. Despite these limitations, a cross-country oil pipeline inspection and maintenance policy formed on the basis of our methodology is an effective tool to mitigate risk. It is cost effective and environment friendly.

The established decision support system helps the pipeline operators to dynamically evaluate pipeline health and to make decisions on types of inspection and a maintenance program for a specific stretch at any time they desire. Therefore, all the pipeline sections will receive attention with respect to health, although inspection and maintenance may be exempted for specific sections during a given period because of the better condition of the pipeline during risk analysis study.

Although the model developed in this chaptr is related to cross-country petroleum pipelines, the similar methodology can be applied to develop a risk-based inspection and maintenance model for any productive system. However, considerable research would be involved in such a study.

Conclusion

Risk is part and parcel of any business. Hence, success of business depends on how effectively risk is managed. Various tools and techniques are available to identify risk, analyze their effect, and develop responses for mitigation. However, the use of specific tools and techniques will depend on application requirement as well as management objectives and perception. Analytic hierarchy process is an effective tool for risk management as it employs a group decision-making process with the involvement of the operators. Risk-based inspection and maintenance not only suggests cost effective operations of facilities, but also integrates inspection and maintenance with safety and environmental management. In today's business there is tremendous importance in successful environmental management along with achieving business throughput. Risk management not only develops means for zero shutdown most economically, but also suggests ways of improving facility performance by addressing environmental and safety regulations. A strong institutional framework is needed for its effective function. The framework will vary from organization to organization depending on characteristics of the operations and projects for which risk is required to be managed.

References

Annual Report of CONCAWE. Brussels (Conservation of Clean Air and Water, Europe), 1994.

Cooper, D. F. and Chapman, C. B., *Risk Analysis for Large Projects: Models, Methods and Cases*. Chichester: John Wiley & Sons, 1987.

Dey, P. K., Tabucanon, M. T. and Ogunlana S. O., 'Planning for project control through risk analysis: a case of petroleum pipeline laying project', *International Journal of Project Management*, Vol. 12. No. 1, 23–33, 1994.

Dey, P.K., Tabucanon, M. T. and Ogunlana, S. O., 'Petroleum pipeline construction planning: a conceptual framework', *International Journal of Project Management*, 14, no. 4, 231–240, 1996.

Dey, P. K., *Symbiosis of Organisational Re-engineering and Risk Management for Effective Implementation of Large-scale Construction Project*. Doctoral thesis, Jadavpur University, 1997.

Dey, P. K. and Gupta, S. S., 'Decision support system for pipeline route selection', *Cost Engineering*, Vol. 41 No. 10,: 29–35, October 1999.

Dey, P. K. and Gupta, S. S., 'Risk-based model aids selection of pipeline inspection, maintenance strategies', *Oil and Gas Journal, PennWell*, Volume 99.28, 39–67, 9 July 2001.

Dey, P. K., 'A risk-based model for cost effective inspection and maintenance', *Journal of Quality in Maintenance Engineering, Emerald, MCB University Press*, Vol. 7, No. 1, 25–43, March 2001a.

Dey, P. K., 'Integrated approach to project feasibility analysis: a case study', *Impact Assessment and Project Appraisal*, Vol 19, No. 1, 135–245, September 2001b.

Dey, P. K., Ogunlana, S. O, Tabucanon, M. T. and Gupta, S. S., 'Multiple attribute decision making approach to petroleum pipeline route selection', *International Journal of Service Technology and Management*, Vol. 2, Nos.3/4, 347–362, 2001.

Dey, P. K., 'Benchmarking project management practices of Caribbean organizations using analytic hierarchy process', *Benchmarking: An International Journal*, Vol. 9, no. 3, 326–356, 2002.

Dey, P. K., 'Analytic Hierarchy Process analyses risk of operating cross-country petroleum pipelines in India', *Natural Hazard Review, American Society of Civil Engineering*, Vol. 4 issue 4, 213–221, 2003.

Dey P. K., Ogunlana, S. O. and Naksuksakul, S., 'Risk-based maintenance model for offshore oil and gas pipelines: A case study', *Journal of Quality in Maintenance Engineering*, Vol. 10. No. 3, 169–183, 2004.

Dyer, R. F. and Forman, E. H., 'Group decision support with the analytic hierarchy process', *Decision Support Syst.*, no. 8, 99–124, 1992.

Golden, B. L., Wasli, E. A. and Harker, P. T., *The Analytic Hierarchy Process: Applications and Studies*. New York: Springer Verlag, 1989.

Hopkins, P. *Ensuring the Safe Operation of Older Pipelines*. International Pipelines and Offshore Contractors Association, 28th Convention. September 1994.

Islei, G., Lockett, G., Cox, B. and Stratford, M., 'A decision support system using judgmental modeling: A case of R&D in the pharmaceutical industry', *IEEE Trans. Engineering Manage.*, vol. 38, 202–209, August 1991.

Jamieson, R. M. *Pipeline Integrity Monitoring*. Pipeline Integrity Conference Aberdeen, Scotland. 29–30 October 1986.

Korpela, J. and Tuominen, M. 'Benchmarking logistic performance with an application of the analytic hierarchy process', *IEEE Trans. Engineering Manage.*, Vol. 43, No. 3, 323–333, August 1996.

Mian, S. A. and Christine, N. D., 'Decision-making over the project life cycle: an analytical hierarchy approach', *Project Management Journal*, Vol. 30, No. 1, 40–52, 1999.

Mustafa, M. A. and Ryan, T. C., 'Decision support for bid evaluation', *International Journal of Project Management*, Vol. 8 No. 4, 230–235, 1990.

Ng, M. F., Tummala, V. M. R. and Yam, R. C. M., 'A risk-based maintenance management model for toll road/tunnel operations', *Construction Management and Economics*, 21, 495–510, July 2003.

Partovi, F. Y., Burton, J. and Banerjee, A., 'Application analytic hierarchy process in operations management', *International Journal of Operations and Production Management*, Vol. 10 No. 3, 5–19, 1990.

Payne, B. L., *California Hazardous Liquid Pipeline Risk Assessment*, in Proceedings of the 1993 International Pipeline Risk Assessment, Rehabilitation and Repair Conference. Huston, Tx: Gulf, 1993.

Pipes and Pipelines International. Oil Industry Pipeline Leakage Survey. January–February 1993.

Rowe, W. D., *An Anatomy of Risk*. New York: John Wiley and sons, 1993.

Saaty, T. L. *The Analytic Hierarchy Process*. New York: McGraw-Hill, 1980.

Saaty, T. L., *Decision Making for Leaders*. New York: Lifetime Learning, 1982.

Saaty, T. L., 'Priority setting in complex problems', *IEEE Trans. Engineering Manage.*, vol. EM-30, 140–155, August 1983.

Tomic, B. *Risk-based Optimization of Maintenance Methods and Approaches, Safety and Reliability Assessment – An Integrated Approach*. New York: Elsevier Science Publishers B V.,. 1993.

Tummala, V. M. R. and Mak, C. L., 'A risk management model for improving operation and maintenance activities in electricity transmission networks'. *Journal of the Operational Research Society*, 52, 125–134, 2001.

US Department of Transportation. Pipeline Safety Regulations. 1 October 1995.

Vargas, L. G., 'An overview of the analytic hierarchy process and its applications', *Europe J. Operat. Res.*, vol. 48, no. 1, 2–8, 1990.

Vaughan, E J. *Risk Management*. New York: John Wiley and Sons, 1997.

15 Managing construction risks and uncertainties

A management procurement and contracts perspective

Adekunle S. Oyegoke, Oluwaseyi A. Awodele and Saheed Ajayi

Chapter summary

This chapter deals with the relationship between construction risks, procurement strategy and contract documents. It is argued that the selection of appropriate procurement strategy and contract documents can mitigate construction risk. In procurement terms, agile specialist approach has been cited as an example of a procurement process that can mitigate risks. Different ways in which construction risks and uncertainties are identified, quantified and managed through the provisions in the contract documents of construction management contract are presented as an example. Moreover, the chosen construction management contract has been chosen as an example due to its popularity in the US and in the UK. In this chapter, key issues and challenges in mitigating risk in construction projects have been identified, these include:

- the risk associated with the procurement selection process
- appropriate distribution of risk and responsibility allocation
- the use of an appropriate contract documents
- transparency principles in procurement processes
- alignment of procurement strategy with the conditions of contract.

Introduction

A construction project is a complex process that involves many stakeholders, long project durations, and complex contractual relationships, which in turn increase the likelihood of project risks. A construction project involves multiple organisations which creates intricacies in inter-firm relationships and management. From a supply chain point of view, there are many players and organisations upstream and downstream in the supply chain. Their involvement makes the construction work prone to a high level of risk and uncertainty in different construction phases. PMI (2004) defines project risk as an uncertain event or condition that, if it occurs, has a positive or a negative effect on project objectives.

Construction risk can be identified along the project management body of knowledge of cost, time, quality etc. It can also be categorised along different events and activities, for instance, risks as a result of design, payment, weather etc. It can be project/site specific, e.g. ground conditions, project location, and project complexity. Others can be those imposed by external factors like the cyclical nature of the economy, government regulation framework, material shortages, etc. Since construction work is not risk free it is important to minimise, share, transfer, or accept risks – but risks should not be ignored (Cox and Townsend 1998).

Project risk should be assigned to a party that is able to manage it, although there is no good procurement route for all projects and at all times. However, what procurement systems should attempt to do is to distribute risks and allocate responsibilities to the stakeholders. This can be in a form of single or multiple points of responsibility, or point of performance responsibility. Therefore, procurement selection is vital in minimising and sharing risks.

In order to manage the relationships between different players and firms, contract documents most especially the conditions of contracts simplify and clarify the rights and responsibilities of the various parties in the event of a dispute. The contract conditions are also aim at classifying and clarifying duties and relationships of the parties, and at avoidance of disputes. Therefore it is vital to align procurement strategy with suitable contract documents in order to mitigate some of the risks.

This chapter demonstrates how construction risks can be minimised through proper scrutiny in selection phase, transparency in procurement processes and procedure, and adoption of suitable conditions of contract/contract documents. An example of how a contract document manages project risk is given in the chapter.

Risks and uncertainty in a project environment

According to Barrie and Paulson (1992), the building industry is characterised as custom-oriented, incentive-dependent and predicated on human factors, which consequently leads to the industry being fragmented. The project environment can be categorised into three levels; industry, project and task/trade levels. Industry level risks are affected by economic, social and political factors which may affect the outcome of a project. This type of risk is managed by the government through directives and guidance, for instance, collaborative procurement, demand management, among others. Project level is a stage where different components/disciplines are interchangeable resulting in many players both in the demand and supply sides of the construction business. Some of the risks in this level will include project initiation, quality of brief, choice of procurement selections among others. The task/trade level is typified by its temporary and multi organisational nature that results in the involvement of tasks/trades organisations in the execution of the projects. These organisations interact with each other and are affected by their internal organisation dynamisms and external contingency factors.

Procurement systems

In a construction project, procurement serves as an organisational system that assigns specific responsibilities and authorities to both people and organisations, and defines the relationships of the various elements (Love et al. 2002). In other words, project procurement establishes the contractual framework that assigns risks and responsibilities, and determines the nature of relationship between the project teams within the duration of their interactions.

The level of risk and uncertainty in a project environment results in two major weaknesses of the prevailing procurement systems: those inherent in the systems and those caused by changes in the environment (Atkin and Pothecary 1994). These weaknesses affect the choices of supply chains for projects: a single contractor (single point of performance responsibility) or multiple suppliers (multiple points of performance responsibilities) that are integrated or fragmented or a mixture of different solutions.

Cox and Ireland (2002) argue that the construction supply chains remain fragmented and highly adversarial due to the conflicting nature of demand and supply. The outright fragmentation of a supply chain contributes to the degree of risks and uncertainty in the project environment. Koskela (2003) proposes a lean supply chain aimed at narrowing down the sources of supply, reducing waste and producing continuous improvements throughout the supply network. The lean concept also requires a fundamental construction supply chain reorganisation and increase in the level of customisation.

Contracting documents

There are many ways in which risks and uncertainties in construction project environments can be managed. Flanagan and Norman (1993) assert that fundamental risks are liabilities and responsibilities that are inherent in any construction project between project stakeholders. They further postulate that standard forms of building contract allocate risks between parties by expressed and implied terms.

In a broader sense, safety management is a risk response development by setting the enhancement steps for opportunities and responses to threats. Risk response control in a form of risk management is insurance and a mutual indemnity by both the parties as well as a bonding (performance, payment and professional indemnity bonds) (Oyegoke 2002). Haltenhoff (1999) listed ways to dispose static risks either by elimination, avoidance, prevention, reduction, assignment, expensing or by retention and management. Figure 15.1 shows that all risk types occur in project resources management in the form of responsibilities. Failures or defaults to carry

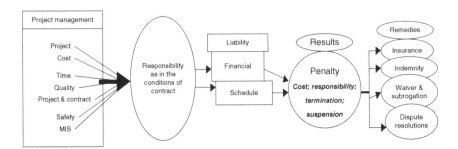

Figure 15.1 Classification and effects of the project risks in relation to project resources management

out the responsibilities will lead to a liability either financially (a direct or consequential loss; claims) or in a schedule recovery or an extension of time, etc. It may also result in a penalty in the form of claims (liquidated and ascertained damages), an additional responsibility, a contract termination or a suspension. Remedies include insurance (injury to third party), a mutual indemnity, a waiver and subrogation and dispute resolutions.

Haltenhoff (1999) argues that the risk management area of knowledge includes surety bonding and insurance in the static risk area, as well as contracting and construction processes and procedures in the dynamic risk area. Hence, this chapter is based on the key premise that risks and uncertainties in a project can be causally linked with contracts and procurement strategies employed.

Figure 15.1 presents the classification and effect of the project risks in relation to project management. Risks in construction projects can occur in managing any of the key areas of project management body of knowledge. The conditions of contract assign rights, duties and liabilities in order to manage the relationships between the stakeholders. Failure of any of the parties to the contract will lead to liabilities which can be financial or project elongation. This can result to termination or suspension of contract if the issue is not adequately addressed. There are a number of remedies that can be used to protect the parties to the contract and the third party.

Proposed Framework

The CMAA contracting documents example

In this section, the authors demonstrate interrelationships between the construction management contracting documents and project risks management. PMI risks management classifications (i.e. risk management planning, identification, qualitative and quantitative analysis, response planning, monitoring and control) are used as a frame of reference.

Table 15.1 Management of project risk based on the CMAA services and practice

| Themes | Project phases | Planning | | Assessment | Analysis | | Management | | Monitoring & control |
		Management planning	Identification		Qualitative	Quantitative	Response planning		
Project management	Pre-design	Organise project team with the owner	Project organisation & Project requirements		Establish a relationship of mutual professional trust and respect	Identify each team member responsibilities; and pre-design project conference	Establish MIS that shows overall status and forecast compared to established plan		Construction management plan and project procedures manuals
	Design	Pre-design project conference	Design documents and project funding		Design document review	Document distribution and contract agreement review	Periodic project meetings		Cost control and time control
	Procurement	On-going consulting activities	Bidding and contracting process		Provision for permits, insurance and labour affidavits	Bid opening and evaluation	Pre-bid meetings, bid opening and pre-award conferences		Monitoring compliance with construction contracts execution
	Construction	Professional planning and project execution	Construction process efficiency		Verify on and off site facilities	Measure compliance of schedule and project/ contract documents	On-site meeting and management reporting		Claims, time and quality management, and management reporting
	Post-construction	Effective project documents transmission	Effective project close-out		Verify documents that relate to move-in or startup	Prepare manuals and assemble record drawings	Preparing final documents e.g. final cost report		Cost, time management, and project/ contract administration

CMAA contracting documents were used for this analysis to demonstrate how it can mitigate risks. Other contracting documents can be analysed in this way. A study by Oyegoke and Kiiras (2001) indicates that there are no significant differences between the standard forms of contracts for different management procurement routes. The differences can only be found on their details; which often depend on the responsibilities of the Construction Management (CM) firm either in consulting or contracting role.

Project management

In the pre-design stage, the project owner is required to assemble and organise a project team/organisation and provide basic project purposes to the team. The Construction Manager then quantifies this process by outlining Construction Management Plan (CMP) as a strategy of fulfilling the owner's requirements. The Construction Manager plans, conducts and documents a pre-design conference, which addresses the CMP with respect to the design phase and establishes a management information system that keeps the team informed as to the overall status of the project. The project team (Construction Manager and design team) writes out a Project Procedure Manual that outlines the responsibilities of the team, levels of authority, the systems, methods, and procedures to be followed for project execution. Both the CMP and PPM (Project Procedure Manual) serve as a control technique.

As the design process proceeds, the Construction Manager periodically reviews the design documents, focusing on the need for clarity, consistency and co-ordination among the contractors. The Construction Manager co-ordinates the distribution of information among all the team members and the transmittal of all the documents to the regulatory agencies. The Construction Manager develops and/or reviews appropriate construction contract agreements and general and supplementary general conditions for the inclusion in the bid documents in consistent with the project requirements. The Construction Manager then conducts periodic project meetings for the purpose of assessing the progress, verifying the adherence to the CMP, documenting the performance, planning for completion and taking the necessary action to resolve problems.

Moreover, the Construction Manager develops cost control procedures in order to monitor and control the project expenditures, both current and projected, within the allocated budget and to modify the master schedule and the milestone schedule to reflect the actual performance. The Construction Manager is involved in on-going consulting activities by making recommendations to the team members regarding constructability, cost, duration, phasing and sequencing of construction. After the project documents have been finalised, the Construction Manager involves in solicitation and pre-qualification of bidders and draws the guidelines by which bidders are evaluated. He also places notices and advertisements, issuance of the addenda, information to bidders, and delivery of the bid documents. It is the duty of the Construction Manager to arrange pre-bid meetings, bid openings and pre-award conferences.

In the construction stage, the Construction Manager verifies that the on-site facilities are provided; this will include the provision of office facilities and site work required for the general access and utilities to all on-site organisations. He then establishes the procedures for planning and monitoring compliance with the project time line that relates to the master and detailed construction schedules. He also establishes procedures on issues such as budget and cost monitoring, payment requests, change orders, claims management, management reporting, quality management as well as owner-purchased materials and equipment.

In the post-construction stage, the Construction Manager prepares and transmits documents connected with the final payment, the organisation of operation and maintenance manuals as well as assembles the record drawings, contractor follow-up, owner move-in or start up, contractor call-back and contractor close-out (CMAA 1999).

Cost management

Cost management begins with a preliminary cost investigation that includes all the cost components. There is a need for the contingency amounts in the budget estimate stage to take care of unforeseen costs. The estimates of construction and project costs are developed into project and construction budgets in the formats based on the work breakdown structures. At the owner's request, the Construction Manager uses cost analysis methods to prepare the cost estimates for conceptual design alternatives. The Construction Manager provides ongoing cost management services to ensure that the budget is adhered to as the design is developed.

A cost verification exercise is prepared at different stages to make sure that the project remains within the construction and project budgets. The Construction Manager prepares a schematic design cost estimate and a preliminary design estimate as well as in-progress and final design document estimates from the working drawings and specifications. While value analysis studies are used for the purpose of optimising value in the project design. In order to maintain compliance with the construction budget, the Construction Manager provides cost monitoring to assist design professionals.

In the procurement stage, the Construction Manager prices in detail all the proposed addenda (supplementary documents), in a manner similar to the final estimate quantification and pricing methodology. All the submitted bids by the contractors are tabulated for the bid analysis. The analysis includes evaluation of all the alternate bids and unit prices, compared with the Construction Manager's final estimate of the construction cost based on the bid documents. After the bids are received, the amounts of the acceptable contractor proposals replace the estimated line-item amounts and become the construction phase budget (CMAA 1999). The Construction Manager has the responsibility to install a budget-control system, which provides for the accurate and timely estimating/tracking of the project costs from the conceptual estimate to the final budget accounting (Haltenhoff 1999).

The construction stage is a phase where the progress payments for the works done are carried out. This can be in two ways: (i) percentage of the completion of the scheduled activities; (ii) percentage of the completion by the division of work. The Construction Manager in conjunction with the contractor(s) determines the schedule of the value for each of the scheduled activities or a schedule of the value for each bid package for the contractor's progress payments. The Construction Manager as a financial controller establishes and implements a change order control system by preparing an estimate of the cost of the change order listing the anticipated labour, materials, equipment, subcontract work, contractor's overhead and profit, as well as any justified impact cost.

The pricing for a change order can be forward pricing (done prior to the start of or during the work) or post-pricing (done at some point in time during or after the work is completed). Therefore, the sum of all the contracts awarded establishes the construction budget. Each construction budget line item is divided by a listing called a schedule of the values (a breakdown of each work-scope into identifiable units of work that are measurable in terms of quantity and cost). The most difficult part of the budget management occurs prior to the start of construction, when construction knowledge, estimating experience, and communication skills are the only tools to accomplish what must be done. The credibility of the budget during the formation and maintenance depends on estimating expertise and timely communication. In the post-construction stage, the Construction Manager summarises the total project costs in the final report, listing all the change orders and identifying any unresolved issues, which may have a cost impact (Table 15.2 refers).

Time management

The Construction Manager prepares the master schedule to identify major components of the project, their sequence and duration, which will then be submitted to the owner for acceptance. After the owner's acceptance, a milestone schedule is prepared. It may be in the form of a bar graph for small projects to networks for large and/or complex projects. The milestone schedule includes the dates for the design professional selection, design professional orientation meetings, anticipated cost/benefit studies to be submitted, and completion dates for the project phases. As the scope of the project is developed during design stage, the Construction Manager makes recommendations for revisions to the master schedule if the need arises.

Table 15.2 Management of project cost risk based on CMAA services and practice

Themes	Project phases	Planning	Assessment	Analysis		Management	
		Management planning	Identification	Qualitative	Quantitative	Response planning	Monitoring & control
Cost management	Pre-design	Preliminary cost investigation	Project and construction budget(s)	Review budgets with established cost limitations	Cost analysis, LCC, energy studies, preliminary cash flow	Allow for design contingency	Monitor cost management plan and prepare cost reports
	Design	Preliminary design estimate	Project estimates	Verify construction and project budgets	Value analysis studies and final design document estimates	Design estimate/cost management	Cost monitoring and reporting
	Procurement	Estimate of addenda	Bid analysis and negotiation	Evaluate alternative bids and unit prices	Prepare bid analysis	Bid negotiation	Monitor pricing methodology in the final estimates of construction costs
	Construction	Trade-off studies on materials, systems, etc.	Project costs overruns	Change order control	Forward and post-pricing	Establish a detailed audit record e.g. for claims	Monitor cost management procedures
	Post-construction	Summarise total project costs	Total project cost	Identify unresolved issue with cost impact	Listing all change order	Identify unresolved issue with cost impact	Monitor project costs and prepare final cost report

All revisions concerning time are submitted to the owner for acceptance. When the design is about to be completed, the Construction Manager develops the pre-bid construction schedule and identifies the major milestones for the inclusion in the bidding documents prior to transmitting the contract documents to the bidders. The Construction Manager and the owner jointly determine how float related to the various paths of activities in the schedule will be managed during the course of construction.

In the procurement stage, the Construction Manager provides a master schedule to the bidders and makes them aware of their scheduling responsibilities and obligation to participate in the schedule development as required by the contract documents. The successful bidder(s) must submit a contractor's construction schedule to the Construction Manager that provides an orderly progression of the work to achieve the completion within the contract time. In the construction stage, the Construction Manager monitors the schedule compliance (construction schedule/ master schedule) in order to provide the accurate report of the actual construction progress as compared with the established schedule. If the network-based project schedules are part of the contractual scope of work, the Construction Manager may use these schedules for developing the progress payment requests, payment certifications and cash flow projections.

It is the duty of the Construction Manager to quantify the time impact of the critical delays encountered during the project and document information regarding time extensions requested, granted or denied. When the contractor is behind schedule due to lack of performance on his part, the Construction Manager directs the contractor to recover the time at the contractor's cost. But if a project is legitimately behind schedule, the Construction Manager and the contractor(s) determine the opportunity for time recovery, by reviewing the critical activities and specific durations of those activities. As the project is near completion or completed, the Construction Manager reviews and evaluates construction claims that remain open or submitted after the substantial completion. In the post-construction stage, the Construction Manager develops an occupancy plan that provides the owner with a smooth and orderly transition into the completed project and facilitates the revenue income or beneficial use as quickly as possible (see Table 15.3).

Quality management

The Construction Manager meets with the owner to clarify the goals and objectives of the quality management programme and then develops a comprehensive project quality management plan, with the direct input from the designers and the owner. In the design stage, the Construction Manager sets out the design procedures in QMP (Quality Management Plan) and gives the opportunity to review the design submittals as they developed. This leads to the approvals by the owner, users, government

Table 15.3 Management of project time risk based on CMAA services and practice

Themes	Project phases	Planning	Assessment	Analysis		Management	
		Management planning	Identification	Qualitative	Quantitative	Response planning	Monitoring & control
Time management	Pre-design	Sequencing, management and implementation of design	Schedule adherence	Prepare master schedule	Highlight key events from master schedule -milestone schedule	Develop construction schedule	Monitor and implement schedules and prepare periodic reports
	Design	Maintaining master schedule	Realistic design phase schedule	Revision of master schedule	Milestone schedule compared with construction schedule	Determine management of float	Monitor milestone and construction schedules
	Procurement	Development of pre-bid construction schedule	Contractor's construction schedule	Clarification of contractor's schedule responsibilities	Milestone schedule compared with contractor's construction schedule	Contractor participation in schedule development	Monitor contractor's construction schedule
	Construction	Monitor schedule compliance	Construction schedule	Ways for contractor's to recover lost schedules	Extensions/impact analysis	Recovery schedules and claim review	Monitor schedule compliance and construction progress
	Post-construction	Develop occupancy plan	Occupancy plan	Determine participant in occupancy plan	Move-in frequency	Obtain city/state/federal reviews and certification	Monitor occupancy plan

agencies and other agencies having a jurisdiction over the project. Design quality control involves checking of the concepts, calculations and material selection procedures to attain the owner's quality standard, which should be in conformance with the quality management plan.

In the procurement stage, the Construction Manager complies with the standard for advertisement and the solicitation of the bids for public agencies and private owners. The Construction Manager participates in all the pre-bid meetings, site visits and addendum preparation and conducts with the owner a pre-award conference with the apparent successful bidder to discuss the terms on costs, conditions, and scope of work. The Construction Manager requires the contractor to provide necessary bonding, insurance and other special requirements as contained in the contract documents.

In the construction stage, the QMP states the specific requirements for the quality control and quality assurance. The Construction Manager verifies the testing and inspection of the contractor's work and ascertains that it is performed in accordance with the contract specification. He also maintains a thorough documentation in the form of reports and record keeping, e.g. the non-conforming and deficient work is recorded in a log removed as a result of an acceptable remedial action by the contractor(s). Towards the end of the project, the contractor(s) may request a semi-final inspection of the work, which will require the contractor to develop a detailed punch list of the outstanding work. The final acceptance of the work shows that the contract is completed with no outstanding items remaining. In the post-construction stage, the Construction Manager reviews and discusses the overall quality management of the project with the owner. He also prepares a final report for the overall project with the recommendations to the owner regarding activities during the course of the work (see Table 15.4).

Project/contract administration

The Construction Manager's contracting structure and the assignment of the contractual responsibilities generate an unprecedented amount of project-related information. The volume of information in a construction management contract requires a multi-level, need-to-know reporting structure as well as an efficient information storage and retrieval system. The Construction Manager develops the communication procedures for recording and controlling the flow of the submittals by the design professional for the approval by the owner. In the design stage, the Construction Manager develops and implements a system for the flow of information to the entire project team members and organises design review meetings to provide the owner's comment and approvals. It is the duty of the Construction Manager to prepare a project cost report to compare the budget for the project to the actual costs incurred and the forecast to complete.

In the procurement stage, before the bidder pre-qualification, the Construction Manager develops a contract scope breakdown where the

Table 15.4 Management of project quality risk based on CMAA services and practice

Themes	Project phases	Planning	Assessment	Analysis		Management	
		Management planning	Identification	Qualitative	Quantitative	Response planning	Monitoring & control
Quality management	Pre-design	Clarify owner's objectives	Quality management organisation	Determine methodology for quality control	Review scope of work and quality control	Develop quality management plan	Quality management, control and assurance
	Design	Manage design process/design procedure	Document control/design quality	Constructability reviews and quality management specifications	Design criteria changes, testing requirement, value engineering etc.	Project review meetings and reports	Verify funding, quality control and assurance
	Procurement	Procurement planning	Contractor's selection	Proposal document protocol and bid opening	Advertisement and solicitation of bids, and bidders list selection	Instructions to bidders and pre-bid conference	Pre-award conference with a successful bidder
	Construction	Preconstruction conference	Construction quality	Report and recordkeeping, and changes in the work	Inspection and testing, checking work quality	Final review, documentation and punch list work	Issuance of progress payment and certificate of final acceptance
	Post-construction	Pre-warrantee check-outs	Operations and maintenance manuals	Review and discuss overall quality management	Quality management assessment with the owner	Quality management plan	Final report and recommendations

local contracting practice must be a major factor and develops the list of potential bidders if open bidding is not exercised. If there is a need for notices and advertisements (open bid), the Construction Manager prepares and places it in trade journals, and newspapers etc. and administers the distribution of the bid documents in co-ordination with the design professionals. The Construction Manager organises the pre-bid conferences and meetings that centre on the explanation of the project requirements to the bidders. He also assists in bid opening and evaluation as well as post-bid interviews. After the successful bidder has been selected, the Construction Manager issues the notice to proceed to the successful bidder on behalf of the owner (see Table 15.5).

After the pre-construction orientation conference, the Construction Manager assists the owner in the transfer and acceptance by the contractor(s) of any purchased materials and equipment. He also monitors the contractor's progress in securing a proof of the insurance, building permits, bonds etc. The Construction Manager prepares the onsite communication procedures, organises project site meetings and establishes contract documentation procedures, field reporting and quality review. On the non-conforming work, he notifies the contractor in writing and informs the owner and design professionals. He must follow-up the non-conforming work in a form of the acceptance, removal and payment until a satisfactory solution is reached.

In the post-construction stage, the Construction Manager co-ordinates the compilation, organisation and indexing of the maintenance manuals and operating procedures and binds them into the document sets. The Construction Manager also co-ordinates all the requirements for spare parts and warranties and regulatory agencies permits to permanently utilise or operate a facility. After the punch list item is completed, the Construction Manager arranges a final inspection of the facility with the inclusion of the owner and design professionals. The Construction Manager prepares a close-out report that will be in a form of the final project history report.

Safety management

An early member of the project team should be the Construction Manager's safety co-ordinator who begins to develop the input from a safety perspective. Some projects may require the expertise of a certified safety professional due to special conditions, which may exist on the jobsite, e.g. asbestos or hazardous waste removal. During the design stage, the safety co-ordinator meets with the design team to achieve an understanding of the scope for the project. The safety co-ordinator is provided with the opportunity to review the drawings and to discuss the specific elements of the project to determine potential safety hazards, which may exist once the project is begun. The Construction Manager provides input for the construction contract documents concerning specific safety devices, equipment, and personal protective equipment.

Table 15.5 Management of project/contract administration risks based on CMAA services and practice

Themes	Project phases	Planning	Assessment			Management	Monitoring & control
		Management planning	Identification	Analysis		Response planning	
				Qualitative	Quantitative		
Project contract/ administration	Pre-design	Enhance communication flow	Project administration and reporting	Establish Communication procedures	Recording and controlling the flow of submittals	Communication procedures	Recording and controlling the flow of submittals
	Design	Systematic flow of information	Design phase progress	Consultation among team member	Project cost compared with project budget	Design review meetings	Schedule management and project cost report
	Procurement	Bidder pre-qualification	Securing bidders	Develop bidders list	Bid opening and evaluation, and review addenda	Pre-bid conferences and meeting, and post-bid interview	Project cost report, cash flow report and schedule maintenance report
	Construction	Pre-construction orientation conference	Efficiency in construction process	On site communication and contract documentation procedures	Quality review/ nonconforming work	Securing proof of insurance, permits, labour affidavits, and bonds	Project site meetings, field reporting, special record keeping etc.
	Post-construction	Requirement for spare parts and warranties	Project maintenance	Maintenance manuals and operating procedures	Move-in/start-up activities	Contractor callbacks	Contract close out, final payment, and close out report

In the procurement stage, the Construction Manager reviews the contractor's safety related submittals and jobsite safety programme. He also develops lines of communication with the agencies responsible for enforcing the compliance of the regulations applicable to the construction of the project. At the pre-construction conference, the Construction Manager's safety representative addresses all the prime contractors on the issues of the emergency response programmes and procedures, safety meeting times and schedules, training requirements, site surveys, etc. In the construction stage, before the performance of all the critical activities, the contractor must perform a job hazard analysis, which outlines the plans and procedures to be followed in order to perform the work in a safer manner. Thereafter, a safety co-ordination meeting takes place between the Construction Manager and the contractor to discuss the job hazard analysis. The Construction Manager also participates as a member of the jobsite safety committee, conducts the periodic safety audits and the evaluation of the status of the programme and of accident frequency and severity (see Table 15.6).

Application of the framework

An agile specialist example

In order to mitigate the level of risks and uncertainty, the agile specialist's approach utilises both positive attributes of integration and fragmentation by involving the agile specialists in design through construction and facility management of their specialist tasks (Oyegoke et al. 2008). Agile specialist approach distinctively adds extra value when compared with traditional and lean supply chain management. A 'lean supply chain' enables material flow from the supply base through the suppliers' tiers towards the contractors and subcontractors (Briscoe et al. 2001; London and Kenley 2001; Vrijhoef and Voordijk 2004).

Key steps that reduce risks in agile specialist approach (Oyegoke et al. 2008) are the in-built transparency in the process through competition between specialist, openness and transparency in selection process, prequalification and involvement of the agile specialist in early design and construction.

(1) An owner identifies his building needs through a rigorous briefing stage.
(2) The owner forms an agile management team of experts.
(3) The agile specialists are augmented with a design sub-team that provides the overall product design, documents and performance specifications.
(4) The instructions to tenderers (ITT) spell out the project information (general plan, performance and technical specifications), the tender format, the selection and evaluation procedures, the rules for disqualification, the latest date for the notification of intention and the tender submission date.

Table 15.6 Management of project safety risk based on CMAA services and practice

Themes	Project phases	Planning	Assessment	Analysis		Management	
		Management planning	Identification	Qualitative	Quantitative	Response planning	Monitoring & control
Safety management services	Pre-design	Owner's commitment	Project safety	Safety management options	Safety programme/organisation/staffing consideration	CM safety coordinator	Certified safety professional or certified industrial hygienist
	Design	Project scope understanding	Design safety	Review drawings with design team	Potential hazards/specific safety devices	Safety input to construction contract documents	Mitigate potential hazards by providing safety devices
	Pre-bid	Pre-bid conference	Performance safety	Drafting guidelines/responsibilities	Safety performance as prequalification	Written safety programme and emergency response coordination	Contract requirements and drafting guidelines
	Pre-construction	Pre-construction conference	Construction safety	Review of safety submittals	Review of contractor's jobsite safety programme	Emergency response programmes and procedures	Communicate with compliance agencies
	Construction	contractor safety enforcement and compliance	Project safety	Safety enforcement and compliance	Job hazard analysis	CM safety training	Safety audits, safety coordination meetings and monthly reports

(5) Tenders are prepared in two parts: technical tenders and price tenders.
(6) The agile management team evaluates the tenders by giving the scores to both the technical and price tenders as well as to the completion times in proportions that justify the complexity of the project. Hence, the economically/technically most advantageous tender is selected.
(7) The agile specialists are selected based on the design/engineering solution, constructability, maintainability, life cycle costs, schedule, stated methods and technical specifications through the closing negotiation with the owner.

The agile specialist approach follows production philosophy in product development, project execution processes coupled with integrated management system and systemic flow of information and communication. In addition, the agile specialist approach shortens the supply chain by eliminating the main contractor (first tier) and allows the agile specialists to complete design, manufacturing and installation of components and systems that decrease product development time. This subsequently leads to economic integration of the agile specialist firms through the complete task execution, and results in a greater level of transparency, total quality management, just-in-time delivery and other logistic issues like better management of inventory, lead-time and reduction in production costs.

The risk can be managed through the integrated upper organisational structure which consists of a management team that is responsible for integrating agile specialists and encouraging systemic flows of information, coordination, communication and overall strategic supply chain management. The integrated upper level is high on formalisation in terms of the document control of project development and building design processes on behalf of the owner (client). This self-directed management team involves high-task certainty, high functional specialisation, clear task descriptions and maximum control. Self-direction also absorbs the impacts of uncertainties that are likely to emerge from within a project environment.

On the lower level of organisation structure (i.e. project implementation), is the fragmented network of agile specialists. This level is decentralised (authority and responsibilities) and high on formalisation. The network reduces uncertainty and risks in a project environment because it has:

1 low-task certainty in preconstruction phases due to the contracting of competing specialists based on the overall design only
2 high-task certainty in a construction phase as a result of the early agile specialist involvement in project development and building design.

The network is very high on functional (packaged) specialisation and involves interdependent tasks that specialists perform concurrently. The uncertainty is further reduced as the approach allows each agile specialist and its self-directed team to perform the given task (package) independently based on the clear package description (function, performance and technical specification) with their own organisational resources under sufficient indirect control. The management team integrates a set of technical engineering and design processes that take place within each agile specialist. After the construction startup, the same team manages the fragmented project implementation, that is, the tasks to be carried out by the specialists.

Discussions

It has been demonstrated that procurement strategy and conditions of contract can be used to mitigate construction project risks. The agile specialist approach provides two layers of managing risk. The approach encourages process transparency and competition among the agile specialist. It is important to align procurement strategy to the type of contract document that will be use. Early contractor involvement is vital and where that is not possible a design audit should be carried out by the contractor to ascertain its compliance with building regulations and standards, and to advice on its buildability.

The analysis shows that the construction management contracts services and practices and the contract documents explicitly identify, quantify, and develop responses and control of project risks. In a broader sense, safety management can be said to be risk response development by setting the enhancement steps for opportunities and responses to threats. Risk response control in a form of risk management is insurance and a mutual indemnity by the both parties, as well as bonding (performance, payment and professional indemnity bonds). Clause 9.1 and 8.1 of A-1 and GMP-1 respectively (Construction Manager's liability insurance) state that the Construction Manager must purchase and maintain an insurance to prevent the Construction Manager from claims under workers' compensation, disability benefits, bodily injury, disease or death of CM employee, or the third party etc. However, the Construction Manager is entitled within a stated limit, the comprehensive general and automobile liability insurance. On the other hand, clauses 9.2 and 8.2 respectively (the owner's insurance) state that the owner must be responsible for purchasing and maintaining its own liability insurance or additional insurance for further protection. Both the owner and Construction Manager will indemnify and hold harmless each other against any claim, demands, suits etc. for which any of them is liable.

This chapter shows that total project resources management assigns responsibilities to project stakeholders through the standard forms of agreement. Figure 15.1 shows that all risk types occur in project resources management in a form of responsibilities. Failures or defaults to carry out the responsibilities will lead to a liability either financially (a direct or consequential loss; claims) or in a schedule recovery or an extension of time etc. It may also result to a penalty in a form of claims (liquidated and ascertained damages), an additional responsibility, a contract termination, or a suspension. Remedies include insurance (injury to third party), a mutual indemnity, a waiver and a subrogation and dispute resolutions.

Both the owner and Construction Manager must provide to each other the copies of all the policies obtained for the project and within 30 and 60 days in A-1 and GMP-1 respectively, each of them must provide a written notice for the cancellation, non-renewal, or endorsement reducing or restricting the coverage. The Construction Manager must indemnify and hold harmless the owner, the design professional, and their employees, agents and representatives (vis-à-vis) from and against player any, and all the claims, demands, suits, and damages for bodily injury and property damage for which the Construction Manager is liable or as a result of his negligent acts or an ommission in carrying out the Construction Manager's services.

The total liability of the Construction Manager arising out of this indemnity for losses that are not insured must not exceed the amount of the total compensation actually paid to the Construction Manager by the owner. Furthermore, the owner shall be responsible for the accuracy and completeness of all the project documents. There is a waiver of subrogation between the owner and the Construction Manager against the contractor, design professionals, consultants, agents and employees of the other, and between the owner and the Construction Manager from the contractor for damages during the construction covered by any property insurance as in the conditions of contract.

Conclusions

The agile specialist approach shows that project risks can be effectively managed through proper risk allocation and responsibility distribution among the project players. It is important to have a thorough project briefing in order to determine the appropriateness of different procurement routes. In management route as demonstrated in the agile approach the project was divided into specialist task and bidding was done through an agile specialist's for different trades or work sections. This approach generated a healthy competition between specialists in design, materials and life cycle costing. Each task should be demarcated from start to end, eliminating problems with task interdependence and schedule adherence. An experienced management firm/contractor should coordinate and managed the project resulting in better prequalification of the specialists, balanced evaluation of tenders, better understanding of technical solutions and their cost and quality implications. This results in reduction in the overall risk and uncertainty in a project environment and subsequently, the attainment of project objectives. The link between procurement strategy and conditions of contract in managing project risks cannot be overemphasised. The procurement route sets out roles, responsibilities and risks while contract clarifies the nature of interaction between the in terms of rights, duties and liabilities. CMAA documents are more explicitly in setting our duties and functions because the Construction Manager is rendering its service in a consultant capacity in parallel to consultant services. The analysis of the conditions of contract shows that most of the items under planning, assessment, and management are interchangeable. This means that the areas of project risk management interact with each other and all project resources management is aimed at managing project risk.

References

Atkin, B. and Pothecary, E. (1994) *Building futures.* London: St George's Press.

Barrie. D. S. and Paulson, B. C. (1992) *Professional construction management.* New York: McGrawHill.

Briscoe, G., Dainty, A. R. J. and Millett, S. (2001) Construction supply chain partnerships: skills knowledge and attitudinal requirements. *European Journal of Purchasing and Supply Management* **7**, 243–255

CMAA (1999) *Standard construction management services and practice.* 3rd edn, McLean: CMAA documents

Cox, A and Ireland, P. (2002) Managing construction supply chains: the common sense approach. *Engineering Construction and Architectural Management* 9 (5/6), 409–418.

Cox, A. and Townsend, M. (1998) *Contracting for business.* 1st edn, London: Thomas Telford Publishing.

Flanagan, R. and Norman, G. (1993) *Risk management and construction.* 1st edn, Oxford: Blackwell Scientific Publications.

Haltenhoff, C. E. (1999) *The CM contracting system; fundamentals and practices.* 1st edn, New York: Prentice-Hall, Inc.

Koskela, L. (2003) Theory and practice of lean construction: achievement and challenges. In Hansson, B. and Landin, A. eds., *Proceedings of 3rd Nordic Conference on Construction Economics and Organization. Lund University,* 239–256. Lund: Lund University.

London, K. A. and Kenley, R. (2001) An industrial organization economic supply chain approach for the construction industry: a review. *Construction Management and Economics,* Vol. **19**, 777–788.

Love, P. E. D., Irani, Z., Cheng, E. and Li, H. (2002) A model for supporting inter-organisational relations in the supply chain. *Engineering, Construction and Architectural Management,* No. **9**, 2–15.

Oyegoke, A. S. and Kiiras, J. (2001) Consulting and contracting perspectives of construction management contracting systems: a case of US, UK and Finland. *Proceeding of 2001 Brunei International Conference on Engineering and Technology, Brunei.*

Oyegoke, A. S. (2002) Risks management in construction management contracting systems – application of PMI risk management principles. *Proceedings of ICEC International Cost Engineers Council 3rd World Congress, Melbourne, Australia. CD-Rom.*

Oyegoke, A. S., Khalfan, M. M. A., McDermott, P. and Dickinson, M. (2008) Managing risk and uncertainty in an agile construction environment: application of agile building specialist model. *Int. J. Agile Systems and Management,* Vol. 3, Nos. 3/4, 248–262.

PMI (2004) *A guide to the project management body of knowledge* (PMBOK Guide), Project Management Institute.

Vrijhoef, R. and Voordijk, J. T. (2004). Improving supply chain management in construction: what can be learned from the electronics industry?. *In Proceedings CIB World Building Congress.* 2 May 2004, Toronto.

16 Risk management practice in a construction project

A case study on the NHS PFI/PPP hospital project in the UK

Farid Ezanee Mohamed Ghazali and
Mastura Jaafar

Introduction

Private Finance Initiative (PFI)/Public Private Partnership (PPP) is increasingly becoming a favourable procurement approach for the UK government when procuring its public projects such as prisons, highways and hospitals where the National Health Service (NHS) is usually the owner of the latter projects, due to direct involvement of the private sector in modernising public service deliveries. By allowing each sector to do what it does best, the intended facilities and services can be delivered in the most cost effective and efficient manner. Gentry and Fernandez (1997) identified the four most influencing factors that lead to the change of procurement approach from the conventional ones to PFI/PPP:

- Ability of private sector companies to provide the facilities/ services required;
- Acceptable degree of project control by public sector clients;
- Availability of financial resources from public funding; and
- Specific legal framework outlined for private investment and regulatory oversight.

Since this type of procurement approach involves long-term operational contracts, any PFI/PPP construction project is subject greatly to risk occurrence. According to Chicken (1994), every construction project is subject to risk. Risk is described as a potential problem that has not yet occurred but, if it does occur could prevent or limit the achievement of objectives defined at the outset of the construction project (Burke 1999). Thus it is essential to manage risks effectively and efficiently through appropriate risk management arrangements for each and every construction project including the PFI/PPP in order to ensure value-for-money deliveries. The PMI (2000) perceives risk management as "a systematic process of identifying, analysing, and responding to project risk. It includes maximising the probability and consequences of positive events and minimising the probability and consequences of negative events to project objectives". The APM (2000) sees risk management as a process that identifies preventive measures to avoid a risk or to reduce its effect on a specific project. Risk management can also be used as a tool to consider risk allocation in contracts and risk transfer to insurers for any construction projects.

Pledger et al. (1990) suggested that risks should be managed in three key distinctive stages, namely risk identification, risk analysis and risk mitigation. O'Reilly (1994) identified that risk management should start with risk identification, which is classified as the *sine qua non* or an indispensable action for any meaningful risk management exercise to ensure a successful project delivery. The APM (2000) then emphasised the importance of the two next upcoming sequential processes in risk management; risk analysis and risk mitigation/management. The risk analysis is split into two "sub-stages"; a qualitative analysis "sub-stage" that focuses on identification and subjective assessment of risks and a quantitative analysis "sub-stage" that focuses on an objective assessment of the risks (APM 2000). The risk mitigation/management involves the formulation of management responses to the main risks. There is also a risk monitoring and control process, which involved the development of contingency and containment plans that comes after risk mitigation/management. However since this process only takes place during project implementation stage and hardly depends upon the three preceding processes, thus the risk monitoring and control is not as important as risk identification, risk analysis and risk mitigation/management.

While the importance of risk management has been clearly highlighted for every construction project, the effectiveness of its implementation in real practice is not known. This chapter explores a real NHS PFI hospital project in the UK as a case study to determine how risk management processes have been applied into the actual construction project. The tools and techniques used to carry out risk identification, risk analysis and risk mitigation/management processes in the case study are discussed. This chapter also analyses the potential implications of the actual results achieved by the case study clients on the three risk management processes in the real project. The significance of this chapter is that it provides a comprehensive overview on how each risk management process is being undertaken in a public sector project that involved long-term operational contracts using the PFI/PPP approach.

Methodology

The selection of the case study is made based on its ability to meet the aim of this chapter, which is to determine how the three key risk management processes have been practised in an actual construction project through the tools and techniques used. Since this type of procurement approach requires long-term operational contracts and hence exposed most public clients to greater risks, the NHS PFI hospital project has been selected as the case study. The selection of a particular NHS PFI hospital project for the case study is also another difficult task as most of the information attached in the contract documents is classified as confidential and not suitable as a case study. Unless the project has already been commissioned and started to deliver its operational services for a number of years, then it can be used as a case study like the one used in this chapter.

Once the case study has been decided, the aim of this chapter is achieved through the following data collection means; literature review, the Full Business Case (FBC) report and interviews. Literature review focuses on establishing the best practice in terms of tools and techniques used within the three key risk management processes in the construction industry. The FBC is a document required from each and every PFI/PPP construction project to ensure appropriate governance is in place. It usually comprises of comprehensive reviews on the undertaken procurement process as well as project affordability and value for money across various procurement strategies prior to formal commitment from the project sponsors such as the NHS for UK hospital projects through contract agreements.

Among the data that have been generated from the FBC includes risks identified, risk analysis results and risk mitigation/management strategies. However due to data confidentiality, not all information can be gathered from the FBC for the case study. Thus, interviews have been used as an additional data collection means that are able to comply with the aim of this chapter. A series of interviews have been conducted with a few of important personnel within the clients' organisation involved in the case study. Nevertheless the key data collection source comes from the contract manager of the project sponsor, which is responsible to make sure that not only contract agreements are delivered successfully during the operational stage but also risks have been managed effectively and efficiently at the pre-contract award stages.

The risk management best practice

The section focuses on establishing best practices for the three key risk management processes, namely risk identification, risk analysis and risk mitigation/management. Best practices here refer to the tools and techniques that are commonly being implemented in the real construction projects for each risk management process. Although there are various mechanisms available in the risk management best practice, the application of each tool and technique specified in the actual project practice is minimal. This is probably due to lack of awareness of the key players on the importance of effective and efficient risk management as well as time and cost saving. The following sub-sections explain the best practices identified for each and every risk management process assessed in this chapter.

Risk identification

According to Smith (2003), all major investment decisions for a construction project are made at the early stage of a project cycle such as the appraisal and sanction stages. Therefore all potential risks that could affect the delivery performance of a construction project ought to be identified at the outset. The ICE (1998) classified risk identification as a process that is used to determine types and sources of risk associated with the investment objectives as well as the key parameters related to those objectives. The risk identification process should start by determining all critical risk events, which could have significant impact on the delivery outcome should they occur in the project.

A number of established techniques have been widely used by the practitioners in the construction industry for risk identification. Amongst them include brainstorming, checklist, interview, the Delphi technique and historical reports. Brainstorming allows participants to use their knowledge and experience not only to identify potential risks together with their associated cost impact, brainstorming also allows participants to propose the most effective ways for managing those risks. All decisions in brainstorming must have the consensus from each participant involved before structuring them into risk register. Hence, it eliminates any possible bias in the decision-making made via brainstorming.

The Delphi technique is a structured group approach that focuses on establishing opinions and thoughts from experts in the industry without the need to conduct a face-to-face meeting with them i.e. survey questionnaire. Historical reports reflect the approach where information is generated through documented report of the past projects. For NHS PFI/PPP hospital projects in the UK where their capital costs exceed £1 million, there is an obligation for the project sponsors to prepare the Post Project Evaluation (PPE) report. The purpose of PPE is to assess impact of a project, programme or policy while it is in operation or at the end of the operational contract. As a result, project sponsors are able not only to improve project performance but also achieve best value for money from public resources, improving decision-making and learning lessons.

Risk analysis

Perry and Hayes (1985) described risk analysis as a process that quantifies the impact of risks in terms of cost, schedule and quality on a particular construction project. Among the important elements that require quantification in risk analysis are the likelihood of occurrence of a particular risk and also its potential consequences on the construction projects. Hayes (1987) has divided risk analysis into three categories; qualitative, quantitative and a combination of both techniques. Qualitative risk analysis is a process that requires the probability occurrence and impact of a risk to be measured through a numerical or qualitative scale that is subjective. The typical range of measuring scale used in qualitative risk analysis is high, medium or low (APM 2004). The Risk Probability-Impact Matrix is another type of assessment available for qualitative risk analysis where risks are classified according to their probability of occurrence and impact based on the judgment made in the matrix table as shown in Figure 16.1 below.

Unlike the qualitative, quantitative risk analysis involves the use of sophisticated computer-based programmes to quantify the cost impact of a risk if it occurs in a construction project. However Smith (2003) said that the choice

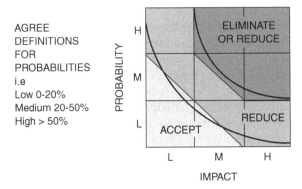

Figure 16.1 The Probability-Impact Matrix (Adapted from www.ttm.co.uk)

of quantitative risk analysis technique would usually be constrained by the experience and expertise of the user as well as the availability of relevant computer software. There are two most commonly-used type of quantitative assessment in risk analysis; probability analysis and sensitivity analysis.

Smith (1999) described probability analysis as a technique that outlines the probability distribution for each risk and then considers those risks in combination. Sensitivity analysis is another quantitative assessment that allows the effect of economic changes in a construction project through "what-if" exploration, which can be achieved by isolating the identified key variables as well as evaluating the effects of incremental changes in the values assigned to the key parameters (Smith 1999). A Spider Diagram, as shown in Figure 16.2, is one of the common techniques for sensitivity analysis. However the Spider Diagram approach does not give a clear indication about the magnitude and extent of cost variation in response to changes in the risk variables (Yeo 1991).

Risk mitigation/management

Risk mitigation is a process that formulates appropriate strategies for managing identified risks effectively and efficiently throughout a construction project. Smith (1999) identified three types of response strategies for risk mitigation/management; namely risk avoidance, risk transfer and risk retention/shared. Risk avoidance is described as an elimination of a possible loss exposure by not undertaking activities that have been identified as having a high likelihood of occurrence during a specific project. This elimination can be achieved by weighing up potential values of the risk-prone activity against the estimated potential loss should the risk occur in the project. Risk transfer is a strategy that involves the relocating of risk ownership to the best able-to-manage party without affecting the number of risks existing in the project i.e. third party transfer. Risk retention/shared is where the project owner absorbs the ownership of a risk and accepts any consequences that may incur as a result of risk occurrence in the actual project.

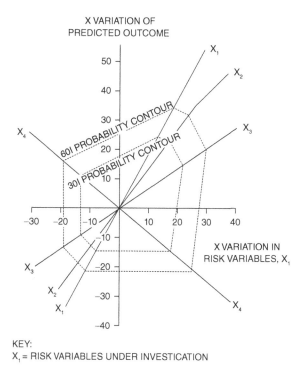

Figure 16.2 Sensitivity Analysis' Spider Web Diagram (Source: *Yeo, 1991*)

Data analysis and results

This section involves a comprehensive overview of the real risk management practice being carried out in an actual NHS PFI hospital project in the UK through case study. The tools and techniques used, including the definite results achieved by the case study organisation in the actual project for their three risk management processes, namely risk identification, risk analysis and risk mitigation/management, are thoroughly described in the upcoming subsections. All the data and information are genuinely absorbed from the case study project's Full Business Case (FBC) report. The FBC is a compulsory document for each and every PFI/PPP hospital project in the UK to support a series of decisions made by the NHS Trust involved including potential benefit realisation and affordability of the public clients to ensure long-term viability of the project. The case study involved the relocation of an NHS hospital in the UK onto a new site due to the poor condition of the existing building and also limited land area for further expansion. In general, the total cost estimated for the relocation scheme was approximately £46 million, which included the procurement of seven facilities through a single PFI contract. However due to confidentiality purposes, the name of the project and also the NHS Trust involved in the case study could not be disclosed.

Risk identification

The procuring NHS Trust implemented only a single risk technique in risk identification which is the brainstorming. During the brainstorming sessions, only three key personnel from the client's organisation were involved in risk identification. They were the Project Director (who was also the head of the project board), the Assistant Director of Finance and the Project Manager for the case study project. Apart from the named personnel, other participants involved in the brainstorming sessions mainly comprised of external consultants, who were appointed by the procuring NHS Trust to provide technical, legal and financial advice to them. Risks were identified under six key risk areas, namely planning, pre-commissioning, design, land purchasing, construction and operation and maintenance. In the actual risk identification exercise of the case study, only three key risks were determined for land purchasing. Table 16.1 outlines most of the key risks identified by the case study clients under each risk area via the brainstorming.

Table 16.1 Risks Identified for the Case Study

Risk Area	Key Risk
Planning	• Procedural Delay
	• Planning Permission Delay (non-NHS sites)
	• Planning Permission Delay (NHS sites)
	• CDM regulations compliance
	• Consultation delay
Pre-commissioning	• Delay of infrastructure improvements
	• Access to sites (non-NHS sites)
	• Availability/ capacity of utilities
	• Access to sites (NHS sites)
	• Environmental conditions
Design	• Trust variation
	• Delays through design phase
	• Design consultants delivery/ life expectancy
	• Fit for purpose
	• Contractor variation
Land Purchasing	• Land purchasing delay
	• Cost of land purchase
	• Identification of site suitability/ acceptability
Construction	• Trust variation
	• Time overrun
	• Cost overrun
	• Phasing/ decanting
	• Weather conditions
Operation and Maintenance	• General legislative change
	• Health specific legislative change
	• Residual value
	• Poor quality of services
	• Maintenance and repair cost overrun

Risk analysis

Besides identifying risk, the brainstorming technique also was used by the procuring NHS Trust to estimate the probability occurrence of those risks, which in actual fact was the qualitative risk analysis. The probability occurrence for each risk identified was estimated based on two circumstances; firstly, what would be the probability of occurrence of a particular risk if the case study project were delivered via public procurement and secondly, if it was procured through the PFI/PPP. Table 16.2 specifies the actual results gathered from qualitative risk analysis of the case study project.

As for the quantitative risk analysis, the quantification of potential cost impact was determined through the estimated Net Present Value (NPV) of the identified risks over a number of possible duration of the operational PFI/PPP contracts. According to the case study's FBC report, the initial duration for the operational PFI/PPP contract was set at 26 years with a possible extension of 15 years and then another ten years but with no further. The procuring NHS Trust estimated the potential NPV of the identified risks over the maximum life expectancy of the proposed assets/facilities where it has been standardised at 60 years for all infrastructure and building projects in the UK. Thus the quantitative risk analysis of the case study was conducted over the following duration of contract options; 26, 41, 51 and 60-year.

Table 16.3 summarises some part of the actual results gathered from the quantitative risk analysis on the 26-year contract option for further analysis later in this chapter.

Risk mitigation

The procuring NHS Trust had used negotiation as the key means to mitigate most of the risks identified to the best able-to-manage parties to ensure they could be effectively and efficiently managed throughout the case study project. According to the contract manager interviewed, the selected contractor/consortium agreed to absorb most of the elemental/project risks with the unitary charge imposed at £8.7 million per annum throughout the initial 26-year of operational PFI/PPP contract. The unitary charges include the costs of construction and facilities management (FM) of the case study project. However the initial unitary charge demand was found to be the affordability ceiling of the procuring NHS Trust. Thus further negotiation had been carried out with the selected contractor/consortium to reduce unitary charge by making several minor changes to the drafted PFI contract. Among the changes made included a modification to sale agreement of the existing hospital assets/facilities from total sale onto a lease arrangement, which subsequently enabled the unitary charge to reduce from £8.7 million to £8.2 million per annum where the latter figure fits with the affordability cash flow of the case study clients. Table 16.4 indicates the risk mitigation/management decisions made in the actual case study project.

Table 16.2 The Qualitative Risk Analysis Results

Risk Area	Key Risks	Public Procurement Probability (%)	PFI/PPP Probability (%)
Planning	• Procedural Delay	80	80
	• Planning Permission Delay (non-NHS sites)	70	30
	• Planning Permission Delay (NHS sites)	30	20
	• CDM regulations compliance	10	10
	• Consultation delay	5	5
Pre-commissioning	• Delay of infrastructure improvements	10	5
	• Access to sites (non-NHS sites)	10	5
	• Availability/capacity of utilities	5	5
	• Access to sites (NHS sites)	5	5
	• Environmental conditions	5	5
Design	• Trust variation	60	10
	• Delays through design phase	50	10
	• Design consultants delivery/life expectancy	25	10
	• Fit for purpose	5	5
	• Contractor variation	5	5
Land Purchasing	• Land purchasing delay	60	30
	• Cost of land purchase	40	20
	• Identification of site suitability/acceptability	25	5
Construction	• Trust variation	90	50
	• Time overrun	75	25
	• Cost overrun	50	20
	• Phasing/decanting	30	5
	• Weather conditions	20	20
Operation and Maintenance	• General legislative change	95	95
	• Health specific legislative change	95	95
	• Residual value	70	30
	• Poor quality of services	50	50
	• Maintenance and repair cost overrun	50	20

Table 16.3 The Risk Analysis Results or 26-Year Contract Option

RISKS 26-Year Analysis	Probability (%)		Risk NPV to NHS Trust PSC (£ '000)	Gross Risk under PFI (£ '000)
	PSC	PFI		
Design				
Trust variations	60	10	219	38
Delays (through design phase)	50	10	92	18
Design consultants delivery/ life expectancy	25	10	393	182
Fit for purpose	5	5	23	24
Contractor variations	5	5	18	19
Construction				
Trust variations	90	50	1097	656
Time overruns	75	25	268	94
Cost overruns	50	20	1135	474
Phasing/decanting	30	5	42	8
Weather conditions	20	20	39	13
Operations and Maintenance				
General legislative change	95	95	283	291
Health specific legislative change	95	95	1201	1236
Residual value	70	30	199	78
Poor quality of services	50	50	299	299
Maintenance and repair cost overruns	50	20	153	100

Table 16.4 The Risk Mitigation

RISKS 26-Year Analysis	Risk with Trust Under PFI	NPV of Risk to Trust PFI (£ '000)	NPV of Risk Transfer (£ '000)
Design			
Trust variations	Absorbed	38	180
Delays (through design phase)	Transferred	0	92
Design consultants delivery/ life expectancy	Transferred	0	393
Fit for purpose	Transferred	0	23
Contractor variations	Transferred	0	18

(continued)

Table 16.4 (continued)

RISKS 26-Year Analysis	Risk with Trust Under PFI	NPV of Risk to Trust PFI (£ '000)	NPV of Risk Transfer (£ '000)
Construction			
Trust variations	Absorbed	656	441
Time overruns	Transferred	0	268
Cost overruns	Transferred	0	1135
Phasing/ decanting	Transferred	0	42
Weather conditions	Transferred	0	39
Operations and Maintenance			
General legislative change	Transferred	0	283
Health specific legislative change	Absorbed	1236	−35
Residual value	Transferred	0	199
Poor quality of services	Transferred	0	299
Maintenance and repair cost overruns	Transferred	0	153

Discussion

The section discusses and analyses the actual process used and results gathered by the procuring NHS Trust in the case study project in order to determine the possible implication and impact from its risk management practice in regards to risk identification, risk analysis and risk mitigation.

The risk identification process

In contrast to the tools and techniques identified from risk management best practice that proposed the use of combination techniques such as brainstorming, checklist, historical reports, etc., the risk identification process in the case study applied just brainstorming as the only tool used. Also the number of key personnel from the case study clients involving the brainstorming was minimal. Besides relying too much on the expertise of outside consultants, the minimal involvement from public clients' personnel in the brainstorming probably due their lack of knowledge and experience on PFI transactions may have had an effect on the value of results achieved in the risk identification process.

Unlike other public sector organisations in the UK such as the Highways Agency (HA) where most of their PFI/PPP projects are centralised and managed by the HQ, each PFI transaction in the NHS is like a new learning curve for an individual NHS Trust. A particular NHS Trust is usually involved only in one major PFI/PPP project throughout its organisational lifecycle. As a result, most of the procuring NHS Trusts have to acquire services from outside experts in terms of legal, financial and technical matters. One of the possible consequences from lack of commitment and involvement from the clients' personnel in the process is that all risks identified fail to address the project aim and objectives. In other words, all the identifications are made based on general construction project risks but not specifically correlating them with the case study project.

The risk analysis process

In terms of qualitative risk analysis, the case study clients had applied brainstorming as the sole technique to establish the probability of occurrence of the risks identified. Although it is capable of providing problem solutions through collective and consensus outcomes from participants that may also minimise bias decisions, brainstorming could also dictate peculiar estimates especially when it is used as the sole technique in qualitative risk analysis like in the case study. Figure 16.3 portrays 13 risks that have been identified from the actual qualitative risk analysis results as shown in Table 16.2 and then singled out as critical risks based on public procurement's probability of occurrence estimates for further discussion and analysis purposes in this chapter.

The classification of "critical risks" is made based on two considerable literature sources that specify various ranges of risk severity. The first source stated that a risk can be classified as critical when the probability of occurrence is estimated at 65% or greater while those that lie between 35% and 65% are considered as medium risks (www.portal-step.com). Risks with probability of occurrence estimated at 35% or less can be considered as low risk.

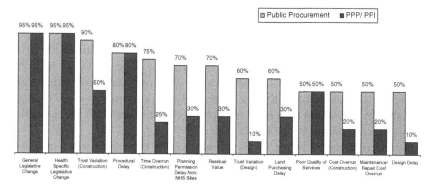

Figure 16.3 The Probability Analysis Results on the Critical Risks

Another source of risk classification stated that risks should be divided under the following categories: 10% or below probability of occurrence – low risk; 20% to 50% probability of occurrence – medium risk; and 50% or more probability of occurrence – critical risk (www.ttm.co.uk). As for the discussion and analysis purposes, risks that were estimated to have 50% or more probability of occurrence in the case study are considered as critical risks.

The probability occurrences of all 13 critical risks identified are expected to be minimised significantly through the PFI/PPP approach except four risks where their probability of occurrence estimates were identical; general legislative change, health specific legislative change, procedural delay and poor quality of services. Both legislative-related risks are beyond the control of project parties and are usually being dictated by politically-motivated alteration of the current policies due to change in the ruling government and/or variation in the political climate from election campaigns. Thus probability of occurrence estimates made for these risks can be categorised as acceptable and justified. The probability of occurrence for both public procurement and PFI/PPP options under procedural delay risk was identically estimated at 80%, which can be considered "risk with great possibility to occur" in the case study project. However these high identical estimates for both procurement options can be reasonably justified based on the fact that most of public projects in the UK are delivered on value for money basis. Hence various comprehensive activities/processes ought to be carried out by the procuring clients prior to project sanction.

Another critical risk estimated with similar 50% probability of occurrence if the case study were to be procured through either public procurement and PFI/PPP was poor quality of services. As mentioned earlier under this discussion and analysis heading, the implementation of brainstorming as a sole technique could dictate peculiar results to the qualitative risk analysis of the case study. Thus the probability of occurrence for poor quality of services via the PFI/PPP had been identified as one of the actual results that seem to be peculiarly estimated. In the PFI/PPP procurement, the concessionaire would generate revenue based on the unitary charge imposed to the public clients either being paid monthly or on an annual basis during project operations. However the payments or "unitary charges" are subject to the performance of the concessionaire and also deliverability of the procured assets/ services to the end user. Any kind of poor quality of services from the concessionaire would lead towards a potential payment deduction. Therefore the 50% probability occurrence of the poor quality of services as estimated in the actual case study seemed peculiar where 10%–20% would be a more reasonable figure for the PFI/PPP.

As for quantitative risk analysis application in the case study, only the assessment of cost impact of risks identified was quantified over the number of contract duration i.e. 26-year via the risk's NPV as shown in Table 16.3. However the procuring NHS Trust did not carry out a sensitivity analysis in the case study, which is an important tool to determine the likely range of variation for elements of the project data (Smith 2003). Had the sensitivity analysis been applied in the actual case study project, the procuring NHS Trust would be able to identify risks that are sensitive towards any sort of economic changes and hence affecting the principal's cash flow. Since it had been carried out during the actual case study's quantitative risk analysis, a "sensitivity analysis model" is developed in this chapter for the purpose of determining potential cost variations of the critical risks with peculiar probability of occurrence estimates over a range of uncertainty values. The "sensitivity analysis model" developed in this chapter is presented in the form of spider diagram. According to Woodward (1995) results gathered from a sensitivity analysis portray a linear property of the sensitivity relationship between the key variables at certain level of change in variation.

Out of 13 critical risks as portrayed in Figure 16.3, six risks were identified to have been peculiarly estimated in terms of their probability of occurrence in the actual case study; planning permission delay, land purchasing delay, cost overrun in construction, poor quality of services, cost overrun in maintenance/repair and residual value. Although the probability of occurrence of two other critical risks, namely design delay and time overrun in construction, were identified as reasonably estimated ones, they had also been included in the sensitivity analysis. The decision to include these risks was made based on their close association with cost overrun in construction where the occurrence of one risk would dictate the occurrence of the remaining two in the case study.

Out of eight critical risks involved in the sensitivity analysis, five of them are pre-sanction risks while the remaining three are operational-related ones that have the probability to occur at commissioning stage. Thus the "sensitivity analysis model" had been divided into two stages; risks at sanction and risks at commissioning. Table 16.5 outlines the classification of risks under different categories for the sensitivity analysis purposes. Prior to undertaking the sensitivity analysis, several reasonable assumptions had been made on facts and figures related to the case study project such as the discounting rate factor and also range of uncertainty values, which were beyond the knowledge of the author.

Table 16.5 Risk Categories for the Sensitivity Analysis

Risks at Sanction	*Risks at Commissioning*
• planning permission delay • land purchasing delay • design delay • time overrun in construction • cost overrun in construction	• poor quality of services • maintenance/ repair cost overrun • residual value

The effect of discounting is bringing in a different range of future cash flows of the project back to its current values. In other words, the implementation of the discounted cash flow would enable project clients to predict how their financial stream is going to look like in the future. Since there is lack of information about the real discounting rate applied in the actual case study project, the author decided to use the 6% of discounting rate suggested by the Office of Government Commerce OGC (2003a) in both sensitivity analysis models. The range of uncertainty values meanwhile depends largely on the extent within which the probability occurrence of the assessed risks is expected to deviate from their actual estimates.

Figure 16.4 portrays the results obtained from the undertaken sensitivity analysis for risks at sanction. The result clearly indicates that cost overrun in construction is the only risk at sanction that is sensitive towards any potential changes in the probability estimate and hence could have a significant impact towards the case study project's cash flow. Nevertheless, other assessed risks at sanction were found to have only a trivial sensitivity level towards any potential changes in their actual probability estimates. In the actual quantitative risk analysis as shown in Table 16.3, the cost overrun in construction was predicted to have 50% probability of occurring at the approximate value of £1.14 million if the case study were to be procured via public procurement. Owing to the high cost impact, any deviation of the probability occurrence of cost overrun in construction in the actual project would have a significant financial impact towards the procuring NHS Trust.

Table 16.6 identifies the range within which the cost impact of cost overrun in construction may vary as a result of an economic change or error in

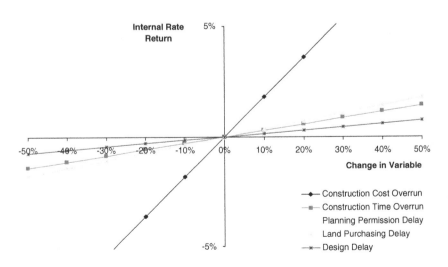

Figure 16.4 The Sensitivity Analysis

Table 16.6 Potential Cost Impact Variation for Sensitive Risks at Sanction

Risk	Assessed Option	Estimated Probability	Estimated Risk NPV (£'000)	Range of Risk NPV Variation (£'000)
Cost Overrun in	PSC	50%	1135	681–2270
Construction	PFI	20%	474	119–593

the probability estimates during the qualitative risk analysis. For instance, should the probability estimate be increased by 20% from the original 50%, then the potential cost impact of cost overrun in construction would increase by extra £454,000 from the initial estimate of £1.14 million. The budget/cash flow availability of the procuring NHS Trust is strictly limited. Thus the most significant implication would definitely be on the procuring NHS Trust's cash flow as a result of this cost impact increment.

The sensitivity analysis was carried out on the three critical risks at commissioning, namely poor quality of services, maintenance cost overrun and residual value in order to determine their sensitivity. Figure 16.5 portrays the sensitivity analysis results for risks at commissioning. Poor quality of services risk was found as the most sensitive risk at commissioning. The other two key risks involved in the sensitivity analysis, residual value and cost overrun in repair/maintenance, were found to be less sensitive than the poor quality of services although their consequences would also have a major impact on the case study project's cash flow.

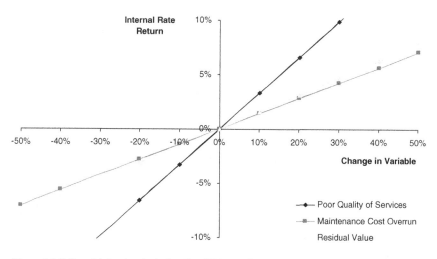

Figure 16.5 Sensitivity Analysis for Key Risks at the Commissioning

Table 16.7 Potential Cost Impact Variation for Sensitive Risks at Commissioning

Risk	Assessed Option	Estimated Probability	Estimated Risk NPV (£'000)	Range of Possible Risk NPV Variation (£'000)
Poor Quality of	PSC	50%	299	180–598
Services	PFI	50%	299	60–419
Residual Value	PSC	70%	199	86–285
	PFI	30%	78	26–78
Maintenance/ Repair	PSC	50%	153	92–306
Cost Overrun	PFI	20%	100	50–100

Table 16.7 outlines the potential range of cost impact variations of sensitive risks at commissioning due to deviation of the actual probability estimates. For example, should the actual probability estimate for poor quality of services deviates by 5%, then there is a probability for the procuring NHS Trust to incur additional cost of £20,000 if neither one of the procurement options were to be used to procure the case study project.

The risk mitigation process

The procuring NHS Trust in the case study implemented negotiation as the only technique to mitigate and manage all the identified risks, to the best able-to-manage parties. Basically in PFI/PPP, most of the elemental/ project risks such as design, construction and finance are transferred to the concessionaire or SPV. However this kind of PFI/PPP flexibility does not mean that all risks should be transferred to the private sector. According to the OGC (2003b), risks that have been categorised as transferable should be non-negotiable while seeking to transfer the absorbed risks would lead towards a poor Value-for-Money (VFM) deal to the principals or project clients. Typically, in most PFI/PPP projects, the mitigation strategies are decided based on the results achieved in risk analysis. From the identified 13 risks with high probability estimates in the case study project, the procuring NHS Trust agreed to absorb three risks while nine other risks had been transferred completely to the SPV. Only one risk was shared between them. The transferred risks were mainly project-specific ones except for general legislative change. Risks absorbed/retained by the procuring NHS Trust, which include the Trust's variations in design and construction are mostly caused or dictated by the principal/project clients.

Conclusion

Lack of knowledge on PFI transaction and also commitment in the brainstorming sessions by the Trust's personnel as well as the use of brainstorming as a single technique for risk identification and qualitative risk analysis have affected the effectiveness of the risk management practice. There were a few disadvantages found in the case study as a result. Firstly, the probability estimates for some risks with high probabilities were peculiarly estimated by the principal. Secondly, the qualitative and quantitative analyses produced some contentious and peculiar results that may dictate future financial implication to the public sector clients. Thirdly, the sensitivity analysis, which assesses cost variation of risks from economic changes, was not conducted in the case study. Otherwise the cost impact of four most sensitive risks in the case study project, namely cost overrun in construction, poor quality of services, cost overrun in maintenance/repair and residual value, could have been identified and managed earlier by procuring NHS Trust for effective and efficient risk management practice i.e. preparing appropriate contingency plan.

References

Association for Project Management (APM) (2000) *A Guide on Project Risk Analysis and Management*, www.eurolog.co.uk/apmrisksig/publications/minipram.pdf.

Association for Project Management (APM) (2004) *Project Analysis and Management Guide.* High Wycombe, 2nd Edition.

Burke, R. (1999) *Project Management Planning and Control.* West Sussex, Wiley.

Chicken, J.C. (1994) *Managing Risks and Decisions in Major Projects.* London, Chapman & Hall.

Gentry B. and Fernandez, L. (1997) Evolving Public-Private Partnerships: General Themes and Urban Water Examples, *Globalisation and the Environment: Perspectives from OECD and Dynamic Non-Member Economies.* Paris, OECD.

Hayes, R.W. (1987) Risk management in engineering construction: implications for project managers: a report of research supported by the SERC specially promoted programme in construction management and prepared by the Project Management Group, UMIST, UK.

Institution of Civil Engineers (ICE) (1998), *Risk Analysis and Management for Projects (RAMP).* London, Thomas Telford.

Office of Government Commerce (OGC) (2003a), The Use of Discounting Factor in Public Sector www.ogc.gov.uk/sdtoolkit/reference/ogc_library/pfi/series_1/andersen/7tech_03_01.html.

Office of Government Commerce (OGC) (2003b), Design and the Procurement Process – Bid Evaluation www.ogc.gov.uk/sdtoolkit/reference/ogc_library/PFI/series_3/technote7a/7atech_10.htm.

O'Reilly, M. (1994) *Risk, Construction Contracts and Construction Disputes.* London: CIRIA publication

Perry, J.G. and Hayes, R.W. (1985) Risk and its management in construction projects, *Proceedings of the Institute of Civil Engineering*, 78(1), 499–521.

Pledger, M., Clark, R.C. and Needler, H.M.J. (1990) Risk analysis in the evaluation of non-aerospace projects, *International Journal of Project Management*, 8(1), 17–24.

Project Management Institute (PMI) (2000) *A Guide to the Project Management Body of Knowledge.* Pennsylvania: PMI.

Smith, N.J. (1999) *Managing Risk in Construction Projects.* Oxford: Blackwell Science.

Smith, N.J. (2003), *Appraisal, Risk and Uncertainty.* London: Thomas Telford.

Time to Market (TTM), *Risk Probability/ Impact Matrix* www.ttm.co.uk/ptiriskmatrix. htm.

Woodward, D.G. (1995) Use of sensitivity analysis in build-own-operate-transfer project evaluation, *International Journal of Project Management*, 13(4), 239–246.

Yeo, K.T. (1991) Project cost sensitivity and variability analysis, *International Journal of Project Management*,9(2), 111–116

Your Project Management Portal: Manage Risk/ Process www.portal-step.com/7. 1ManageRisk.htm.

17 Managing risk management knowledge

Eduardo Rodriguez and John S. Edwards

Highlights

- Identifies knowledge as a way to mitigate risk
- Links the concepts of risk management and knowledge management
- Demonstrates the value of using risk management and knowledge management for project development
- Explains how risk management has evolved over the years, to current developments in enterprise risk management and risk analytics
- Gives examples of how knowledge management can support risk management

Introduction

Risk appears in the design and implementation of strategy and the execution of projects. It affects multiple stakeholders. The purpose of risk management can be generalized as keeping aware of risk and designing potential solutions when risk appears. Knowledge of the most common risks has provided the input for the design of products in insurance companies that help other organizations to avoid or mitigate losses. Equally, financial institutions have used their knowledge in developing products that help to deal with the variation of the markets in order to compensate, protect, and mitigate possible losses.

The "management" part of the term risk management means the capability to put the resources of the enterprise to control, communicate, train, adapt the organization to a culture of managing and living with risk. It is vital to identify and evaluate risks so as to know what to do when an adverse event is a reality. Knowledge is involved as a means to reduce risk, to evaluate risk and to find ways of protecting stakeholders against risk. In particular, knowledge creation from data analytics is one of the crucial aspects of enterprise risk management as it is described in the updated COSO (2017) Enterprise Risk Management framework.

The effectiveness of risk management has been criticized (Taleb et al. 2009). In this chapter some knowledge gaps are indicated. These include the need to improve knowledge in risk management for example in:

model development, interpretation of the model outcomes, the design of information systems, and in attaining experience and understanding it. This chapter introduces knowledge management concepts and indicates some of the points to analyze in the relationships between risk and knowledge management.

Finally, some applications of knowledge management are indicated, using examples drawn from the financial services sector. Experience in the form of accumulated knowledge is crucial in management practice. It seems that organizations have repeated the same errors many times, the good times have not been analyzed in depth and the failure of accumulation of knowledge of the results of similar strategies that affect the market keeps everyone playing with the same strategies. As Francis Bacon said: "Truth emerges more readily from error than from confusion" (Cardwell 1994).

The rest of the chapter is structured in four parts. The first part introduces some of the concepts of knowledge management; an understanding of knowledge in the setting of knowledge management; the processes of knowledge management; and the components of a knowledge management system. The second part indicates possible links between risk management and knowledge management showing the role of knowledge in the evolution of risk management. The third part covers "ways to know" in risk management, incorporating both qualitative and quantitative approaches. The final part gives four examples of the ways in which knowledge management processes can support the practice of better risk management.

About knowledge management

Wiig (1993) indicates that "Knowledge consists of truths and beliefs, perspectives and concepts, judgements and expectations, methodologies and know-how." This view is complemented by others that include the concept of reasoning in the definition, such as Beckman (1997): "Knowledge is reasoning about information and data to actively enable performance, problem-solving, decision-making, learning, and teaching."

Among the main authors that have had influence on the literature of knowledge management (KM) are Nonaka and Takeuchi (1995), Davenport and Prusak (1998), and Alavi and Leidner (2001).

Nonaka and Takeuchi (1995) concentrated on the interaction between two knowledge types: tacit and explicit knowledge. Tacit knowledge is represented by experience, beliefs, and technical skills accumulated in people's minds. Explicit knowledge is knowledge expressed in documents, data and other codified forms. The interactions between people correspond to the conversion between tacit and explicit knowledge from the individual to the organizational level. The dynamic is expressed in the SECI Model of knowledge creation as comprising four processes.

- Combination is a conversion of explicit knowledge to explicit knowledge.
- Internalization is to pass from explicit to tacit knowledge; this is how to learn to work on the solution of the problem through action.
- Externalization: the tacit knowledge is converted into explicit knowledge. This is presented through different means: methodologies, models, metaphors, concepts etc.
- Socialization is the step from tacit to tacit knowledge. This means the conversion of experience and practice into new experience and practice keeping the bases of human relationships.

The *ba* or spaces for knowledge creation are also different according to the SECI model processes: originating *ba* for socialization, interacting *ba* for externalization, cyber *ba* for combination and exercising *ba* for internalization.

Davenport and Prusak (1998) mentioned four knowledge elements in their definition of knowledge which complement the SECI model. First, the sources of knowledge are: experience, values, context, and information. Second, people are considered the original repository of knowledge from information and experience. Third, processes and procedures act as means to retrieve, describe, and apply knowledge. The fourth element refers to the organization as the setting for knowledge management.

Alavi and Leidner (2001) not only support the idea that different entities, processes, resources, and assets are required to achieve a sustainable competitive advantage based on knowledge assets, but also present knowledge as a competitive factor. This understanding can contribute to building a knowledge-sharing, creation, and application infrastructure.

Knowledge management, then, is ". . . the systematic, explicit and deliberate building, renewal, and application of knowledge to maximize an enterprise's knowledge–related effectiveness and returns from its knowledge assets" (Wiig 1997). Similarly, Beckman (1997) indicates that: "KM is the formalization of and access, to experience, knowledge, and expertise that create new capabilities, enable superior performance, encourage innovation, and enhance customer value." What is common in these two definitions is the methodical access to experience – knowledge in order to develop enterprise capabilities.

Alavi and Leidner (2001) describe knowledge management as a process, with four sub-processes, that identifies and leverages the collective knowledge of the organization in order to compete (von Krogh 1998). Equally, they state that KM requires more than IT; it requires the creation of a means to share knowledge, information processed by individuals and adapted to be communicated.

The sub-processes presented by Alavi and Leidner (2001) are as follows:

- Knowledge creation: Organizational knowledge creation involves developing new content and replacing the content already in place. The knowledge creation is related to the organization's social and collaboration capacity to grow knowledge and to validate it (Nonaka 1994).
- Knowledge storage and retrieval: This process refers to the reality of the need to manage organizational memories; knowledge is created and at the same time forgotten. There are different forms of keeping organizational memories: through databases, information systems, and networks of individuals. There is a difference between individual and organizational memories. The first is developed based on personal experience and observations, while the second can be in documents, databases, systems to support decisions, etc.
- Knowledge transfer: This process takes place ". . . between individuals, from individuals to explicit sources, from individuals to groups, between groups, across groups and the group to the organization." Knowledge transfer channels may be formal or informal, personal or impersonal. These channels may be supported by technology and each category may have a different solution.
- Knowledge application: This process is associated with competitive advantage development and for that there are three mechanisms to create capabilities: directives, organizational routines, and self-contained task teams. Technology can be involved in the application of knowledge which supports knowledge integration and knowledge application by providing access to and updates of directives, organizing, documenting, and automating routines (Alavi and Leidner 2001).

Finally, the knowledge management processes require a knowledge management system (KMS) as support. Alavi and Leidner (2001) identified the KMS as the "kind of information systems applied to managing organizational knowledge." However, the transition of an information system into a KMS requires several components that take into consideration the system design stage. It is helpful to regard all of these as part of the KMS, not just the technology. One component is data architecture which includes data in multiple formats, structured and unstructured. Another component deals with knowledge attributes of the KM sub-processes.

Thus the KMS components (Lehaney et al. 2004; Davenport and Prusak 1998; Malhotra 1999; Edwards et al. 2003) can be summarized as follows:

- People interactions: KM and knowledge acquisition are subject to perceptions and agreement. These human interactions require two subsystems:
- Technology acting as support and the way to enable the KM function
- Organizational structures.

KMSs continue to evolve and there are now several combinations of KM-tools and KM-practices in order to support the KM processes (Cerchione and Esposito 2017). Currently the attention developing around knowledge creation from data (Edwards and Rodriguez 2016) provides a wide spectrum of analytics techniques and methods to use in risk management.

Why link Risk Management and Knowledge Management?

As Marshall et al. (1996) said: "Risk Management is frequently not a problem of a lack of information, but rather a lack of knowledge with which to interpret its meaning". This raises an important point: how do we convert a risk management information system into a KMS?

The evolution of risk management (RM) demands different knowledge. Once, "what to know" was the risk in a specific asset, liability, or contingency, and learning was based on the experience of the specific risk. Risk management in finance more recently is tied to the product development, investment strategies and hedging strategies for portfolios. When the view is integral, holistic and based on analysis across the organization (becoming ERM, Enterprise Risk Management) the demand for new knowledge is even higher, for example the possibility to study multivariate loss distributions.

To illustrate this different knowledge, Nwogugu (2005), referring to the financial markets, pointed out that when modelling and quantification processes are involved risk assessment and decision-making need a combination of methods based on belief systems, time series models and the analysis of multiple factors that consider trading rules, market structure, new information announcements, macroeconomic trends, etc. in general. The "rigid" models and their outputs are not enough to explain risk behaviours and decision making. Knowledge of the market conditions and factors affecting the inputs of the models is also needed.

In another direction, Wu et al. (2018) study risk knowledge in the context of the environment and society, indicating the importance of additional factors related to products for risk perception and decision making for investing money in purchasing protection to the risk exposure.

These two examples are an indication of how understanding risk knowledge is moving from the traditional view in financial markets, to the society risks and decisions related to risk perceptions and their relationships to finding protection using different instruments (derivatives, insurance or devices).

Knowledge in Risk Management evolution

Achieving a systematic way to prevent, assess, control, and manage adverse events has been an evolving challenge as society has evolved. The required knowledge for using the proper methods to avoid, reduce, accept, spread, and transfer risk was originally related to finding solutions for risk affecting tangible assets and life contingency risks. Recently, the focus of risk analysis has been on financial-related risks and converting risk management into a competitive advantage, because investment revenues of organizations and individuals had been affected by market fluctuation and the complexity of financial products. Equally, risk management knowledge has been observed as a way to build a sustainable competitive advantage. Based on the identification of new risks, new knowledge was required and new organizations for risk management emerged.

Risk managers, regulators, boards, and executives are interested in analyzing the organization as a whole and developing knowledge to understand the enterprise risk. ERM represents new knowledge challenges demanding a high level of effort. For instance, the search for the description of the loss distribution and the search for tools to know of individual risks and the integrated risk are important tasks that each time require more and more sophisticated knowledge to accomplish them.

To begin to understand the risk knowledge dynamic we look at the processes of knowledge management from the SECI model in a risk management setting, as follows:

- Socialization: social interaction among risk management employees, shared risk modelling experience and qualitative assessment
- Combination: merging, categorizing, reclassifying and synthesizing the risk modelling process and qualitative assessment
- Externalization: articulation of best practices and lessons learned in the risk modelling process and qualitative assessment
- Internalization: learning and understanding from discussions and mathematical modelling review and qualitative assessment.

Developing "ways to know" is part of the daily work in risk management. In the next section there is an illustration of the dynamics of the KM process using three examples of "ways to know" in risk management.

Some "ways to know" in Risk Management

This section presents examples of the "ways to know" in risk management using risk analytics such as: risk modelling, risk classification, and development of analytics capabilities through designed metrics and tools. These ways to know are complemented by a review of the experience in RM and how to adapt KM processes in the improvement of the RM process.

Risk modelling process

The risk modelling process is considered as a support to the decision-making process in order to pursue a strategic planning process and strategy implementation. The issues with ERM related to the risk modelling process are: to be more efficient and effective in order to get better solutions for risk issues, to extend the experience, results, solutions to more problems, to use technology in a better way, to organize an integral risk information system and to improve the decision-making process. All these challenges are associated with work across the organization done with similar tools, different types of risk and similar modelling processes (mathematical and conceptual) (Panjer 2006).

Additionally, as Buchanan and O'Connell (2006) said "companies must be able to calculate and manage the attendant risks." The decision-making process involves resources and people. In particular, computers, information, and minds are connected in order to develop solutions to decision problems. Creation of knowledge through operational research, decision analysis tools, data structures and methods of sharing knowledge and collaboration are all part of the support systems. In some cases, the systems are designed to make decisions based on algorithms, rules and predefined models (prescriptive analytics).

Machines have limitations, as Turing (1948) presented in the sense of capacity of thinking. Using human knowledge in the creation of models, augmented by the machines' support, can contribute to the reduction of uncertainty develop the capacity for better modelling process and risk management. In particular, the risk modelling process requires risk knowledge and management of the RM organizational structure to coordinate the knowledge required.

In mathematical terms, risk management is a decision-making process under uncertainty. Models support the transition from uncertainty to risk understanding. A model (Klugman et al. 1998) in the insurance risk management world is defined as "a simplified mathematical description which is constructed based on the knowledge and the experience of the actuary combined with data from the past." The description of the modelling process includes the steps of modelling: first selection, then model calibration, fit validation, new model selection, comparison of models and identification of application for the future.

The risk modelling process is part of a risk management system and a risk measurement process; but risk modelling is principally a subset of mathematical modelling where means are designed to access past experience, conceptualize, put into operation, develop, and discover relationships between variables using risk knowledge. Risk modelling is a process that plays a very important role in RM. Quantifying risk is part of the definition of knowing risk. In a modelling process there are multiple questions to answer, such as: what resources does the organization need? What does the organization want? What does the organization measure? What is the impact? What has been the experience? What are the errors? Where is the failure based on lack of knowledge management?

Risk modelling knowledge is based on the measure of variability. Risk modelling supports RM value added to the organization given the search to maximize expected profits, which are exposed to potential variability which can transform them into losses (Oldfield and Santomero 1997). Risk modelling knowledge contributes to the main points or pillars considered in the Basel II Accord: capital allocation, separation of the operation and credit risk, and alignment of regulatory and economic capital.

Risk modelling knowledge could facilitate knowledge transfer, reducing difficulties of the diverse "languages" spoken inside the organization related to risk and the application of risk expertise to solving different problems (Dickinson 2001; Warren 2002; Shaw 2005). Additionally, risk modelling can provide organizations the knowledge for defining problems, variables and their relationships, and supporting tools to facilitate the search task if there is a high volume of knowledge available (Alavi and Leidner 2001) or to organize the search tools that are crucial in RM (Simoneau 2006).

Risk classification

The risk classification problem is one of the most important subjects in insurance and reinsurance companies, as much for underwriting as for actuarial calculation of premiums, reserves, and ruin problems. Scoring methods are used in bank loans analysis. The process requires identifying good or bad prospects, the probabilities of monetary losses on a portfolio and the survival of loans to maturity. Risk classification and credit analysis have a probability of misclassification, which must be minimized. However, in risk classification there is a prior problem, this is to define the classes or groups used to separate risks.

The knowledge required is in several areas with different barriers to overcome, for example in insurance:

- Risk can be classified relative to frequency. For example bad or good drivers, winter season or not. Besides the intensity of the risk can change: it is possible to have a small number of large claims. The strategy of the company can be given as a mix between number and intensity. This, in terms of the model, means a different set of variables for making decisions about the risk in each group.
- Classification methods require complete data. In insurance, sometimes data is missing, as variables important to the claim process may be difficult to find in a file. For example, car parts that are expensive can impact on the claim costs in automobile insurance, but this amount of detail is usually not kept on record.
- Loss distributions are usually heavy tailed, that is they depart from the normal distribution and in many cases are mixed probability distributions (two or more distributions describing the phenomenon).
- The time effect is very important: today the risk is good but tomorrow it may change.

- Type of variable: in insurance, many are nominal and it is necessary to create dummy variables to include them in models.
- Contracts in insurance cover different risks simultaneously. This means some groups could be good for one risk but bad for another.
- Difference between collective and individual contracts. The analysis of collective contracts is similar to that of a company.

Developing analytics: metrics and tools for risk management

One of the most popular and widely used metrics is the VAR (Value at Risk). There are weaknesses with this metric, as there are in other metrics, but the point is to use them in two ways: acknowledging their limitations and modelling constraints, and putting various metrics together in order to see the whole picture of the risk map.

Regarding this, The Economist (2010) wrote: "So chief executives would be foolish to rely solely, or even primarily, on VAR to manage risk." The point here is that VAR is a good tool when there is liquid security, over short periods and under normality of market behaviour, but not under other market conditions or the attributes that are in place when some derivatives are designed and put in the market.

With the trend towards using Big Data analytics to study risk, it is possible to obtain more awareness, and to prepare organizations to develop methods to mitigate risks or protect themselves. Having board members with appropriate understanding of analytics and knowledge creation methods is crucial for converting data into actions (Yoost and Mathaisel 2016). The relevant aspects of the board's oversight are related to policies in technology, adaptation to regulation, achievement of customer profitability and improvement of operations performance indicators.

Other metrics include CVAR (Conditional VAR), Loss Given Default (LGD), probability of default (PD), Asset – Liability Management (ALM) indicators, Capital Asset Pricing Model (CAPM) and related betas, indexes of stock exchange houses etc (Crouhy et al. 2001). Additionally, the experimental design for validating results includes stress testing, a review of the models with samples that are dedicated for model creation and model training, and validation using samples from a different time and a different sample to those originally used for model creation and training.

From the tools perspective, those used to "know" in risk management include:

1   Transition matrices of risk levels:
    These are very important in credit risk for analyzing the Expected Losses and the evolution of the portfolio (See Figure 17.1).
    The importance of the transition matrix is based on exposure, that is moving between risk levels. This is observed in two dimensions: the

				ProbabilitiesTransition Matrix period of time A to B						
		N	**Sum**			**N**	**Sum**		**N**	**Sum**
LOW TO	LOW	85.1%	77.2%	**MED TO**	LOW	1.4%	1.8%	**PRI TO** LOW	0.7%	0.2%
	MOD	0.9%	2.5%		MOD	4.4%	16.1%	MOD	1.1%	0.6%
	MED	3.3%	5.8%		MED	82.4%	65.7%	MED	1.5%	2.1%
	HI	0.5%	1.2%		HI	2.7%	4.9%	HI	10.2%	16.4%
	CRI	0.0%	0.1%		CRI	0.0%	0.0%	CRI	4.4%	0.9%
	PRI	0.0%	0.3%		PRI	0.1%	0.2%	PRI	69.8%	67.5%
		N	**Sum**			**N**	**Sum**		**N**	**Sum**
MOD TO	LOW	5.4%	5.2%	**HIG TO**	LOW	0.7%	0.6%	**CRI TO** LOW	0.3%	0.0%
	MOD	79.9%	62.4%		MOD	1.5%	2.7%	MOD	0.3%	0.0%
	MED	3.9%	9.0%		MED	4.2%	3.9%	MED	0.6%	0.3%
	HI	0.9%	2.2%		HI	75.6%	69.1%	HI	4.4%	3.7%
	CRI	0.0%	0.0%		CRI	0.6%	1.3%	CRI	51.8%	36.6%
	PRI	0.1%	0.3%		PRI	0.9%	3.3%	PRI	9.6%	10.9%

Figure 17.1 Transition matrices by risk classification

number of points of exposure (N) and the value that represents that exposure (Sum). From this analysis, more elements of risk management appear to be analyzed such as concentration and quality of credit decisions.

2   Risk maps and early warning systems

Figure 17.2 represents two different ways to display risk maps. The first indicates ranges of probability of default and ranges of value paid in compensation for adverse event occurrence and in each case the total amount of exposure that the organization has in place.

Exposure ranges

Rank Prob. Default	0-100k	100k-1M	1M-10M	10M-100M	10M+
0.0000-0.0022	1	4	1	0	10
0.0022-0.0032	1	9	6	-	0
0.0032-0.0042	2	11	4	0	4
0.0042-0.0055	4	16	20	0	6
0.0055-0.0069	4	25	9	2	7
0.0069-0.0090	7	26	12	4	16
0.0090-0.0119	11	38	10	5	0
0.0119-0.0167	16	48	14	14	18
0.0167-0.026	22	54	8	2	73
0.026+	58	68	24	20	-

	G1	G2
PD group 1	161	50
PD group 2	371	92
Exposure group 1	36%	11%
Exposure group 2	83%	20%
PD is probabiulty of default and G1, G2 the observed groups		

Total loss in each range ($million) / **Exposure by Ranges**

Probability Rank	<$A	A-B	B-C	C-D	D+
	Low		Med	High	
0.0000-0.0022	13	7	2	15	24
0.0022-0.0032	15	6	5	40	67
0.0032-0.0042	109	2	16	8	48
0.0042-0.0055	39	19	8	8	60
0.0055-0.0069	59	21	14	21	100
0.0069-0.0090	27	9	15	11	131
0.0090-0.0119	81	15	15	37	69
0.0119-0.0167	68	10	21	25	106
0.0167-0.026	55	16	17	23	120
0.026+	79	21	28	30	84

Distribution		
Low PD and Low Claim Paid	235	13%
Low PD and Med Claim Pid	100	5%
Low PD and High Claim Paid	391	21%
ModPD and Low Claim Paid	177	10%
Mod PD and Med Claim Paid	86	5%

Figure 17.2 Example of two risk maps based on probability of default and ranges of compensation payments

The purpose of the second map is to review where the amount paid as compensation for the occurrence of adverse events is concentrated, based on the combination of probability of default and ranges of exposure to risk. It also shows the benefit of conventions for identifying what is good and what is not (for example visual codes such as "traffic lights" to differentiate where to take or not take action).

3 Exposure management tools

There are several tools that can be used in exposure management in order to know the current state of the portfolio and forecast it in future. Figure 17.3 shows how by using reports, typically interactive, risk knowledge can improve. The purpose is to show what is happening in different dimensions, such as risk segments, economic sectors or regions through time for the relationship between price and expected losses. The example shown here indicates additionally the option of analyzing concentration and changes in the concentration. This allows the development of knowledge of risk transformation and awareness to make decisions about risk mitigation and protection.

			Change in Risk (Expected Loss/Exposure)				Change in Pricing (Premium/Expected Loss)			
			Value	Quarter	6 Months	12months	Value	Quarter	6 Months	12 months
Exposure	Risk levels	A								
		B								
		C								
		D								
		E								
		F								
		G								
		Total and Changes by period								
Exposure	Geographic Regions	R1								
		R2								
		R3								
		R4								
		R5								
		R6								
		R7								
		R8								
		R9								
		R10								
		Total and Changes by period								
Exposure	Segments	**SEG1**								
		SEG2								
		SEG3								
		SEG4								
		SEG5								
		SEG6								

Figure 17.3 Example of a tool for following the relationship between expected losses and pricing

In the context of risk analytics there are multiple developments providing possibilities to improve the ways to know in risk management. Digital technologies, the abundance of data and improvements in modelling and computational capacity all offer new opportunities. Wu, Olson and Dolgui (2017), in an editorial of the journal *Engineering Applications of Artificial Intelligence*, indicate the need to develop systems to deal with the high levels of uncertainty in current complex organizational systems. They call for the use of artificial intelligence tools to create knowledge from data and to support risk control. As an example of this, Ivanov, Dolgui and Sokolov (2018) provide a set of applications of analytics techniques to use for areas such as disruption risks. This includes the adoption of "ways to know" through the process of gathering and analyzing data in current manufacturing processes, control of the supply chain and in particular controlling risk related to operations and strategy. The methods supplement the risk time series analysis using simulation and development of exposure control systems and the use of sensors producing big data to be analyzed through analytics techniques.

How to improve knowledge capacity based on RM experiences, successes and failures?

Development of risk management has been based on the accumulation of experience, dealing with crises, successes, failures and creation of products. Yet risk management is not just based on modelling capacity: knowledge is

Table 17.1 Examples of knowledge problems and steps for solution

Knowledge Problems	Some steps taken to manage knowledge
• Acting under different regulations • Dealing with several kinds of models • Managing organizations and cultures for different risk types • Satisfying consumer market • Preparing to act before an adverse event and after the fact • Improving risk control process in expansion • Improving risk management system functionality and controls	• Developing capacities for prediction and classification • Search for understanding and use of information rather than information itself • Mining for learning from RM experiences • Managing the cost of integrating risk analyses, control, and risk policy creation, deployment and application • Searching for understanding of the interactions with external customers and the solutions provided by the financial institutions • Searching for building a Risk Knowledge Portal in order to connect many sources of experience (content integration), explicit and tacit knowledge, measurement process, and management of operations at an acceptable cost

required in other areas. This section indicates how knowledge is involved in risk management. Knowledge problems in risk management have come from many different sources (see Table 17.1).

The analysis of the risk modelling process identifies some of the needs for knowledge and systems to manage it (KMSs). The requirements that Crouhy et al. (2001) proposed for risk information systems include managing data globally using distributed database technology. These authors indicate that a "risk management system approach is not simply an aggregate of applications, data, and organization; instead, it is born out of an IT vision."

The architecture for risk management needs to gather the key information that is supplied by different areas, in a data warehouse. The design of the KMS has to take into account that data can be both static and dynamic, and to provide adequate access to all the users. Additionally, it is necessary to take into consideration the fact that there are different functions in risk management with specific needs, such as trading operations that require systems to support the monitoring of trades, prices and the decision-making process through models. King (2006) argues that the KMS is an enabler of knowledge sharing.

Risk management information systems can have KMS components that support risk control based on current resources used, such as: expert systems in credit decisions, data mining for risk classification and predictive modelling, and system components to enable interaction and collaboration in the risk management context. Collaboration in this sense takes the form of the design and development of systems to gain knowledge and problem-solving capacities that involve multiple people, disciplines and teams.

Nowadays, new computational capabilities are providing a scenario where risk analysis can come from the use of immense sources of Big Data. Work on each type of risk with analytics tools can provide support for many decisions, and preparedness for events that can hit the performance of organizations and societies. Rau-Chaplin et al. (2014) indicate how Big Data and the use of new technologies allows the development of solutions for the analysis of vast datasets in specific risks. In particular the use of parallel computing power through MapReduce allows handling the huge volume of data, organizing and processing it for input into the analytics modelling process.

Applications: how to observe the practice of KM working for RM

The following four examples indicate means to improve the implementation of knowledge management in support of risk management. The answer to the question "why know" is fourfold: the search for performance improvement, increase in the firm's value, developing capabilities to compete, and improvement of the decision-making process. Organizations have tried communities of practice, data mining tools, collaboration tools, and knowledge portals, amongst other ways, to support the four KM processes presented earlier, as follows:

- Knowledge creation: In RM, new risk implies finding new ways to measure it and identifying the potential effects that it could have. Acquisition, synthesis, fusion and adaptation of existing risk knowledge are all part of the ways to understand new and current risks (Hormozi and Giles 2004; Chaudhry-SAS 2004; Dzinkowski 2002).
- Knowledge storage and retrieval: RM actions and methods require codification, organization and the representation of risk knowledge. They include the activities of preserving, maintaining and indexing risk knowledge (Basel II Accord, 2004).
- Knowledge transfer: ERM is a multidisciplinary task and thus an inter-departmental development. ERM and its holistic view of risk across the organization requires risk knowledge dissemination and distribution in order to support individuals, groups, organizations and inter-organizations to develop RM capacity (Desouza and Awazu 2005; Spies et al. 2005).
- Knowledge application: Risk knowledge can be converted into a competitive advantage for financial institutions willing to adopt best practices, and develop products and methods for risk control (Gibbert et al. 2002).

The first example is related to the need for the influence of knowledge in the modelling process and knowledge creation. The second example is related to the use of knowledge portals for risk knowledge storage, retrieval, dissemination and control. The third addresses the value of means of communication and knowledge sharing in the risk management organization. The fourth considers the possible improvement in the value of the firm because of the importance of knowledge management variables in the investor's decision.

Knowledge creation: risk modelling process

Earlier, risk modelling was presented as a "way to know" in risk management. Development of models is a process that requires data analysis and mathematical and statistical tools to create answers to problems. In some cases the outcome of the models is used to build a metric or risk map, in others to classify and to evaluate risks.

In risk modelling, knowledge creation includes identification of assumptions, conceptualization process, identification and selection of techniques to use, selection of processes, development collaboration, methods of solution, prototyping models and testing.

A knowledge management approach to the risk modelling process follows seven steps (Rodriguez and Edwards 2008b):

1   Studying traps, errors and constraints of the process, understanding flows of information to produce knowledge and how to use these flows in risk modelling. Risk modelling processes (Mladenic et al. 2003) comprise decisions that come from humans and machines, in particular people's minds interact with computers in search of a problem solution.

2 Identifying the enablers to convert risk knowledge from tacit to explicit knowledge and vice versa.
3 Understanding flows of information to produce risk knowledge.
4 Understanding risk knowledge organization.
5 Searching for KM technologies and techniques to support the risk modelling process.
6 Designing the risk knowledge management system to support risk modelling.
7 Connecting organizational performance metrics and the risk modelling process: outcomes and resources.

Risk knowledge creation appears in the risk modelling process itself and with the outcomes of the process. There are many different components of knowledge in the risk modelling process to store and retrieve such as: documents, selection of the proper data, new data created required in another model, creation of taxonomy and metadata and selection of structured and unstructured data. The means for storage and retrieval include data warehouses, specialized modelling systems and content management systems.

Risk modelling knowledge can be transferred through presentations, portals, meetings, discussions, collaboration activities, content management design, distribution, testing, and reporting. Application of risk modelling knowledge is represented by decisions, business processes, and models in other areas of the organization such as: impact analysis, evaluation, new developments, new strategic and tactical decisions.

Business understanding and its connection to knowledge creation using the modelling process is crucial for obtaining results in organizations. Nalchigar and Yu (2018) proposed a framework to put analytics into action in a business context. They explain how the modelling process is not only related to studying the variables and their values but also to building a conceptual model to tackle the problem of putting the analytics system into operation. Operation of the analytics system means understanding the selection of algorithms/models, contribution of the analytics algorithms to indicators such as precision, managing the lack of knowledge/expertise in the algorithms to use or to generate insights for achieving strategic goals.

Finally, modelling and creation of metrics also support the "way to know" in risk management. Analytics work requires information systems support, information interpretation and communication of meaning. The enterprise performance evaluation sub-system design answers questions such as: how to measure, to interpret and to discover directions of the organizational performance connecting risk metrics with risk modelling and the analytics process itself. Davenport et al. (2005) presented analytics as a strategic capacity that the organization can develop for competitive advantage. Barcelo (2015) explained the relationship of knowledge creation and analytics and how models are developed. The main points to take away are that the modelling process does not end, it is ongoing development; and it is looking for useful results even though they are not perfect. The possibility to validate models provides insights in solutions to problems using multiple techniques and data types. In particular Barcelo (2015) observed: "A descriptive model explains what happens, but curiosity, which perhaps is one of the factors that make us humans, is not satisfied only with description; it needs to explain how things happen and, if possible, why. This calls for predictive models. In other words, science usually aims to be predictive."

Knowledge storage and retrieval: a portal for country risk

People's ideas and previous experience with information systems were the input for organizing and prioritizing a Risk Knowledge Management System design project (Rodriguez and Edwards 2006). The objectives of the project were:

- Manage explicit knowledge transfer in order to support business activities and transactions.
- Increase business volume through increasing cross selling, improving customer orientation, and developing country and sector cluster analysis.
- Improve Satisfaction Index: internal – external improvement, the customer can have an integral view.
- Improve efficiency and reduce cost of processes: process coordination and development.
- Improve productivity as a result of process improvement.
- Improve innovation through new solutions which are potentially new services.
- Reduce information overload. This means more focus on the quality than on the quantity of data.
- Organize, categorize and identify the information and knowledge silos that need to be shared and analyzed: from marketing research, from external sources, from lessons learned by the market developers and other sources
- Support the relationships between knowledge issuers and users.

The conception of this KMS for risk management was based on the four pillars represented by alignment, integration, connection and coordination.

- Alignment is represented by the balanced interaction of the components of the KMS infrastructure and the consolidation of knowledge to address the customer's needs according to different segments.
- Integration is related to the application of KM processes to the business processes in a company division or the organization as a whole. This integration preserves the strategy and the processes of pre-selling, selling and post-selling products and services of the organization.

- Connection is the creation of links between knowledge issuers and the kinds of knowledge which have been required for solving transactions or customers' questions.

- Coordination is represented by the work flow and organization of stages for knowledge creation from multiple groups that have to provide their input to be included in a single document that has multiple users.

The KMS design is related to business processes and the purpose is to take actions connecting KM processes with business processes. Table 17.2 indicates examples of actions taken.

Knowledge transfer: importance in RM

Even though there are benefits from ERM concepts there are organizational issues to address as well. Lam (2003) indicated that even having a good RM practice for each risk, there are many difficulties in consolidating information and supplying guidelines to the board and senior management in order to answer strategic questions. Lam also noted that benefits of ERM are based on the concept of integration: integration of risk organization, integration of risk transfer practices and integration of risk practices with business processes. Besides, he indicated that based on the preparation of the organization for ERM, the expected benefits are: "increased organisational effectiveness, better risk reporting, and improved business performance." The question is: are these methods producing a better understanding of risk policy at the top of the organization?

Rodriguez and Edwards (2010) formulated and tested five hypotheses in order to relate KM concepts to risk knowledge sharing. A survey obtained responses from 121 RM staff in financial services. The responses were analyzed using multiple regression and stepwise regression. The results are shown in Table 17.3.

Four of the five hypotheses were supported: only the web channel functionality was not significantly associated with risk knowledge sharing.

Additionally, Rodriguez, Edwards and Koenig (2010) identified some points to keep in mind to strengthen the communication of risk knowledge. Knowledge, experience and feedback in an organization have a flow in both directions: top-down and bottom-up. The value of working on increasing the board's understanding is shown through the capacity of the board to communicate risk appetite and policy integration. This means a positive step of alignment as Kaplan and Norton (2006) indicate. However the means of communication used appears not to have a positive effect on the board understanding of risk policy, suggesting that different approaches for communication and knowledge transfer are needed. Probably, a more systematic and ongoing approach is required, based on specific needs and requirements in the board role and with a continuous learning process.

Table 17.2 Actions based on KM processes that support RM in the country risk portal example

KM process	Action in Risk Management
• Focusing on a KM strategy that mobilizes knowledge transfer to the exporters using the web channel	• Review of contradictions of document content, similar content from different authors/ no updating process/ different messages
• Bringing in experts and practitioners to talk about their experiences	• Standardization of the content, reports, data and information structures in order to share content and to get consistent results
• Implementing a cluster of explicit knowledge for countries (called country portal)	• Increment of capacity to provide independent answers to different knowledge consumers or to create undifferentiated solutions
• Participating in other projects to coordinate efforts such as Business Intelligence, connectivity of international operation and sector portal	• Inclusion of attributes of the business processes related to people's expertise in order to select people using an expertise locator
• Developing prototypes for the **KM** strategy based on international business in order to solve the knowledge needs on the exporter's value chain	• Review of the value of the documents from the users' perspective
	• Adjustment of document formats and presentations for identification and extraction of pieces and sections of those documents, fundamentally for sharing
• Developing portal strategy and portal solutions for specific audiences of bankers, brokers, customers and ultimately providing online transaction support to them	• Structuring of naming conventions and taxonomy for storing and retrieving documents
	• Ungrouping of some knowledge that is part of the same document and is required independently
	• Creation of links between different pages in order to find the correct document
	• Changing of the practice of users and issuers to understand that knowledge is based on the capacity for sharing it
	• Organization of data in order to give access to different software
	• Production of content in two different languages
	• Creation of prototypes for the definition of solutions and for communication with the programming group

Table 17.3 Hypotheses supported in the investors and KM study

Hypotheses	Results
People	
H1: Organizational capacity for work coordination is positively associated with the perceived quality of risk knowledge sharing	Supported
H2: The perceived quality of communication among groups is positively associated with perceived quality of risk knowledge sharing	Supported
Process	
H3: The perceived quality of the risk control process is positively associated with the perceived quality of risk knowledge sharing	Supported
Technology	
H4: The web channel functionality is positively associated with the perceived quality of risk knowledge sharing	Not Supported
H5: The risk management information system functionality is positively associated with the perceived quality of risk knowledge sharing	Supported

Knowledge application: value of KM for investors

Knowledge can be considered as a factor that reduces risk (Dickinson 2001). The existence of silos of knowledge can adversely influence the knowledge transfer process, and business units can require education in how to transfer experiences, taking into consideration that the pace of change can reduce the value of experience in some fields. The point is to review how knowledge leverages the results of the organization and the value of the organization itself.

Rodriguez and Edwards (2008a) studied the relationships between knowledge management variables and investment attractiveness. The research investigated three questions:

1   Is there a difference in risk level between companies with different levels of knowledge?
2   Are intangible assets directly associated with investment risk level?
3   What are the variables that can identify groups of similar companies based on risk, knowledge and structural capacity?

The differentiation of the groups of companies based on the combination of risk and knowledge management variables suggests a differentiation among organizations describing knowledge and risk. Low and high levels of the KM index appear related to low and high risk levels respectively. The analysis of the variables suggests: first, that the correlations are moderate and there are indications of relationships between risk variables and knowledge management variables such as innovation, years in control and quality of management. Second, that not all the clusters include risk variables and when the cluster does include risk variables they appear with variables such as years in business of the company, historical record in the Fortune 500 list, people management, quality and innovation.

Third, if we consider risk levels on a scale from 1 to 4 (with 1 the lowest) it is possible to say that levels 2 and 3 are not well differentiated but 1 and 4 are. High scores in people management, revenues, profits, and years of expertise are associated with a high score in the risk index. On the other hand, the low levels include low levels of history (Fortune 500 list), low profit scores and low assets. Variables like people management, percentage of intangibles, years of expertise, quality management appear in the splits describing risk levels, which shows their relevance. However, the variable categories are not clear indicators of risk classification. Percentage of intangibles score is not a discriminatory factor anywhere in the sample; it only appears in the decision tree describing medium, moderate and high risk levels in combination with moderate levels of people management and profits. However, the risk index appears correlated to the percentage of intangibles score.

Rodriguez and Edwards (2007, 2008a, 2008b, 2008c) cited several observations from the literature that indicate the basis for proposing that KM processes add to the firm's value. For example, Johnson et al. (2002) studied the relationship between knowledge, innovation, and share price. They found that the knowledge-based enterprise obtains firm value based on human capital, research and development, patents and technological assets. The evaluation of companies is based on intangible assets when the financial statement of the organization is more based on intellectual capital than on the valuation models of traditional accounting practice. Johnson et al. (2002) stated that it is difficult to find how research and development (R&D) influences the share price because of the different ways in which R&D affects organizational performance. Similarly, goodwill is considered significant in the share value for non-manufacturing companies but not for manufacturing organizations.

Nohria and Stewart (2006) wrote that during the twentieth century management emphasized risk and that "Uncertainty and doubt push the boundaries of management as we know it . . . the flight from uncertainty and ambiguity is so motivated, and the desire to reduce what is fundamentally unknowable to probabilities and risk so strong, that we often create pseudo uncertainty." Miller and Bromiley (1990) examined the risk factors used for different measures affecting strategic risk management. The factors are related to the risk of income stream, stock returns, and strategic risk.

The measures of risk and the association with performance are not clear and in some cases contradictory results have been presented in different studies because of the measures used for risk and its attributes. Miller's and Bromiley's (1990) results show a negative association between performance and risk: "Performance reduced subsequent income stream uncertainty for high performers and increased income stream risk for low performers."

There are several examples (see Table 17.4) of relationships between the investor's psychology and the perception of risk. Perception of risk can affect decisions, and in turn can be affected by knowledge and information about reputation, business longevity and perception of organizational competitiveness. Brand equity or brand value plays an important role in the investment decision. Brand equity includes the human factors associated with the brand evaluation.

Motameni and Shahrokhi (1998) identified several perspectives on brand equity valuation and their associated measures. Factors such as style, culture, and attitude are those that people need to keep the value of the brand. Jucaityte and Virvilaite (2007) went on to introduce the influence of the consumer into the traditional view of the economic value of the brand. The influence of the consumer is represented by psychographical and behaviourally-oriented metrics for brand evaluation.

Mizik and Jacobson (2008) presented the pillars of the brand evaluator model developed by Young and Rubicam (a leading marketing communications agency). They showed that the most important for brand valuation was to identify the financial impact of the perceptual brand attributes. They concluded that the stock return is associated with three elements: energy, perceived brand relevance, and financial performance measures.

Finally, Edwards and Rodriguez (2019) indicate how several applications of knowledge of risk management can be implemented in the real world with an implicit risk (in particular to risk assessment and risk control) that is bias. Alternatively, the issue can come from the capacity for dealing with bias and the ways to find and avoid it in the knowledge creation process. In KM and RM applications the presence of bias is part of the possible inputs but the crucial weaknesses arise from ignorance of the bias and the inappropriate use of knowledge creation data and techniques.

Discussion

The issue of a consolidated management practice for risk management and knowledge management is a controversial one. On the one hand, the two disciplines and practices are considered independent, with many technical components which are not possible to use jointly. On the other hand, practice shows that the processes in both disciplines have many points in common which can be used to gain synergy in a firm.

Bowman (1980) describes the paradox of having a negative association between corporate return and risk when there is a positive relationship,

Table 17.4 Studies identifying of the value of intangibles

The value of intangibles in a firm – risk mitigation

- Catasus and Grojer (2003) demonstrated that credit decisions take into consideration the value of intangibles and that it is possible to get access to capital markets if a company is intensive in intangible assets.
- Guimon (2005) observed the value of the intellectual capital report in the credit decision process and presented a wide set of references where investment decisions and cost of capital are based on intangibles disclosure.
- Patton (2007) studied metrics for knowledge in organizations where traditional project management may not be most effective. The concept of intangible assets as skills and their strategic value imputes an important role to intellectual capital and intellectual property in creating competitive advantage.
- Hillestad (2007) examined ratios on the Oslo Stock Exchange and found that intangible assets add value to increase equity and to produce better equity ratios.
- Fiegenbaum and Thomas (1988) examined a broad list of studies of risk-return association. They found that individuals mix risk averse and risk seeking behavior when there is a search for a target return.
- Lim and Dallimore (2002) analyzed the information that is needed by, and valuable for, investors. They presented a difference between the relationship with the disclosure of the intellectual capital in manufacturing and services, showing that better disclosure of intellectual capital is related to better understanding of the health of the company.
- Brammer et al. (2004) studied UK companies in the "list of the best" in Management Today and found that in the short term, reputation led to a higher return.
- Mulligan's and Hastie's (2005) work on the impact of information on investment decisions found that when information is used for risk judgement, negative (i.e. adverse) information has a high impact.
- Wong et al. (2006) studied the behavior of individuals regarding selling securities. Investors, whether winners or losers, decide to keep or to sell according to attributes of the people involved in the trading process.
- Orerler and Taspinar (2006) did research into utility and risk and the decision-making process. Their conclusion is that the psychological component of the investment decision, in terms of how much risk to take, is important in the process and there is more risk tolerance for the decision when there is more knowledge or control.
- Baker and Wurgler (2007) used two main assumptions: first, investors follow their beliefs about future cash flows and investment risk that are not supported by information or facts and, second, that making decisions against people's feelings can be risky and costly. They concluded that it is possible to measure the sentiment of investors and that it is possible that the results affect the stock market.
- Desai et al. (2008) presented a relationship between the consumer's perception of risk when they manage business and the years in business for the organization. Operational and strategic risks are associated with years in business in the way of sustainability of the competitive level of the organization in the market.

from the portfolio point of view, between return and risk. The meaning is that companies with lower risk and higher return have a higher share price, reducing the return for the owner or buyer of the stocks This result has inspired many articles in order to understand whether the better the organization's knowledge, the better the risk. The risk will have consequences for the investor's decision depending on the return appetite. In the final analysis, the integration of KM and RM could contribute to the creation of a more sustainable competitive advantage, because:

1 Strategy requires information interpretation in order to produce knowledge so as to control the enterprise's risks.
2 The strategy formulation process and the purpose of wealth growth need to be analyzed under a framework of uncertainty, trends, and complexity. The control of result variance is an important strategic goal.
3 Good results in a company require the combination of a clear business definition, understanding of risk, and the development of people's capacity for productivity improvement.
4 There is a need for understanding of integrated knowledge and risk assessment tools that support more efficient and effective management.
5 The knowledge practice is based on merging people, technology, business processes and leadership with the creation, storage, retrieval, transfer and application of knowledge. This merger looks for the identification, classification, transference, hedging, planning, and evaluation of enterprise risk.

Conclusion

In conclusion, this chapter presents risk and knowledge management as two disciplines with common ground, where knowledge management has great potential to support the improvement of RM. The four examples presented in the Applications section serve as "concept demonstrators" for the validity of the five claims in the Discussion section. We feel that they are sufficient to suggest that the integration of KM and RM might help to avoid failures in risk management and improve the core competences of the organization, in response to the issues that Marshall et al. (1996) identified as stemming from dysfunctional culture, unmanaged organizational knowledge, and ineffective controls. "Hence, the evidence is consistent with the proposition that effective risk management encourages valuable firm-specific investment by essential stakeholders and allows the firm to exploit opportunities as well as guard against downside exposure" (Andersen 2008). Big data, analytics, and artificial intelligence offer new opportunities but also new difficulties. The challenge is to mobilize the professional risk processes and people's knowledge to develop means to coordinate their expertise and actions for solving complex risk problems, supported by appropriate use of technology.

References

Alavi M and Leidner D 2001, "Review: Knowledge Management and Knowledge Management Systems: Conceptual Foundations and Research Issues", MIS Quarterly, Vol.25, No.1, pp 107–136.

Andersen TJ 2008, "The Performance Relationship of Effective Risk Management: Exploring the Firm-Specific Investment Rationale", Long Range Planning, Vol.41, No.2, pp 155–176.

Baker M and Wurgler J 2007, "Investor Sentiment in the Stock Market", Journal of Economic Perspectives, Vol.21, No.2, pp 129–151.

Barcelo J 2015, "Analytics and the art of modelling", International Transactions In Operational Research, Vol.22, pp 429–471.

Basel II: Revised International Capital Framework, 2004, www.bis.org/publ/bcbsca.htm.

Beckman T 1997, "A methodology for knowledge management", International Association of Science and Technology for Development (IASTED) AI and Soft computing Conference, Banff, Canada.

Bowman E 1980, "A risk/return paradox for strategic management", Sloan Management Review, Vol.21, pp 17–31.

Brammer S, Brooks C and Pavelin S 2004, "Corporate Reputation and Stock Returns: Are Good Firms Good or Investors", Faculty of Finance, Cass Business School, City University, London.

Buchanan L and O'Connell A 2006, "A brief history of decision making", Harvard Business Review, Vol.84, No.1, pp 33–41.

Cardwell D 1994, The Fontana History of Technology, Fontana Press, Hammersmith, UK.

Catasus B and Grojer J 2003, "Intangibles and credit decisions: results from an experiment", European Accounting Review, Vol.12, No.2, pp 327–355.

Cerchione R. and Esposito E., 2017, "Using knowledge management systems: A taxonomy of SME strategies", International Journal of Information Management, Vol.37, No.1, pp 1551–1562.

Chaudhry A 2004, "CRM: Making it simple for the Banking Industry". Paper 180, SAS Institute-SUGI 29.

Crouhy M, Galai D and Mark R 2001, *Risk Management*, McGraw-Hill, New York.

COSO 2017, Committee of Sponsoring Organizations of the Treadway Commission, Enterprise Risk Management – Integrating with Strategy and Performance.

Davenport T and Prusak L 1998, Working Knowledge, Harvard Business School Press, Boston.

Davenport T, Cohen D and Jacobson M 2005, "Competing on Analytics", Working Knowledge Research Center, Babson Executive Education.

Desai P, Kalra A and Murthi BPS 2008, "When Old Is Gold: The Role of Business Longevity in Risky Situations", Journal of Marketing, Vol.72, pp 95–107.

Desouza K and Awazu Y 2005, "Maintaining knowledge management systems: A strategic imperative", Journal of the American Society for Information Science and Technology, Vol.56, No.7, pp 765–768.

Dickinson G 2001, "Enterprise Risk Management: Its Origins and Conceptual Foundation", The Geneva Papers on Risk and Insurance, Vol.26, No.3, pp 360–366.

Dzinkowski R 2002, "Knowledge for all: Knowledge sharing at the World Bank", Financial Times Mastering Management Online, June. www.ftmastering.com/mmo (last accessed November 11 2008).

Edwards JS, Handzic M, Carlsson S and Nissen M 2003, "Knowledge management research and practice: vision and directions", Knowledge Management Research and Practice, Vol.1, No.1, pp 49–60.

Edwards JS and Rodriguez E 2016, "Using knowledge management to give context to analytics and big data and reduce strategic risk", Procedia Computer Science, Vol.99, pp 36–49.

Edwards JS and Rodriguez E 2019 (2019), "Remedies against bias in analytics systems", Journal of Business Analytics, DOI: 10.1080/2573234X.2019.1633890.

Fiegenbaum A and Thomas H 1988, "Toward risk and the risk-return paradox: Prospect theory explanation", Academy of Management Journal, Vol.31, No.1, pp 85–106.

Gibbert M, Leibold M and Probst G 2002, "Five styles of customer knowledge management, and how smart companies use them to create value", European Management Journal, Vol.20, No.5, pp 459–469.

Guimon J 2005, "Intellectual capital reporting and credit risk analysis", Journal of Intellectual Capital, Vol.6, No.1, pp 28–42.

Hillestad C 2007, "An analysis of financial ratios for the Oslo Stock Exchange", Economic Bulletin, Vol.78, pp 115–131.

Hormozi A and Giles S 2004, "Data mining: A competitive weapon for banking and retail industries", Information Systems Management, Vol.21, No.2, pp 62–71.

Ivanov D, Dolgui A, and Sokolov B 2018, "The impact of digital technology and Industry 4.0 on the ripple effect and supply chain risk analytics", International Journal of Production Research, published online 28 Jun 2018.

Johnson L, Neave E and Pazderka B 2002, "Knowledge, innovation and share value", International Journal of Management Reviews, Vol.4, No.2, pp 101–134.

Jucaityte I and Virvilaite R 2007, "Integrated model of brand valuation", Economics and Management, Vol.12, pp 376–383.

Kaplan R and Norton D 2006, Alignment, Harvard Business School Press, Boston.

King W 2006, "Knowledge sharing", in Encyclopedia of Knowledge Management, Schwartz D (editor), Idea Group Reference, Hershey, PA, pp 493–498.

Klugman S, Panjer H and Willmot G 1998, Loss Models From Data To Decisions, John Wiley and Sons Inc, New York.

Lam J 2003, Enterprise Risk Management, John Wiley & Sons, Hoboken, NJ.

Lehaney B, Clarke S, Coakes E and Jack G 2004, Beyond Knowledge Management, Idea Group Publishing, Hershey, PA.

Lim L and Dallimore P 2002, "To the public-listed companies, from the investment community", Journal of Intellectual Capital, Vol.3, No.3, pp 262–276.

Malhotra Y 1999, "Beyond hi-tech hidebound knowledge management: strategic information system for the new world of business", Working Paper, Brint Research Institute.

Marshall C, Prusak L and Shpilberg, D 1996, "Financial risk and need for superior knowledge management", California Management Review, Vol.38, No.3, pp 77–101.

Miller K and Bromiley P 1990, "Strategic risk and corporate performance: An analysis of alternative risk measures", Academy of Management Journal, Vol.33, No.4, pp 756–779.

Mizik N and Jacobson R 2008, "The financial value impact of perceptual brand attributes", Journal of Marketing Research, Vol.XLV, pp 15–32.

Mladenic D, Lavrac N, Bohanec M and Moyle S 2003, Data Mining and Decision Support: Integration and Collaboration, Kluwer Academic Publishers, Boston.

Motameni R and Shahrokhi M 1998, "Brand equity valuation: A global perspective", Journal of Product & Brand Management, Vol.7, No. 4, pp 275–290.

Mulligan E and Hastie R 2005, "Explanations determine the impact of information on financial investment judgements", Journal of Behavioral Decision Making, Vol.18, pp 145–156.

Nalchigar S and Yu E 2018, "Business-driven data analytics: A conceptual modelling framework", Data and Knowledge Engineering, Vol.117, pp 359–372.

Nohria N and Stewart T 2006, "Risk, uncertainty and doubt", Harvard Business Review, Vol.84, No.2, pp 39–40.

Nonaka I 1994, "A dynamic theory of organisational knowledge creation", Organization Science, Vol.5, No.1, pp 14–37.

Nonaka I and Takeuchi H 1995, The knowledge-creating company: how Japanese companies creates the dynamics of innovation, Oxford University Press, New York.

Nwogugu M 2005, "Towards multi-factor models of decision making and risk: A critique", Journal of Risk Finance, Vol.6, No.3, pp 267–274.

Oldfield G and Santomero A 1997, "Risk management in financial institutions", Sloan Management Review, Vol.39, No.1, 33–47.

Orerler E and Taspinar D 2006, "Utility function and risk taking: An experiment", The Journal of American Academy of Business, Vol.9, No.2, pp 167–174.

Panjer H 2006, Operational Risks Modelling Analytics, John Wiley and Sons, New York.

Patton J 2007, "Metrics for knowledge-based project organizations", SAM Advanced Management Journal, Winter, pp 33–43.

Rau-Chaplin A, Yao Z and Zeh N 2014, "Efficient data structures for risk modelling in portfolios of catastrophic risk using MapReduce", Procedia Computer Science, Vol.29, pp 1557–1568.

Rodriguez, E 2006, "Application of knowledge management to enterprise risk management: Country Risk Knowledge (CORK)" In Proceedings of 3rd Knowledge Management Aston Conference, Edwards J (editor), pp 190–202, The OR Society, Birmingham.

Rodriguez E and Edwards JS 2007 "Knowledge management applied to enterprise risk management: Is there any value in using KM for ERM?" In Proceedings of 8th European Conference on Knowledge Management, Remenyi D, (editor), pp 813–820. Academic Conferences Limited, Reading, UK.

Rodriguez E and Edwards JS 2008a, "Risk and knowledge relationships: An investment point of view", In Proceedings of 9th European Conference on Knowledge Management, Remenyi D, (editor), pp 731–742 . Academic Conferences Limited, Reading, UK.

Rodriguez E and Edwards JS 2008b "Before and after modelling. Risk Knowledge Management is required" ERM Symposium 2008 Casualty Actuarial Society – PRMIA. Society of Actuaries, Schaumberg, IL.

Rodriguez E 2008c "A bottom-up strategy for a KM implementation at EDC Export Development Canada". Chapter 10 in Making Cents Out of Knowledge Management, Liebowitz J, (editor), Scarecrow Press Inc., Lanham, MD.

Rodriguez E and Edwards JS 2010, "People, technology, processes and risk knowledge sharing", Electronic Journal of Knowledge Management, Vol.8, No.1, pp139–150.

Rodriguez E, Edwards JS and Koenig D 2010, "The board of directors, executives and risk knowledge management" in Proceedings ICICKM 7th International Conference on

Intellectual Capital, Knowledge Management & Organisational Learning, Remenyi D, (editor), pp 304–405, Academic Conferences Limited, Reading UK.

Shaw J 2005, "Managing all your enterprise's risks", Risk Management, Vol.52, No.9, pp 22–30.

Simoneou L 2006, "Enterprise search: The foundation for risk management", KM World, Nov 1, 2006.

Spies M, Clayton AJ and Noormohammadian M 2005, "Knowledge management in a decentralized global financial services provider: A case study with Allianz Group", Knowledge Management Research and Practice, Vol.3, No.1, pp 24–36.

Taleb N, Goldstein D and Spitznagel M 2009, "The six mistakes executives make in risk management", Harvard Business Review, Vol.87, No.10, pp 78–81.

The Economist 2010, Number-Crunchers Crunched, Special Report on Financial Risk, February 11.

Turing A 1948, "Intelligent Machinery", Reprinted in "Cybernetics: Key Papers". Evans CR and Robertson ADJ. Baltimore: University Park Press, 1968. p. 31.

Von Krogh G 1998, "Care in knowledge creation", California Management Review, Vol.40, No.3, pp 133–153.

Warren B 2002, "What is missing from the RMIS design? Why enterprise risk management is not working", Risk Management Magazine, Vol.49, No.10, pp 30–34.

Wiig K 1993, Knowledge management foundations: Thinking about thinking. How people and organisations create, represent and use knowledge, Schema Press, Arlington, TX.

Wiig K 1997, "Knowledge management: Where did it come from and where will it go?" Expert Systems With Applications, Vol.13, No.1, pp 1–14.

Wong A, Carducci B and White A 2006, "Asset disposition effect: The impact of price patterns and selected personal characteristics", Journal of Asset Management, Vol.7, No.3, pp 291–300.

Wu D, Olson DL and Dolgui A 2017 "Editorial: Artificial intelligence in engineering risk analytics", Engineering Applications of Artificial Intelligence, Vol.65, pp 433–435.

Wu X, Hu X, Qi W, Marinova D and Shi X, 2018, "Risk knowledge, product knowledge, and brand benefits for purchase intentions: Experiences with air purifiers against city smog in China", Human and Ecological Risk Assessment, Vol.24, No.7, 1930–1951.

Yoost DA and Mathaisel BF 2016, "Board oversight of the risks in using big data and advanced analytics", The RMA Journal, Vol.98, No.5, pp 38–42.

18 Infrastructure development in the UAE

Communication and coordination issues amongst key stakeholders

*Moza T. Al Nahyan, Amrik S. Sohal,
Brian Fildes and Yaser E. Hawas*

Highlights

- Highlights the importance of management processes (especially coordination and communication amongst stakeholders) in risk management in construction projects.
- Applies stakeholder theory to assessing transportation construction projects.
- Identifies factors that contribute to successful and unsuccessful project outcomes.
- Provides guidance for the development of a risk management tool.
- Offers recommendations for improving the management of transportation infrastructure projects in the UAE.

Introduction

Construction projects, as reported by Ellis and Thomas (2003), have many unique features such as a long time period from start to finish, complicated processes, financial constraints and dynamic organization structures. Such organizational, financial and technological complexity generates enormous risks. The diverse interests of project stakeholders on a construction project further aggravate the changeability and complexity of the risks. While risks cannot be eliminated, successful projects are those where risks are effectively managed, of which early and effective identification and assessment of risks is essential.

Large-scale infrastructure construction projects involve many stakeholders and because of the nature of the work involved, these projects are difficult to coordinate and manage (Gil and Beckman 2009) and hence present many risks that relate to slow decision-making, errors in design, utility relocation, poor quality of the work and budget overrun. Furthermore, Guo, Chang-Richards, Wilkinson and Li (2014) highlight that "complexities and uncertainties are endemic in large infrastructure construction projects" (p.815).

In the case where investments amount to $1 billion or more, these projects are referred to as megaprojects and by their nature have high risks associated with them. There are many studies reporting that megaprojects perform poorly, particularly with respect to costs and time. Some relatively large-size projects have not met their delivery times and do not follow the schedule plan; hence actual cost exceeds the budget by varying rates. This is mostly attributed technically to excessive construction variations, redesigns, sudden changes in the clients' requirement and lack of proper planning and design.

Much of the previous research on this topic has been conducted in developed economies. This study is based on a rapidly developing economy – the United Arab Emirates (UAE). The UAE is a constitutional federation of seven emirates: Abu Dhabi, Dubai, Sharjah, Ajman, Umm al-Quwain, Ras al-Khaimah and Fujairah. The complexity of each local government differs according to its size and population, and each emirate follows a general pattern of municipalities and departments. The relationship between the federal and local governments is laid down in the constitution, and it allows for a degree of flexibility in the distribution of authority.

In the last 30 years, the significant changes in economic, business and social activities in the UAE were accompanied by a tremendous increase in transportation demand and as such have led to a corresponding increase in infrastructure project investments to accommodate this demand, alongside an increase in the risks involved in completing such projects. We believe that effective coordination and communication amongst the various stakeholders involved can mitigate many of the risks experienced in completing megaprojects. In this chapter we report the results of a study examining coordination and communication amongst key stakeholders involved in transportation infrastructure projects in the UAE.

The challenges in (or the risks associated with) managing infrastructure projects in the UAE are many. Coordination between federal and local governments, labor quality, changing policies, inconsistency of standards, lack of specialized management, lack of definitive procedures and quality controls to measure the degree of success or failure of infrastructure projects in respect to time, cost and quality of construction have been identified.

A review of local media was undertaken to identify the reported drawbacks and public concerns of the existing implementation system of transportation infrastructure projects in the UAE. Zaneldin (2005) in particular refers to road construction claims whilst Kazmi (2005) highlighted the rise of compensation claims from the boom in construction projects. The leading real-estate companies such as Nakheel and Emaar were reportedly unfazed by project delays (Ditcham 2006; 2007). Contradicting contractor and client opinions however did report concerns about the causes and amount of delay of mega transportation construction projects in UAE (Ahmed 2007; Nazzal 2005; Anonymous 2005). Causes of delay mentioned in the media include lack of coordination, slow decision-making, errors in design, utility relocation, and sudden changes in governmental requirements or modifications of quality standards of the work. Several of the newspaper articles addressed the general public's concerns over infrastructure development. The main concerns are the excessive delays in project delivery, the increasing trend of cost overrun and increasing customer dissatisfaction expressed, especially frustration of road users.

Given the concerns expressed above, that primarily refer to the need for risk management in construction projects, a study was designed to examine the management of transportation infrastructure projects in the UAE in order to understand how management processes (especially coordination and communication amongst stakeholders) impact on reducing risks and hence improve project outcomes. In this chapter, we focus on the following five questions:

1   How are successful and unsuccessful transportation infrastructure projects in the UAE assessed by key stakeholders?
2   How is management practice in the UAE generally assessed by the key stakeholders?
3   How do stakeholders communicate the various activities necessary for completion of a major project? Is communication effective and how it can be improved?
4   How effective is coordination among the stakeholders and how can this be improved?
5   What do stakeholders consider as important in improving project success indicators?

The aim is to identify areas where improvements can be made so that transportation infrastructure projects in the UAE can be completed more successfully.

The remainder of this chapter is structured as follows. The next section reviews literature on the management of construction projects, stakeholder theory and management concerns relating to communication and coordination. Next, the research methodology is described followed by the results. The final section of the paper presents the conclusions and suggestions for future research.

Literature review

The literature on management of construction projects is diverse, tackling several issues including the management of stakeholders relationships and expectations (Macharis 2005; Newcombe 2003; Elias et al. 2000; 2002; 2004), the strategic design and implementation of projects (project management) (Cleland 1995), contract management and theory of organization (Turner and Simister 2001), measures of success as it relates to the project type (Ernzen et al. 2004), and the root causes and effects of delay in infrastructure projects (Ellis and Thomas 2003).

Other important aspect of this literature include methods to achieve accurate and effective project scheduling (Mattila et al. 2003), the development of cost estimation tools (Kyte et al. 2004), innovative project planning (Hegazy 2004), best practices (Kingsley et al. 2004), and the role of IT and technologies for project management (Nono and Tarnoff 2004).

The project stakeholders (the main participants of the construction project) have been traditionally reported in the literature as the client, the architect and the contractor (Macharis 2005; Newcombe 2003; Elias et al. 2000; 2002; 2004). The interactions and interrelationships between these participants largely determine the overall performance of a construction project. The performance of these participants is also interdependent. Hence, in order to perform effectively, a reciprocal requirement exists, whereby each participant requires the other participants to perform their duties effectively and in harmony with each other. The performance of individual participants remains important because overall project performance is a function of the performance of each participant.

The problem of delays in the construction industry is a global phenomenon. Project delays were highlighted in several countries (such as Saudi Arabia, Nigeria, Thailand, Malaysia, Hong Kong, Australia, the USA, etc) as a major concern but for several causes. The research on these aspects is divided between studying the causes of delays or effects of delay on the overall project measures of success (Ellis and Thomas 2003). A common research methodology in identifying the causes of project delays or cost overrun is the use of questionnaire surveys or interviews with contractors, consultants and clients (Elias et al. 2000; 2002; 2004). Among the reported causes are finance and payment arrangements, poor contract management, resources supply, inaccurate estimation and overall price fluctuations (Ellis and Thomas 2003). From the viewpoint of the engineers, cash problems during construction, the relationship between subcontractors and the slow decision-making process of the owner were the main causes of delay (Ellis and Thomas 2003). However, the owners agreed that the design errors, labor shortages and inadequate labor skills were important delay factors. In brief, contradicting opinions are reported by different stakeholders. From a management perspective, reported reasons included poor site management and supervision, inadequate coordination among various stakeholders, poor communication, and low speed of decision-making involving all project teams (Ernzen et al. 2004). In general, previous research indicate that the majority of management problems are client related, while the majority of the technical problems are contractor related (Macharis 2005; Newcombe 2003).

Communication: Literature on construction management widely emphasized the role of communication in effective management. Zwikael et al. (2005) examined project management practices in Japan and Israel and concluded that various types of management styles, scope and time management impact on improving technical performance of projects, while communication and cost management impact on improving overall success measures of projects. Soetanto and Proverbs (2002) suggested that the use of communication effectiveness models to predict satisfaction levels by contractors and clients at the earliest possible stage in the project life cycle was also important.

To minimize defective designs, and subsequent overrun of cost and schedules, Zou et al. (2007) reported that the design team needs to be effective communicators to ensure a successful outcome. A framework for enhancing communication and knowledge sharing in large-scale projects is proposed by Jackson and Klobas (2007). Stewart (2007) further stressed that strategic implementation of innovative information and communication technologies are essential for the long-term survival of construction firms. The lack of communication among parties was reported among the ten most important causes of project delay by Sambasivan and Soon (2007). Chen and Partington (2006) indicated that a good project manager possesses the ability to effectively communicate with different people at different levels.

Coordination: Coordination is recognized as a key issue in successful transportation infrastructure projects. The implementation of such projects entails several stages, and the stakeholder network is more complex than other infrastructure projects. The coordination of the various stakeholders is among the key success factors of such projects. Timmermans and Beroggi (2000) stressed the importance of coordination between organizations with diverse objectives. The concept of international coordination for transportation infrastructure projects is addressed by Short and Kopp (2005), where they reported that despite advances in planning technologies, the European Union's ability to converge to "best practice" is questionable due to lack of the international coordination and unified standards. Chen and Partington (2006) emphasized the ability of project managers to coordinate activities on site. Lack of coordination among project participants was also identified as a key risk factor in Australian projects (Zou et al. 2007).

Stakeholder theory

Freeman (1984) describes stakeholder theory as the understanding of stakeholders' relationships, the processes for dealing with these stakeholders, and the transactions to achieve the project deliverables with satisfactory stakeholders. A detailed description of the literature on stakeholder theory was provided by Elias et al. (2000). In the context of transport policy, Banville et al. (1998) defined stakeholders as those people who have a vested interest in a problem by affecting it or/and being affected by it. In their chapter, a framework for the introduction of the concept of stakeholders was introduced. The concept is applied for the evaluation of transport related strategic decisions.

Construction management particularly focuses on planning the complex array of activities required to deliver a successful construction project, such as a road or a building (Morris 1997). Vinten (2000) argued that a crucial skill for managers of construction projects is being able to manage construction stakeholders' expectations. Failure to address stakeholder expectations has resulted in countless project failures (Bourne and Walker 2005); primarily because construction stakeholders tend to have the resources and capability to stop construction projects (Lim et al. 2005). Successful completion of construction projects is therefore dependant on meeting the expectation of stakeholders (Cleland 1995). Newcombe (2003) listed stakeholders as clients, project managers, designers, subcontractors, suppliers, funding bodies, users, owners, employees and local communities. In construction projects, the interests of stakeholders can vary over the life of a project (Friedman and Miles 2002), and the reasons for these changes include learning, changing values, and specific experiences (Elias et al. 2004). Zwikael et al. (2005) suggested the use of cost overrun, schedule overrun, technical performance and stakeholder satisfaction as the primary performance indicators of projects.

Mitchell et al. (1997) used three attributes in measuring the importance of a stakeholder; namely legitimacy – the moral or legal claim a stakeholder has to influence a particular project; power – their capacity to influence the outcome of a given project; and urgency – the degree to which their claims are urgent or compelling. A stakeholder topology is summarized in Figure 18.1 below, where possessing all three attributes is categorized as highly important (definitive stakeholder), two factors as medium (dominant, dangerous or dependent stakeholder), and one factor as low (dormant, discretionary or demanding). Any individual or organization possessing none of the above factors in a project is regarded as a non-stakeholder.

Stakeholder management capability is tied to understanding of the stakeholder topology of the project, the ability to deal with other stakeholders, and the effectiveness of the decisions to achieve the project objectives. Although stakeholder theory has been subjected to criticisms (e.g. Fassin 2008), others such as Oliverio (2007), Ackermann and Eden (2011), and Jensen and Sandstrom (2011) claim it has considerable merit in clarifying dynamic processes in project management today.

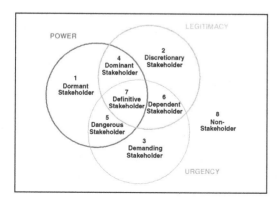

Figure 18.1 Stakeholder topology (Mitchell et al. 1997)

Methodology

Following the stakeholder concept outlined by Mitchell et al. (1997), the stakeholders involved in transportation infrastructure projects in the UAE were initially identified and a preliminary assessment made of their importance (see Figure 18.2). Sponsors/clients and government departments were identified as having the highest degree of influence in project decision-making. Project management firms, consulting firms and contractors were identified as having medium degree of influence in project decision-making. Other stakeholders (such as those highlighted in light grey colour) have low influence on the project execution. These include the media; community groups; insurance firms; labor, legal and political entities; special interest groups (such as environmental protection agencies, traffic safety societies and research institutes); and customer or road users.

For the purpose of this study, stakeholders having high or medium influence were included as they play a significant role during the planning and execution of the project. Furthermore, the progress of construction projects is affected by the quality of the managerial decisions made by these stakeholders. The five groups of stakeholders involved in this study are described in more detail below:

1   *Sponsors/clients:* Sponsors (such as the Council of Ministries or the Executive Council of Abu Dhabi) are responsible for the allocation of the federal funding, or for funding projects within the Emirate of Abu Dhabi. These entities possess the attributes of interest (power, legitimacy and urgency) and are regarded as a highly important and influential group. Clients on the other hand are those governmental agencies who initiate and are responsible for the undertaking of a project. Clients could be federal, state or private entities. An example of federal clients is the Ministry of Public Works (MPW). An example of a state client is the Municipality of Abu Dhabi.

2   *Governmental departments:* These (non-clients) includes local municipalities, police, utility companies, etc. They may be involved during the various stages of the project, but mostly during the coordination activities undertaken by the consultants during the planning and design phases, and by the contractors during the execution (construction) phase.

3   *Management firms:* These are third party entities responsible for coordinating and managing all project activities on behalf of the client and sponsors. Responsibilities including reviewing all documents and design drawings by consultants, approving payments, monitoring progress, etc.

4   *Consulting firms:* Responsibilities of these firms include the design, scheduling, tendering and supervision of all construction activities by the contractors. Some consulting firms get involved in earlier stages of project planning and scoping on behalf of clients (especially if the client organization has no internal planning department).

5 **Contractors:** Responsibilities include the scheduling of the activities of the construction phase and the execution of all civil work. Some contractors might get involved in the earlier design stages, but this is quite rare.

Data collection

The research methodology chosen was qualitative and legitimate for studies such as this one. It has the advantage of being easily implemented, comprehended by decision-makers and UAE culture, and is amenable to various project sizes and management forms. According to Mason (2005), "qualitative research is characteristically exploratory, fluid and flexible, data-driven and context-sensitive". Qualitative research aims at producing "rounded and contextual understandings on the basis of rich, nuanced and detailed data" (Mason 2005: p3). According to Wood (2006), methods of qualitative research involve observations, interviews, sampling, examining, written materials, questionnaires, focus group and case studies. Denzin and Lincoln (2000) argued that qualitative implies an emphasis on processes and meanings, which are not (and cannot be) rigorously examined or measured in terms of quantity, amount, intensity or frequency. Nevertheless, the approach is a useful technique for examining reasons for the success or otherwise of construction projects. Silverman (1993) states that there is no standard approach among qualitative researchers and claims that four primary methods are used namely, observations; analyzing texts and documents; interviews; and recording and transcribing.

Field work involved undertaking 20 interviews representing the five key stakeholders as discussed above. A snowballing approach was adopted in

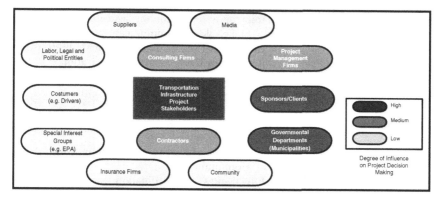

Figure 18.2 Stakeholders involved in transportation infrastructure projects in the UAE

this research to ensure the randomness in recruiting interview participants. Representatives of five sponsors/clients, three governmental departments, two management firms, five contractors, and five consultants were interviewed. All interviewees held senior positions within their organizations and played a significant role in transportation infrastructure projects. Most had been involved in mega/fast-track projects identified by the Ministry of Public Works.

Participants whom agreed to participate were initially contacted by telephone and were briefed on the research project. Request for an interview was made and if granted, the interview date, time and place were arranged. The interview itself comprised ten sections with a total of 58 specific questions, intended to explore the interviewee's opinion on the key research questions. A typical interview lasted for approximately 90 minutes which was sufficient to achieve the targeted depth to each of the detailed research questions.

Table 18.1 Samples of sets and tree nodes for NVivo coding

Sets	Tree Nodes
Stakeholders and Importance	Important Stakeholders Affecting Success
	Judging Stakeholder Importance
	Most Important Stakeholders in Critical Stages
Successful Project Indicators	Project Completed successfully (proportion)
	Interviewee Definition of Project Success
	Organization or Department Success Measures
	Interviewee Agreement with Organization Success Measures
	Reasons for Interviewee Selection of Most Important Success Measures
Communications	Interviewee's Opinion on Criticalness of Stakeholders Communication
	Commonly Used Communication Method(s)
	Reasons for Using Such Communication Method(s)
	Effectiveness of Communication within Organization
	Effectiveness of Communication among Stakeholders
	Interviewee's Opinion on how to Improve Communication
Coordination	Effectiveness of Coordination among Stakeholders
	Interviewee's Opinion on how to Improve Coordination

Data analysis and results

Data analysis

To analyze the data collected, NVivo was used. NVivo is a proprietary desktop software package for the organization and analysis of complex non-numerical unstructured qualitative data. It is primarily used by qualitative researchers working with very rich text-based and/or multimedia information, where deep levels of analysis on small or large volumes of data are required. The software allows users to classify, sort and arrange thousands of pieces of information; examine complex relationships in the data; and combine subtle analysis with linking, shaping, searching and modelling.

About 50 "parent" tree nodes were introduced; each tree node corresponds to a key interview question. The parent tree nodes were then grouped into sets; each set addresses a specific interview issue (e.g. communication, decision-making, etc). Table 18.1 illustrates the sets used and the parent tree nodes. Further coding was carried out by introducing "child" tree nodes to each parent node. For instance, a child node to the "stages influencing project success" parent node included nodes of budget, construction, planning of resources and materials, contracting, design, planning, scheduling and scoping.

Profile of the interviewees

Table 18.2 below presents a profile of the 20 interviewees. All interviewees held senior positions within their organization and played a significant role in transportation infrastructure projects. Those interviewed included general managers, directors/vice presidents, chief engineers, senior architects, and other senior executives. The majority had been in their current position for between one and five years. Almost two-thirds had 20 or more years in their perspective industries with a broad range of experiences covering planning, scoping, design, scheduling, tendering and construction. It is important to mention here that the majority of those interviewed were males with just two females involved who held key senior management positions. All interviewees generously volunteered their time and willingly shared their experiences in their respective fields of specialization. Names of individual interviewees and companies are disguised to ensure confidentiality.

Most of the interviewees had been involved in "mega, fast-track" transportation infrastructure projects in the UAE with substantial developments currently underway. The majority, if not all, national projects are considered mega and are described with the term "fast-track" to reflect the eager desire of the government to accomplish such projects in the shortest time-frame possible.

The projects discussed in the interviews are considered the most important national infrastructure projects in general and transportation in particular. Such focus on transportation projects results from the desire to support the other ongoing infrastructure developments in UAE and the

Table 18.2 Key stakeholders' profile

	Organization Type	Interviewee's Position	Tot. Yrs of Experience	Field of Specialization	Yrs of Exp. in Current Job	Current Job Responsibilities	Gender
1	**Sponsor / Clients**	Executive Manager For Urban Planning & Housing	20	Construction	1	Planning	F
2		General Manager	20+	Management Engineering	5	Management – Planning	M
3		Director of Road Department	7	Mix of Many	1	Design–Management–Engineering	F
4		Senior High Way Design Eng.	10–15	Mix of Many	3	Road Design	M
5		Technical Advisor & SCDIA Committee Member	20+	Management Engineering		Management – Technical Advisor	M
6	**Government Agencies**	Head of Executive Department	5	Construction – Project Management	4	Project Management	M
7		Strategy & Policy Division Director	–	Business development & Tourism	2	Management	M
8		Director of Internal Roads & Infrastructure	5–10	Construction	1	Planning – Construction	M
9	**Management Firms**	Senior Project Manager	20	Mix of many	12	Road Project Management	M
10		Project Manager	20+	Construction	1	Construction Eng. – Project Mngt	M

#	Category	Title	Experience	Field	No.	Specialization	Gender
11	**Consultants**	Senior Liaison Eng.	20+	Construction – Design – Management	2	Construction	M
12		Contracting Consultant	20	Tendering & Contracting	2	Contracting	M
13		Project Manager - Transport Planning Section	10–15	Planning	8	Management. – Planning	M
14		Vice President	20+	Full construction program	4	Management	M
15		Senior Architect	5–10	Construction – Contracting – Design	–	Mix of Many	M
16	**Contractors**	Chief Operating Officer	20+	Construction	1	Management	M
17		Project Director	20+	–	1	Business Development	M
18		Chief Engineer (Roads)	20+	Full Construction Program	18	Construction	M
19		Contracts Manager	20+	Contracting & Management	20+	Construction	M
20		Construction Manager	20+	Construction & Management	5	Quality assurance	M

growing population of the country. Examples of current mega, fast-track projects include the following:

- The Dubai–Al Fujairah highway connecting the east and the west coasts of the country;
- Abu Dhabi airport expansion project to accommodate a substantial increase in passenger travel;
- Various RTA (Road Traffic Authority) highway developments within the emirate of Dubai to support new residential developments such as Al Nakhla and Burj Dubai;
- New road development in the emirate of Abu Dhabi connecting Al Sadeyat and Al Reem Islands to the main land and to facilitate the flow of transportation within the island of Abu Dhabi;
- Zayed Donation Residence Project in northern emirates to support settlement of nationals.

The above-mentioned projects range from five to eight years in duration and vary in terms of their status.

Major findings

Successful and unsuccessful projects

Overall, the majority of the interviewees (12 interviewees out of the 16 who answered this question) agreed that over 70% of the projects had been completed successfully. Two specific questions addressed issues relating to unsuccessful projects: (i) how the specific department/organization defined unsuccessful projects, and (ii) how unsuccessful projects were measured by the department/organization. Generally, the majority of the interviewees indicated that they "fully agreed" with the project measures used by their organization, with only a few indicating that they "agreed to some extent".

Department/Organization Definition of Unsuccessful Project. Unsuccessful projects were defined in a number of different ways. Overall, interviewees identified "cost overrun" (identified by eight interviewees), "time overrun" (seven interviewees) and "poor quality" (six interviewees) as the most common definitions used for unsuccessful projects. Only one interviewee identified "not achieving stakeholders expectations" and "commuters dissatisfaction" as definitions for unsuccessful projects.

Measures Used by Departments/Organizations. The responses show a variety of measures used to describe unsuccessful projects. Overall, interviewees identified "poor quality" (seven interviewees), "cost overrun" (six interviewees) and "time overrun" (six interviewees) as the most common measures used.

FACTORS CAUSING UNSUCCESSFUL COMPLETION OF PROJECTS

Interviewees provided a variety of responses with regard to the factors that cause unsuccessful completion of projects. The most common responses relate to unqualified stakeholders. "unqualified/bad contractors", "unqualified consultants" and "unqualified engineers" were mentioned seven times as the major factors causing unsuccessful completion of projects. Another set of responses relate to lack of coordination mentioned six times and included responses such as "lack of coordination with local governments", "governmental process" and "poor coordination with utility firms". Other common responses relate to "availability of resources"/"materials procurements" – mentioned six times – and "bad design" – mentioned four times – and "price increments" – mentioned four times.

In examining more closely the reasons for project time overrun and cost overrun, interviewees gave a very wide range of responses. The responses given for time overrun cluster around five major reasons, namely:

- Variation in or un-reviewed design and schedules – mentioned ten times;
- Improper planning, scoping or cost estimates – mentioned seven times;
- Human resources issues – mentioned seven times (responses included "availability and reliability of human resources", "varieties of work labor skills" – employees having different cultural background, experience and skills, and "insufficient resources – labor/staff");
- Availability of materials – mentioned six times;
- Availability of qualified contractors and consultants – mentioned five times.

With respect to project cost overrun, the most common response, mentioned 16 times, relates to material cost increases/inflation. The next most common responses were project variations/rework (mentioned six times) and un-reviewed design/schedule (mentioned four times and changes in stakeholder requirements (mentioned four times).

Management practice in the UAE

Two questions were asked with respect to management practice in the UAE. The first question asked interviewees to rank management practice as "excellent", "very good", "good" or "fair". The second question asked for a reason for the response given to the first question.

One-half of the interviewees regard management practice in the UAE as "good" whilst one-quarter indicated this was "excellent" (mentioned by three interviewees) or "very good" (mentioned by two interviewees). Three interviewees identified management practice as "fair" and two of them did not provide a clear answer to this question.

A range of reasons was given for the above responses to the ranking of management practice. These cluster around the following three major reasons:

- Poor human resources – mentioned five times (responses included "unqualified managers", "still learning", "engaging best qualifications" – hiring appropriately qualified staff, "workforce variations" and "availability of resources");

- Lack of use of international design and quality standards – mentioned four times;
- Lack of a continuous improvement culture – mentioned three times.

Two interviewees clarified that the double track megaprojects that are currently taking place in the UAE are overloading all entities and hence affecting negatively on the ranking of management practice in the UAE. Poor human resources were also identified above as a key management concern in the implementation of infrastructure projects in the UAE.

Communication in project management

A series of questions focused on communication, namely how critical communication is to the success of projects, the common methods used for communication, effectiveness of communication within the organization and across the stakeholders, and how communication can be improved. The responses are summarized below.

Criticalness of communication. Except for one interviewee providing no clear answer and another saying that it was "Critical to some extent", all of the other interviewees indicated that communication is "Very critical" (five interviewees) or "Critical" (13 interviewees) to the success of the project.

Communication methods used and reasons for their use. Interviewees indicated using multiple methods for communication within their organization/department. The most commonly used methods are "meetings with minutes" (mentioned 13 times) and "written documents" (also mentioned 13 times). Although "e-mail" was used reasonably frequently (mentioned nine times), the use of "e-meetings" was only mentioned once. "Communication by telephone" was mentioned six times whilst "site visits' was mentioned only once.

Five major reasons identified from the range of responses given for using such communication methods are:

- The method is fast and efficient (mentioned six times);
- The decision made is documented (mentioned five times);
- It is the officially accepted method (mentioned five times);
- The method used assures reaching a contractor/consultant (mentioned four times);
- It allows personal interaction and flow of ideas (mentioned three times).

Interestingly, "confidentiality" was mentioned only once as the reason for using the method of communication.

Effectiveness of communication. Generally, there was agreement amongst the 20 interviewees that communication within their organization/department is "effective" (14 out of 20). Two of the interviewees indicated that it was "effective to some extent", one saying that it was "not effective", and three provided no clear answer. There was less agreement amongst the interviewees on the effectiveness of communication across the different stakeholders involved: seven indicating it was "effective", six indicating it was "effective to some extent" and four saying that it was "not effective".

Suggestions for improving communication. Interviewees made a number of suggestions as to how to improve communication methods. One set of responses relates to establishing communication committees (mentioned three times) that meet regularly (mentioned twice) and having a chair to avoid wasting time (mentioned once). "Making electronic communication official" and "create a manual or system procedure for communication" were both mentioned twice as suggestions for improving communication. An interesting recommendation by one interviewee was to have a stronger commitment to involve other stakeholders in stages of the project, and give more authority for those involved in communication.

Coordination amongst stakeholders

Two questions focused on coordination issues, namely the effectiveness of coordination amongst stakeholders and methods to improve coordination:

Effectiveness of coordination amongst stakeholders. Overall, one-half of the interviewees indicated that coordination amongst the stakeholders was "effective" whilst another seven interviewees indicated that it was "effective to some extent". The remaining three interviewees indicated that coordination amongst the stakeholders was "not effective".

Methods to improve coordination amongst stakeholders. Interviewees provided many suggestions with respects to improving coordination amongst stakeholders. The majority of responses fall into the following areas:

- Introduction of coordination regulations/model – mentioned six times;
- The introduction of new technology – mentioned four times (responses included "develop a GIS-based master plan with federal and local governments involved", "introduce e-government in all aspects", "introduce electronic communication" and "merge the stakeholder network via software technology");
- Early involvement of stakeholders –mentioned four times (responses included "clarify project requirements from beginning", "encourage local authorities to share plans at early stages of project", "involve all stakeholders in early stages" and "briefing meetings with the right people");
- Introduction and legalization of coordination committees/introduction of documented procedures – mentioned three times.

Other suggestions included "information exchange amongst agencies", "increasing the level of trust and professional capabilities", "enhancing problem understanding" and "enhancing decision-making and giving authority".

Improving project success indicators

Interviewees were asked to make suggestions as to how the various project success indicators (project quality, time overrun, cost overrun and increasing stakeholders' satisfaction levels) could be improved. These are discussed below.

Improving quality of projects. From the responses given, the following four ways of improving quality of projects are identified:

- Using competent staff/stakeholders – mentioned seven times (responses included "more qualified staff", "qualified contractors", "better consultants and contractors" and "training staff");
- Adoption of standards – mentioned four times (responses included "adopt international standards" and "adopt optimal standards to suit environment";
- Improved coordination – mentioned three times (responses included "better coordination amongst stakeholders" and "good coordination in planning stage");
- Improved design – mentioned twice.

Reducing project time overrun. Five major suggestions identified from the range of responses given by the interviewees are:

- Proper planning and monitoring, and proper design – mentioned eight times;
- Qualified staff/stakeholders – mentioned five times ("more qualified staff", "staff training" and "qualified contractors");
- Increase coordination and cooperation – mentioned four times;
- Correct outsourcing/timely procurement – mentioned four times;
- Increase financial and human resources – mentioned four times.

Reducing project cost overrun. From the responses given, the following four ways of reducing project cost overrun are identified:

- Qualified staff/equipment – mentioned seven times (responses included "proper equipment and staff", "more qualified staff" and "staff training");
- Improved project management – mentioned five times (responses included "meeting time schedule", "minimize interruptions" and "Proper project management");
- Control over materials – mentioned four times ("control material cost", "looking for different material options and designs", and "secure materials ahead of time");
- Enhance design – mentioned four times;
- Coordination with and meeting stakeholder expectations – mentioned four times.

Interestingly, use of value engineering and new construction technologies was each mentioned only once. "Slowing down activities" was also mentioned once as a measure for reducing project cost overrun.

Increasing stakeholder satisfaction. A large number of suggestions were made by the interviewees. From these responses, three major ways for increasing stakeholder satisfaction are identified:

- Improved communication, coordination, support and involvement of stakeholders – mentioned six times (responses included "having better communication & coordination", "better stakeholder involvement in planning and scoping", "getting governments to talk more", "more support from higher levels (the Cabinet)" and "better coordination between clients and municipality");
- Meeting project objectives (budget, quality and time) – mentioned four times;
- Good design – mentioned three times.

Interestingly, addressing safety and environmental issues, and minimizing claims was each mentioned only once by the interviewees.

Discussion

This chapter set out to examine the management of transportation infrastructure projects in the United Arab Emirates. Specifically, it examines issues relating to communication and coordination. Using the stakeholder theory (Freeman 1984), 20 key stakeholders having a high or medium influence on project decision-making were identified and interviewed. The key findings from this study are summarized below.

- A significant proportion of the projects are considered to be unsuccessful and interviewees indentified a range of factors contributing to poor quality, time overrun and cost overrun in transportation infrastructure projects in the UAE. These factors relate to variations in design and schedules; improper planning, scoping or cost estimates; availability and reliability of human resources; availability of materials; and availability of qualified contractors and consultants. These findings confirm the media reports (Ahmed 2007; Anonymous 2005; Nazzal 2005).
- Generally, management practice is considered to be good in the UAE construction industry. Reasons given for poor management relate to unqualified staff, lack of international design and quality standards and lack of a continuous improvement culture.
- The majority of those interviewed consider communication amongst the stakeholders to be "very critical" or "critical" in project success and there was general agreement that it was effective within their organization. However, there was less agreement on the effectiveness of the

communication amongst the stakeholder groups. These results support the findings of Sambasivan and Soon (2007).

- There was a somewhat equal split amongst the 20 interviewees with respect to the effectiveness of coordination amongst the stakeholders – around half of them said it was "effective" and the other half saying it was "effective to some extent" or "not effective". These results also support the earlier findings of Zou et al. (2007)

The factors identified in this study contributing to unsuccessful completion of transportation construction projects represent major risks for the key stakeholders involved. Management of these risks is critical for successful project outcomes. However, these risks must first be clearly understood so that appropriate mitigation strategies can be developed. A risk modeling approach can be undertaken to estimate the project risk associated with the management issue or management aspect namely communication and/or coordination. It is not our intention in this chapter to develop a risk management tool for application in the context of the UAE transportation infrastructure construction industry. However, we provide some guidance for the development of such a tool. The following dimensions should be considered:

Importance of the management aspect. Communication, coordination or any other aspect of management is likely to vary during the various stages of a transportation construction project. That is, the importance of each management issue varies from one construction stage to another. It must also be noted that different stakeholders have different levels of influence on the project as it moves through the different stages. Hence, the risk modeling approach must capture the importance of each management aspect for each project construction stage by each of the key stakeholders.

Effectiveness of the management aspect. This dimension can capture the quality of the management aspect. Again, effectiveness will vary from one construction stage to another. This can be simply categorized as "not effective", "effective" or "very effective". Where aspects of management are not effective, project risk will increase. The effectiveness of each management aspect will be influenced by the use of appropriate technologies (e.g. communication technologies), the level to which authority is delegated and the existence of competent staff at various levels.

Importance of each key stakeholder. As mentioned above, the importance of each stakeholder involved in the transportation construction project needs to be taken into consideration. Figure 18.1 presented earlier in this chapter illustrates the need to consider the "power", "urgency" and "legitimacy" of each key stakeholder and how these influence each stage of the construction project. Hence, a highly influential stakeholder is likely to have a significant impact on the outcomes of the project and this must be taken into consideration in the risk modeling approach.

Specific stage of the construction project. The final dimension to consider is the various stages of the construction project (e.g. planning, design, tendering, scoping, scheduling and implementation). Each stage of a construction project has its unique characteristics and the interaction of the first three dimensions (importance of management aspect, effectiveness of management aspect and stakeholder importance) will play out very differently in each of the construction project stages.

Developing a risk management model incorporating the four dimensions discussed above will require further research. The results of this research should be of significant benefit to the key stakeholders and operational managers in their decision-making concerning transportation infrastructure project management.

It is clear from the findings reported in this chapter that there is considerable opportunity to improve the management of transportation infrastructure projects in the UAE. In addition to the risk modeling approach suggested above, based on the many suggestions made by the interviewees in our study, we also offer the following recommendations for improving the management of transportation infrastructure projects in the UAE:

1 Issues relating to human resources were mentioned several times by a number of the interviewees. Unqualified managers, engineers, contractors and consultants appear to be a major cause of the many problems experienced. In addition, insufficient labour and the different cultural background, experience and skills that they have also present many problems. There is a clear need for a more strategic approach to human resource management and development. A culture of continuous improvement can help overcome many of the issues relating to communication and coordination and effective human resource management is essential in this respect. Attracting, motivating and retaining the best people is always a challenge and this is an area that requires further examination as to exactly what should be done in the context of the UAE.

2 A much more disciplined approach to establishing policies and standardized procedures for project approvals, for accommodating variations necessary after the projects have been initiated and to allocating the appropriate resources for completing each phase of the project. The need for early involvement of key stakeholders is absolutely vital in this respect, with decision-making taking place through appropriate committees.

3 In addition to establishing a culture of continuous improvement, the adoption of international design and quality standards can also help in improving project success, especially improving client satisfaction. Increasing authority to lower levels will also improve the quality and speed of decision-making.

4 Improving communication amongst the key stakeholders through establishing communication committees, more use of emails, weekly meetings and site visits, early definition of project requirements,

frequent presentations, establishing database systems, and electronic knowledge sharing.
5   Creating a policy or standardized procedures to enforce better coordination among stakeholders particularly in early stages of a project lifecycle.

Conclusion

Effective communication and coordination amongst the key stakeholders in construction projects has been identified as a major concern in past research. This chapter adds to the literature on this topic by presenting evidence from transportation infrastructure projects completed in the United Arab Emirates. The key learning from this study are:

- A more strategic approach to human resource management and development is necessary.
- A more disciplined approach to establishing policies and standardized procedures for project approvals is required.
- There is a need to adopt international design and quality standards.
- A variety of means for improving communication amongst the stakeholders is crucial.

References

Ackermann, F. and Eden, C. (2011) "Strategic Management of Stakeholders: Theory and Practice", *Long Range Planning*, Vol.44, pp.179–196.

Ahmed, A. (2007) Article on "Al Ittihad interchange work is delayed, admits contractor" published in Gulf News on the 24 November, 2007 http://archive.gulfnews.com/articles/05/06/09/168413.html.

Anonymous (2004) "Area of United Arab Emirates", *UAE Statistic Book*, 2004.

Anonymous (2004) "Annual Census Book for Emirates of Abu Dhabi – 2004", *Department of Economics and Planning*, 2004.

Anonymous (2005) "60pc of road project completed", *Gulf News*, 15 December, http://archive.gulfnews.com/articles/02/08/19/60952.html.

Banville, C., Landry, M., Marte, J-M. and Boulaire, C. (1998) "A stakeholder approach to MCDA", *System Research*, 15, pp. 15–32.

Bourne, L. and Walker, D.H.T. (2005) "Visualising and mapping stakeholder influence", *Management Decision*, 43 (5), pp. 649–660.

Chen, P. and Partington, D. (2006) "Three conceptual levels of construction project management work", *International Journal of Project Management* 24, pp. 412–421.

Cleland, D. (1995) *Project Management Strategic Design and Implementation* Singapore: McGraw Hill.

Denzin, N.K. and Lincoln, Y.S. (2000) Introduction: "The Discipline and Practice of Qualitative Research" *Handbook of Qualitative Research*, Thousands Oaks, California: Sage Publications.

Ditcham R. (2006) Article on "Nakheel unfazed by project delay" published in Gulf News on the 26 October, 2006 http://archive.gulfnews.com/articles/06/10/25/10077316.html.

Ditcham R. (2007) Article on "Burj Dubai hits snags" published in Gulf News on the 2 January, 2007 http://archive.gulfnews.com/articles/07/02/01/10101017. html.

Elias, A.A., Cavana, R.Y. and Jackson, L.S. (2000) "Linking stakeholder literature and system dynamics: Opportunities for research", *Proceedings of the international conference on systems thinking in management*, Geelong, Australia, pp. 174–179.

Elias, A.A., Cavana, R.Y. and Jackson, L.S. (2002) "Stakeholder analysis for R&D project management", *R&D Management*, 32 (4), pp. 301–310.

Elias, A.A., Jackson L.S. and Cavana, R.Y. (2004) "Changing positions and interests of stakeholders in environmental conflict: A New Zealand transport infrastructure case", *Asia Pacific Viewpoint*, 45 (1), pp. 87–104.

Ellis, R.D. and Thomas, H.R. (2003) "The Root Causes of Delays in Highway Construction", TRB 82nd Annual Meeting, CD-ROM Proceedings.

Ernzen, J., Williams, R. and Brisk, D. (2004) "Design build vs. design – bid – build: Comparing cost and schedule", *TRB 83rd Annual Meeting, CD-ROM*.

Fassin, Y. (2008), "Imperfections and shortcomings of the stakeholder model's graphical representation", *Journal of Business Ethics*, Vol. 80, No. pp. 879–888.

Freeman, R.E. (1984) *Strategic Management: A Stakeholder Approach*. Boston: Pitman Publishing.

Friedman, A.L. and Miles, S. (2002) "Developing stakeholder theory", *Journal of Management Studies*, 39 (1), pp. 1–21.

Gil, N. and Beckman, S. (2009) "Introduction: Infrastructure meets business, building new bridges, mending old ones", *California Management Review*, 51 (2), Winter, pp. 6–29.

Guo, F., Chang-Richards, Y., Wilkinson, S. and Li, T.C. (2014) "Effects of project governance structures on the management of risks in major infrastructure projects: A comparative analysis", *International Journal of Project Management*, Vol. 32, pp. 815–826.

Hegazy, T. (2004) "Execution Planning: A New Innovative Dimension of Infrastructure Management System", TRB 83rd Annual Meeting, CD-ROM proceedings.

Jackson, P. and Klobas, J. (2007) "Building knowledge in projects: A practical application of social constructivism to information system development", *International Journal for Project Management*, doi:10.1016/j.ijproman.2007.05.011.

Jensen, T. and Sandstrom, J. (2011) "Stakeholder theory and globalization: The challenges of power and responsibility", *Organization Studies*: Sage Pub co. www. egonet.org.os.

Kazmi, A. (2005) Article on "Compensation claims ride the boom in construction projects" published in Gulf News on the 24 November. http://archive.gulfnews. com/articles/05/05/10/164180.html.

Kingsley, G., Cochran, J.A., Wolfe, P. and Crocker, J. (2004) "Best Practice in Consultant Management at State Departments of Transportation", TRB 83rd Annual Meeting, CD-ROM Proceedings.

Kovacs, G.L. and Paganelli, P. (2003) "A planning and management infrastructure for large, complex, distributed projects – beyond ERP and SCM", *Computers in Industry*, 51, pp. 165–183.

Kyte, C.A., Perfater, M.A., Haynes, S. and Lee, H.W. (2004) "Developing and Validating a Highway Construction Project Cost Estimation Tools", TRB 83rd Annual Meeting, CD-ROM proceedings.

Lim, G. and Lee, H. (2005) "Formulating strategies for stakeholder management: a case based reasoning approach". *Expert Systems with Applications*, 28, pp. 831–840.

Macharis, C. (2005) "The importance of stakeholder analysis in freight transport" *European Transport*, no. 25–26, pp.114–126.

Mason, P. (2005) "Visual data in applied qualitative research: Lessons from experience", *Qualitative Research*, 5 (3), pp. 325–346.

Mattila, K.G., Gronevelt, R.A. and Bowman, M.R. (2003) "Project Scheduling Accuracy", TRB 82nd Annual Meeting, CD-ROM proceedings.

Mitchell, R.K., Agle, R.R. and Wood, D.J. (1997) "Toward a theory of stakeholder identification and salience: Defining the principle of how and what really counts", *Academy of Management Review*, 22 (4), pp. 853–886.

Morris, R. (1997) *Early Warning Indicators of Corporate Failure: A Critical Review of Previous Research and Further Empirical Evidence*, Ashgate, Aldershot.

Nazzal, N. (2005) Article on "RAK pavement work delayed" published in Gulf News on 5 December, http://archive.gulfnews.com/articles/01/11/12/32271.html.

Newcombe, R. (2003) "From a client to project stakeholders: a stakeholder mapping approach", *Construction Management and Economics*, 21, pp. 841–848.

Nono, P.K. and Tarnoff, P.J. (2004) "Improved software cost and schedule estimation during early project definition and proposal phases with use case modelling and function point analysis", TRB 83rd Annual Meeting, CD-ROM, 2004.

Oliverio, M. (2007) "Agency Theory", Encyclopedia of Business and Finance, 2nd edition. Detroit: Macmillan Reference.

Sambasivan, M. and Soon, Y.W. (2007) "Causes and Effects of Delays in Malaysian Construction Industry", *International Journal of Project Management*, 25, pp. 517–526.

Short, J. and Kopp, A. (2005) "Transport Infrastructure: Investment and Planning. Policy and Research Aspects", *Transport Policy*, 12, pp. 360–367.

David Silverman (1993). *Interpreting Qualitative Data.* 5th Edition. London: Sage Publications Ltd.

Soetanto, R. and Proverbs, D.G. (2002) "Modelling the satisfaction of contractors: the impact of client performance", *Engineering, Construction and Architectural Management*, 9 5/6, pp. 453–465.

Stewart, R.A. (2007) "IT enhanced project management in construction: Pathways to improved performance and strategic competitiveness", *Automation in Construction*, 16, pp. 511–517.

Timmermans, J.S. and Beroggi, G.E.G. (2000) "Conflict resolution in sustainable infrastructure management", *Safety Science*, 35, pp. 175–192.

Turner, R. and Simister, S.J. (2001) "Project contract management and a theory of organization", *International Journal of Project Management*, 19, pp. 457–464.

Vinten, G. (2000) "The stakeholder manager", *Management Decision*, 28 (6), pp. 377–387.

Wood, P. (2006) "Quantitative Research", Faculty of Education, University of Plymouth, UK.

Zaneldin, E. (2005) "Construction Claims in the United Arab Emirates: Causes, Severity and Frequency", Paper presented at the 6th Annual Research Conference, UAE University, April 2005.

Zou, P.X.W., Zhang, G. and Wang, J. (2007) "Understanding the key risks in construction projects in China", *International Journal of Project Management*, 25, pp. 601–614.

Zwikael, O., Shimizu, K. and Globerson, S. (2005) "Cultural differences in project management capabilities: A field study", *International Journal of Project Management*, 23, pp. 454–462.

19 Supply chain risk assessment in automotive new product development

Atanu Chaudhuri

Highlights

- Supply chain risk assessment
- Automotive new product development
- Identifying vulnerable sub-systems
- Failure modes with highest risk.

Introduction

New Product Development (NPD) can be costly and complex when it requires a high degree of component integration or when a new technology is being developed and included in a new product and when the company develops the product with multiple partners (Kim and Wilemon 2003). In fact, increasing technological complexity in the automobiles coupled with a widely dispersed supply chain has increased the supply chain risks for new product development in the automotive industry.

This not only increases risks of delayed launches of products and overshooting the budget for development projects but can also lead to serious quality lapses resulting in recalls. Thus, while automotive original equipment manufacturers (OEMs) continuously strive to reduce overall development time it may result in curtailed problem solving during NPD which may potentially result in product recalls (Bates et al. 2007). Many recalls in the automotive industry have been attributed to the OEM's failure to develop complete understanding of the performance implications of supplier designed and manufactured components and sub-assemblies. For example faulty airbags supplied by a Japanese manufacturer Toyota has led to the recall of almost 17 million vehicles worldwide manufactured by at least ten carmakers including Toyota, General Motors and Honda (Sharman and Inagaki 2014). While recalls may be one of the serious consequences of inadequate risk management in automotive product development, others maybe budget overruns, delays in launches, other quality related problems (not necessarily resulting in recalls). All of these also impact the financial performance of the company. Risk management and particularly those involving suppliers is usually not at the top of mind for automotive executives involved in a NPD programme. But, increasing involvement of suppliers in development and the key role played by them in the eventual commercialization of the product coupled with the risks involved in such involvement make risk management an imperative for automotive NPD projects. But, it is not a trivial exercise and requires a systematic and scientific approach.

Section two demonstrates briefly on the review of recent articles. Section three provides a step-by-step approach to manage risks in NPD projects. Section four presents data analysis and results and last, section five discusses the implications of the proposed approach to risk management for NPD projects and concludes the study.

Literature review

Blackhurst et al. (2008) developed a methodology for calculating part and supplier specific risk indices to analyze critical parts and suppliers. Such indices can be used by the firm to proactively develop risk mitigation strategies to identify potential disruptions before they occur and take corrective actions. Conducting supply chain risk assessment concurrently with product characteristics assessment is essential (Faisal et al. 2006). Such an approach will help in designing supply chains to mitigate product design and supply chain risks and enhance supply chain responsiveness (Khan, Christopher and Creazza 2012). Daultani et al. (2017) used a Bayesian network model to develop an integrated risk modeling approach across multiple functions in a company. Chaudhuri et al. (2013) suggested that firms should assess supply chain risks during new product development and developed methodology to identify the most vulnerable sub-systems and suppliers. Segismundo and Augusto Cauchik Miguel (2008) used Failure Mode Effect Analysis (FMEA) to manage technical risks during the decision-making process in NPD but they did not consider supply chain risks, which need to be considered during NPD. Hence, we can conclude that limited research exists for supply chain risk assessment in NPD.

Method

The four sub-systems of an automobile, which are assessed for vulnerability are powertrain, chassis, control systems including steering and suspension. Supply chain risk assessment during automotive NPD will involve the following steps:

- Identifying the main factors and sub-factors for vulnerability assessment
- Identifying the most vulnerable sub-system
- Identifying the most vulnerable supplier(s)
- Listing and prioritizing failure modes for the most vulnerable suppliers
- Planning for corrective action.

It will first identify the main factors and sub-factors for vulnerability assessment and then determine the most vulnerable sub-system from the point of view of supply chain risks considering the above main and sub-factors. Once the vulnerable sub-system is identified, a similar approach can be followed to identify the most vulnerable supplier or suppliers for that sub-system. A detailed root cause analysis can then be conducted to identify possible failure modes for vulnerable suppliers followed by prioritization of those failure modes. This will finally help in planning for corrective actions for the vulnerable suppliers.

Identifying the most vulnerable sub-system

The factors identified for vulnerability assessment can be degree of supplier involvement, manufacturing process complexity, supply chain complexity and manufacturing capacity. The above main factors and the different sub-factors for each of them are shown in Figure 19.1. Each company performing this exercise can decide to choose the factors most relevant for them. For the purpose of the illustration the sub-factors have been considered as shown in Figure 19.1.

The steps involved in identifying the most vulnerable sub-system are as follows:

- Identify the main factors and the corresponding sub-factors or drivers for risks
- Preparing a pair comparison table for the main factors
- Determining relative weights for the main factors
- Preparing pairwise comparison tables for the different sub-systems on each sub-factor
- Determining vulnerability score of each sub-system on each sub-factor and aggregating those for the respective main factors
- Determining the overall vulnerability score for each sub-system considering the main factors and their relative weights.

The team responsible for conducting the risk assessment exercise can decide to compare the main factors pairwise using a scale for example 1, 3, 5, 7 and 9. On that scale '1' can indicate equal importance, '3' slightly higher importance, '5' moderately higher importance, '7' much higher importance and '9' very high relative importance.

Vulnerability assessment of suppliers

As the next step of the exercise, the vulnerability of the suppliers involved in the two most vulnerable sub-systems will be assessed. If needed, such an exercise can be conducted only for the strategic suppliers using the same approach as demonstrated for the vulnerability assessment of sub-systems.

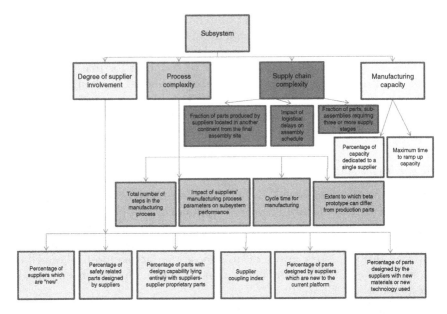

Figure 19.1 A hierarchical structure of parameters for vulnerability analysis of subsystems

For the sake of brevity, the details of the vulnerability assessment for the suppliers are not shown in this chapter.

The OEM and the concerned supplier can then list the failure modes for the supplier, identify the possible effects of the failure modes and assess the severity, frequency of occurrence and possibility of detecting those failure modes using suitable scales. This will help in calculating risk priority number (severity*occurrence*detection). The failure modes with high RPN values can then be prioritized for further root cause analysis and possible corrective actions.

Data analysis and results

First, the pairwise comparison of the main factors involved in the vulnerability assessment of the sub-systems (as shown in Figure 19.1) was conducted.

In the next step the above pairwise comparison matrix is normalized so that the column sums become 1 to generate the normalized pairwise comparison matrix which is shown in Table 19.2 below. Thus 1 divided by the column sum 1.7 in Table 19.1 results in 0.58 in Table 19.2.

In the normalized pairwise comparison matrix, the row sums are added to generate the relative weights for the main factors. Thus, 0.58 + 0.45 + 0.69 + 0.28 equals 1.99 which divided by the corresponding column sum of 4 gives the relative weight of 0.50 for degree of supplier involvement.

Table 19.1 Pairwise comparison matrix for the main factors

	Degree of supplier involvement	Process complexity	Supply chain complexity	Manufacturing capacity
Degree of supplier involvement	1	5	3	5
Process complexity	0.2	1	0.2	5
Supply chain complexity	0.3	5	1	7
Manufacturing capacity	0.2	0.2	0.1	1
Column sum	1.7	11.2	4.3	18

Table 19.2 Normalized pairwise matrix for main factors

	Degree of supplier involvement	Process complexity	Supply chain complexity	Manufacturing capacity	Row sum	weight
Degree of supplier involvement	0.58	0.45	0.69	0.28	1.99	0.50
Process complexity	0.12	0.09	0.05	0.28	0.53	0.13
Supply chain complexity	0.19	0.45	0.23	0.39	1.26	0.31
Manufacturing capacity	0.12	0.02	0.03	0.06	0.22	0.06
Column sum	1	1	1	1	4	

Then pairwise comparison for each sub-system is conducted for each of the sub-factors to generate scores of the sub-systems on each sub-factor for a main factor which are then added to generate scores of each sub-system on each main factor.

Table 19.3 shows that data for each sub-system for each of the sub-factors of 'degree of supplier involvement'. Based on this input, pairwise comparison matrices are generated for each sub-system for each sub-factor. One such pairwise comparison matrix for 'percentage of safety related parts designed by suppliers' is shown in Table 19.4. Following the same procedure as described above for generating relative weights for main-factors, relative weights of sub-systems for 'percentage of safety related parts designed by suppliers' is generated which results in a weight of 0.17 for powertrain, 0.21 for chassis, 0.50 for control systems including steering and 0.13 for suspension. Similar procedure is followed to generate weights for each sub-system on each sub-factor of 'degree of supplier involvement', 'manufacturing process complexity', 'supply chain complexity' and 'manufacturing capacity'. The input data for 'manufacturing process complexity', 'supply chain complexity' and 'manufacturing capacity' are shown in Tables 19.5, 19.6 and 19.7 respectively while the scores for all the sub-systems on the main factors are shown in Table 19.8.

Table 19.3 Input data for pairwise comparison of each sub-system for 'degree of supplier involvement'

	Percentage of safety related parts designed by suppliers	Percentage of parts with design capability lying entirely with suppliers i.e. supplier proprietary parts	Fraction of parts designed by suppliers with new materials or new technology used	Fraction of parts designed by suppliers which are new to the current platform	Coupling index	Percentage of suppliers which are 'new'
Powertrain	0.2	0.3	0.3	0.2	389	8
Chassis	0.25	0.35	0.2	0.1	262	10
Control systems including steering	0.6	0.7	0.5	0.4	271	15
Suspension	0.15	0.45	0.1	0.15	195	5

Table 19.4 Pairwise comparison matrix of sub-systems for 'percentage of safety related parts designed by suppliers'

	Powertrain	Chassis	Control System	Suspension
Powertrain	1	0.80	0.33	1.33
Chassis	1.25	1	0.42	1.67
Control System	3	2.40	1	4
Suspension	0.75	0.60	0.25	1
Sum	6	4.8	2	8

Table 19.5 Input data for pairwise comparison of each sub-system for 'manufacturing process complexity'

	Total number of steps in the manufacturing process	Impact of supplier's manufacturing process parameters on sub-system performance	Overall cycle time for manufacturing (minutes)	Extent to which beta prototypes can differ from production parts
Powertrain	16500	9	98	5
Chassis	10800	3	46	5
Control System	7200	9	15	3
Suspension	9600	5	20	3

Table 19.6 Input data for pairwise comparison of each sub-system for 'supply chain complexity'

	Fraction of parts produced by suppliers in another continent from the final assembly site	Impact of logistical delays on assembly schedule	Fraction of parts, sub-assemblies requiring three or more supply stages or tiers
Powertrain	0.4	7	0.4
Chassis	0.2	9	0.1
Control System	0.3	5	0.15
Suspension	0.2	3	0.2

Table 19.7 Input data for pairwise comparison of each sub-system for 'manufacturing capacity'

	Percentage of capacity dedicated to a single supplier	Maximum time to ramp up capacity
Powertrain	40	16
Chassis	30	10
Control System	40	12
Suspension	60	8

Table 19.8 Scores of sub-systems on main factors

	Degree of supplier involvement	Manufacturing process complexity	Supply chain complexity	Manufacturing capacity
Powertrain	1.40	1.58	1.13	0.58
Chassis	1.24	0.93	0.67	0.39
Control System	2.45	0.78	0.66	0.50
Suspension	0.91	0.71	0.54	0.53

Thus, considering the weights on the main factors shown in Table 19.2, the final vulnerability scores of the sub-systems will be 1.29 for powertrain, 0.97 for chassis, 1.56 for control systems including steering and 0.75 for suspension. A sample calculation is shown below:

$$\text{Vulnerability score of powertrain} = 1.40*0.5+1.58*0.13+1.13*0.31+0.58*0.06=1.29$$

Thus, control systems including steering and powertrain emerge as the two most vulnerable sub-systems for the point of view of supply chain related risk for NPD. Hence the automotive OEM may decide to focus on the suppliers for these two sub-systems to prioritize their efforts for supply chain risk mitigation for their NPD project.

For the sake of brevity, the details of the vulnerability assessment for the suppliers are not shown in this chapter. The OEM and the concerned supplier can list the failure modes for the most vulnerable supplier(s), identify the possible effects of the failure modes and assess the severity, frequency of occurrence and possibility of detecting those failure modes using suitable scales. This will help in calculating risk priority number (severity* occurrence*detection). The failure modes with high RPN values can then be prioritized for further root cause analysis and possible corrective actions.

A sample FMEA conducted for a tier I supplier of drive train is shown in Table 19.9 for illustration. The failure modes with highest RPN values are control softwares for engine, automatic transmission having glitches, supplier lacking knowledge of interfaces with other assemblies/sub-assemblies, impact of drive train components on emission performance not fully understood and increase in raw material costs. A further root-cause analysis can be conducted on the above prioritized failure modes to facilitate identification of mitigation actions. A sample root cause analysis is shown in Figure 19.2 below. The root cause analysis shown is only representative and by no means exhaustive. It may provide an indication of how such root cause analysis for prioritized failure modes can be conducted.

Some mitigation actions for the above identified root causes are shown in Table 19.10 below.

Thus, control software related glitches should be addressed by following suitable project management processes to provide adequate time for

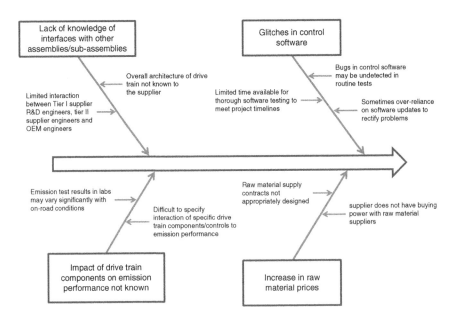

Figure 19.2 Root cause analysis of prioritized failure modes

Table 19.9 Failure mode effect analysis for a drive train component supplier

Failure mode	Failure effect	Severity	Occurrence	Detection	RPN
Manufacturing process not proven	Delay in product launch	5	3	3	45
Specification limits/tolerances not appropriately set	Additional experimentation/increase in costs/delay in product launch	5	3	3	45
Product technology not proven on commercial scale	Additional experimentation/increase in costs/delay in product launch	7	1	5	35
New material not proven	Additional experimentation/increase in costs/delay in product launch	5	3	1	15
Impact of drive train components on emission performance not fully understood	Emission targets not met	9	5	7	315
Control softwares for engine, automatic transmission having glitches	Quality problems, safety issues	9	5	9	405
Increase in raw material cost	Margin reduction	5	7	7	245
Increase in manufacturing cost	Price increase/loss of customers	5	7	5	175
Increasing in logistics cost including inventory	Margin reduction	3	5	3	45
Supplier lacking knowledge of interfaces with other assemblies/sub-assemblies	Quality problems	7	7	7	343
New manufacturing processes required to meet quality requirements taking longer time to prove	Delay in product launch	5	3	5	75
Supplier not aware of lifecycle requirements of products and its variants	Delay in product launch	5	9	3	135
Supplier not able to ramp up capacity	Delay in product launch	5	3	5	75
Logistical difficulties	Delay in product launch	3	3	3	27
Manufacturing process technology not proven	Quality problems	5	3	5	75
Machine reliability issues at supplier	Quality problems	3	1	3	9
Human errors during production process at supplier	Quality problems	3	3	3	27

Table 19.10 Root causes and mitigation action

Root causes	Mitigating action
Limited time available for thorough software testing	Develop overall project plan considering inputs from suppliers regarding software testing time; Follow critical chain project management principles to identify bottlenecks and schedule tasks accordingly
Failure to detect bugs in control software	Focus will be to minimize possibility of bugs and develop appropriate mechanisms to detect them during the testing cycle
Over-reliance on software updates	Minimize reliance on updates. Software updates to be approved only in new versions of the model
Limited interaction between Tier I supplier R&D engineers, tier II supplier engineers and OEM engineers	Organize workshops at project initiation phase and plan regular review meetings; if needed have supplier resident engineers work with OEM engineers for some part of the project
Overall architecture of drive train not known to the supplier	At project initiation phase organize workshops with key suppliers and share overall architecture and how design of individual parts by each supplier will impact overall drive train and the performance of the car
Difficulty in specifying interaction of specific drive train components/controls to emission performance	Experts from OEM and suppliers and also academic researchers can work together can specify the interaction of specific drive train components/controls to emission performance
Varying emission test results between lab and on-road conditions	Emission tests should be conducted under wide ranging on-road conditions and design should be optimized considering performance requirements
Inappropriate raw material supply contracts	Explore opportunities to reduce weight and hence raw material use; design appropriate shared savings contract with supplier
Lack of buying power for supplier for raw materials	OEM collaborating with supplier to enter into joint raw naterial contract with raw material supplier

testing without delaying the project, by focusing on minimizing chances of bugs and resorting to updates only when absolutely necessary and preferably for future models and new versions of the existing models.

Lack of interaction can be addressed by having project initiation workshops with suppliers and regular reviews, and possibility of having resident supplier engineers if needed. Emission performance related risks can be mitigated by forming a team of engineers and researchers, by conducting tests under varying conditions with the focus on optimizing emission performance without sacrificing overall product performance. Risks related to raw material supplies can be addressed by exploring opportunities to reduce material content, designing appropriate contracts and also collaborating with the supplier to enter into joint raw material contracts to enhance overall buying power.

Discussion and Conclusion

This chapter demonstrates a process for identifying the most vulnerable sub-system of an automobile from the point of view of supply chain risks and how causes of vulnerabilities for the most vulnerable supplier can be identified and mitigation actions planned. Insights from such an analysis will provide valuable insights during the development stage of the product so that supply chain risks can be managed.

There can be multiple approaches to address risks in automotive NPD. The choice of the suitable approach will depend on the context. For example if an automotive OEM after identifying the drivers of risk realize that they are all not independent and/or their hierarchy cannot be properly ascertained, the approach outlined to assess the vulnerability may need to be modified by first assessing the relationships between various drivers of risks using an approach like interpretive structural modeling (ISM). ISM may also allow determining the relationships of the risk drivers to NPD and corporate performance measures which will be useful in creating a business case for supply chain risk management in NPD. Once the relationships are ascertained, the company may use an approach like Analytic Network Process (ANP) to identify the most vulnerable sub-system and most vulnerable suppliers and then follow the similar approach as outlined in this chapter.

Implementing the risk management approach will also require a cross-functional approach. A preferred structure will be to make a cross-functional team consisting of people from R&D, manufacturing and supply chain responsible for this exercise with inputs from the development programme leader. If already a cross-functional team is involved in the development programme, some members from that team can work on this exercise. This will also require frequent interactions with the vulnerable suppliers and working closely with them to mitigate the risks. It is also important to assign responsibilities for mitigation and monitoring of those risks to specific individuals and teams with defined expectations for the people who are accountable, the people who need to be consulted and who to be informed regarding the risks.

Future research can be directed at developing a process for collaborative risk management during new product development involving multiple functions within a company as well as suppliers (Chaudhuri and Dani 2015).

References

Bates, H., Holweg, M., Lewis, M. and Oliver, N. (2007). Motor vehicle recalls: Trends, patterns and emerging issues. *Omega*, 35(2), 202–210.

Blackhurst, J. V., Scheibe, K. P. and Johnson, D. J. (2008). Supplier risk assessment and monitoring for the automotive industry. *International Journal of Physical Distribution & Logistics Management*, 38(2), 143–165.

Chaudhuri, A., Mohanty, B. and Singh, K. (2013). Supply chain risk assessment during new product development: a group decision making approach using numeric and linguistic data. *International Journal of Production Research*, 51(10), 2790–2804.

Chaudhuri, A. and Dani, S. (2015). Building a case for collaborative risk management with visibility of risks across supply networks: Investigating the effect of supply network characteristics. 20th International Symposium on Logistics, University of Bologna, July 5–8.

Daultani, Y., Goswami, M., Vaidya, O. S. and Kumar, S. (2017). Inclusive risk modeling for manufacturing firms: A Bayesian network approach. *Journal of Intelligent Manufacturing*, 1–15.

Faisal, M. N., Banwet, D. K. and Shankar, R. (2006). Mapping supply chains on risk and customer sensitivity dimensions. *Industrial Management & Data Systems*, 106(6), 878–895.

Khan, O., Christopher, M. and Creazza, A. (2012). Aligning product design with the supply chain: A case study. *Supply Chain Management: An International Journal*, 17(3), 323–336.

Kim, J. and Wilemon, D. (2003). Sources and assessment of complexity in NPD projects. *R&D Management*, 33(1), 15–30.

Segismundo, A. and Augusto Cauchick Miguel, P. (2008). Failure mode and effects analysis (FMEA) in the context of risk management in new product development: A case study in an automotive company. *International Journal of Quality & Reliability Management*, 25(9), 899–912.

20 To investigate the feasibility of predicting, identifying and mitigating latent system failures in a UK NHS paediatric hospital

Anthony Sinclair

Theoretical background

"Errors are planned actions that fail to achieve their desired consequences without the intervention of some chance or unforeseen agency. Error types include slips and lapses, where the actions do not go according to plan and mistakes where the plan itself is inadequate to achieve its objectives".[1] In addition to error types, which are performance related, there are error forms, which are evident at all levels of human performance and are a product of the cognitive processes of long-term memory. Reason's *generic error modeling* system or GEMS, itself derived from Rasmussen's skill-rule-knowledge classification of human performance.[2] Rasmussen (1983) presents an integrated picture of error mechanisms operating at the three levels of performance:

1   Skill-based performance is sensory-motor in nature, the ability to correlate visual stimuli, motor coordination with space and depth perception to carry out pre-determined tasks, that take place "without conscious control as smooth, automated and highly integrated patterns of behaviour".[3] These are automated routine actions requiring little conscious attention.[4]
2   Rule-based activities would include accuracy checking dispensed medicines. The medicine has been assembled and labeled and finally requires an operative to check that the correct medicine has been dispensed for the selected patient. Rule-based activities require some conscious cognitive activity and some automated behavior. It takes the form of, *if* this happens *then* take this action.
3   Knowledge-based activities have no routines or rules available for their solution and require active conscious thought to derive a solution.

Groenoweg[5] described three stages that are traversed in order to bring about intentional actions. These are

1   Planning the action
2   Mental storage of the action in the memory and
3   Execution of the action.

Errors can occur at each of these stages generating planning mistakes, storage lapses or execution slips. Groeneweg (2002) goes on to explain that slips and lapses may occur as a result of a lack of attention, for example owing to an internal or external distraction, or because of over attention (too many or prolonged checks), which often occurs after a period of reduced attention and double capture slips, where the control of an action is captured by a stronger habitual sequence. Perceptual confusion errors often cause slips resulting in the correct action being carried out on the wrong but similar-looking object. Two other error types also have relevance and these are reduced intentionality, where there is a delay between formulation of an action and execution of the action as a result of a failure in prospective memory. Environmental capture refers to a situation where something in the surroundings distracts an operator from their intended task, for example a telephone ringing, resulting in the failure to carry out the intended action.

Errors at the skill-based level are mainly due to monitoring failures as a result of inattention to the task in hand. Rule-based activities are solved by applying rules, that is an "if . . . then" problem solving approach and errors arise from the application of *bad* rules or the *misapplication* of good rules. Knowledge-based mistakes arise from the concept of *bounded rationality*, a term coined by Herbert Simon,[6] that is, the rationality of individuals is limited to the information they have at their disposal, the cognitive limitations of their minds and the time they have available to solve the problem. In particular, where knowledge available for the required cognitive processing is inadequate or missing it is known as cognitive under specification.

In addition to the three error types, described above, error forms occur at all performance levels and originate in all cognitive processes and are explained later in this paragraph (Reason 1990). He argued that the area in which cognitive functions are carried out in the brain is an area called the *working memory*. In order to function, the working memory draws knowledge from our long-term memory or knowledge base. To retrieve information appropriate to a particular problem-solving situation, two heuristics are employed. These are *similarity-matching*, using the degree of likeness between events or objects, and the second is known as *frequency gambling*, that is selecting events or objects that have occurred before; these are known as cognitive primitives and are thought to process information automatically without conscious effort. When cognitive operations are *underspecified*, that is insufficient information is available to carry out a task, memory defaults to the most appropriate response that has occurred most often previously. A lack of attention to the task in hand, inadequate processes or failure to comply with existing processes, insufficient, incomplete or inadequate knowledge or information all contribute to error causation.

Error types and error forms are only part of the error causation picture. In the accident causation model known as the *Swiss Cheese* model developed by Reason (1990), and used widely in the NHS in the UK, there is a chance that a barrier or defense in the dispensing process will identify the error before the medicine reaches the patient. There is also a chance that the error won't be detected by the barrier and will pass through to cause an accident.

Each slice of cheese represents a barrier or defense put in place to prevent an accident from occurring. The holes represent weaknesses in each barrier but what causes error types and error forms to occur?

One error type can give rise to multiple substandard acts (errors) that may not be repeated in the same way or same order again. However, controlling the effects of Basic Risk Factors (BRFs) would prevent substandard acts from occurring. There are 11 BRFs.[7] These are outlined in Table 20.1.

Table 20.1 Basic risk factors (BRFs) – Groeneweg (2002)

Basic risk factor	Description
Design	Ergonomically poor design of tools, equipment, offices (not user-friendly).
Hardware	Poor quality, condition, suitability or availability of materials: tools, equipment and components.
Maintenance	None or inadequate performance of maintenance tasks and repairs.
Housekeeping	None or insufficient attention given to keeping the workplace clean or tidy.
Error Enforcing Conditions	Unsuitable physical conditions and other influences that have a disadvantageous effect on human functioning.
Procedures	Inadequate quality, insufficient availability of procedures, instructions and manuals.
Training	None or insufficient competence or experience among employees (not sufficiently suited/inadequately trained).
Communication	None or ineffective communication between the various locations, departments or employees of a company, or with the official bodies.
Incompatible Goals	The situation in which employees must choose between optimal working methods according to the established rules on one hand, and the pursuit of production, financial, political, social or individual goals on the other.
Organisation	Shortcomings in the organisation's structure, philosophy, processes or management strategies, resulting in inadequate or ineffective management of the hospital department or clinical area.
Defence	None or insufficient protection of people, material and environment against the consequences of the operational disturbances

Interruptions

Noise or disruption and interruptions merit attention. Boehm-Davis[8] defines an interruption as the suspension of one stream of work prior to completion, with the intent of returning to and completing the original stream of work and a disruption being a momentary lapse of attention on the primary task without a requirement to engage in the distracting task.

Research has shown that interruptions consume about 28% of the knowledge worker's day.[9] It was found that interrupted work environments in which complex intellective tasks are performed leads to lower quality decisions and decreased efficiency. Furthermore, even helpful interruptions that are those that facilitated the completion of simple tasks, were perceived negatively by decision makers. Interruptions are disruptive, just how disruptive depends upon several factors that Trafton[10] summarises as interruption complexity; similarity of the interrupting task to the primary task, how closely the interrupting and primary tasks are related, control over interruption engagement, and the availability of retrieval cues in the primary task.

To recover well from an interruption requires pre-planning and preparation. Boehm et al. (2009) outline two such approaches. The first is aimed at decreasing the amount of information that needs to be remembered upon resumption of the task, by breaking from the task in hand at a suitable boundary or stage in the task enabling the operative to resume work at the next stage or step in their task process. The second strategy is to increase the probability of recalling the origin task (prospective memory), for example by using approaches to increase internal memory (such as rehearsal) or through the creation of *external memory*, such as leaving notes.

Distractions are commonplace in healthcare environments and have been identified as a major contributor to medication errors. The Medmarx report[11] lists distractions as the number one contributing factor in error causation in each of its data reports published between 2002 and 2006. There have not been many studies that have looked at methods to reduce distractions in hospitals.[12] This is not the case in the aerospace industry however, where accident investigation of a number of airline accidents by the National Transportation Safety Board, USA, such as the Eastern Airlines flight 212, concluded that distractions were a significant error causation factor, and hence the Federal Aviation Authority Sterile cockpit rule (1981) (Flight crewmember duties, rule FAR 121.542 / FAR 135.100—Flight)[13] Section (b) of which states that

> no flight crewmember may engage in, nor may any pilot in command permit, any activity during a critical phase of flight which could distract any flight crewmember from the performance of his or her duties or which could interfere in any way with the proper conduct of those duties. Activities such as eating meals, engaging in nonessential

conversations within the cockpit and nonessential communications between the cabin and cockpit crews, and reading publications not related to the proper conduct of the flight are not required for the safe operation of the aircraft.

This *proscribes* both direct distractions, i.e. conversations, or indirect distraction i.e. a colleague eating food, from being made during critical flight activities. In addition to leading to increased error rates, distractions also have been found to increase anxiety and in situations of high workload situations, decrease performance (Speier 1999).

A number of theories and mechanisms have been put forward to explain why distractions are so disruptive and to suggest strategies to mitigate the effects of such disruptions. Research carried out by Oulasvirta and Saariluoma[14] suggests that interruptions are disruptive not because short-term memory becomes overloaded, but rather because information stored in the long-term working memory is poorly encoded. Trafton and Monk[15] outline factors that influence recovery from interruptions. The length of the interruption, similarity between the primary and secondary task, that is the task the operator has been called away to perform, the opportunity to rehearse prior to resuming the primary or main task and the existence of environmental cues, for example a clear marker being left as to where to resume, all affect task resumption performance. Significantly, training in the primary task itself is not by itself sufficient to reduce the impact of task disruption. In addition, training in *how to resume* has been found to be beneficial.[16]

Multiple task management

Another issue to consider is how the brain manages input presented to it simultaneously. Pashler[17] outlines three possible models, which are capacity sharing, bottleneck or task-switching, and cross-talk models. The first model suggests that people have a finite processing capacity which is shared out between tasks and people carry out multiple tasks quite ably until one task becomes more difficult than the others. When this happens, capacity is switched to the more difficult task at the expense of the less difficult tasks. The bottleneck or task-switching model suggests that parallel processing may be possible until two or more tasks require access to the same processing mechanism or mental operation at the same time then a bottleneck results and one or more of the tasks will become delayed. The third model is that interference may arise when multiple tasks require the same sensory inputs in order to be processed, it could be easier to perform more than one task if they both require the same sensory inputs or harder if they require the same sensory inputs. A further complication is that it is thought to be more difficult to perform multiple tasks if they require the same sensory inputs.

Speier[18] outlines two types of interference. Capacity interference is when the number of incoming cues are too numerous for a decision maker to process and structural interference occurs when a decision maker must attend to two inputs that require the same physiological mechanisms, such as two different visual signals. Rubinstein et al.[19] indicate that multi-tasking may be less efficient if a person is switching from a familiar task to an unfamiliar task.

Memory

This leads on to a consideration of memory overload and memory capacity. Baddeley and Hitch[20] proposed a three-part model consisting of a central executive that is thought to control access to the subsidiary elements of the working memory system. The other two parts of the theory were a phonological loop that stores speech-based sounds for a limited period and is itself thought to be made up of two parts: the phonological store can store sound coded items and an articulatory control process that allows sub vocal repetition of the items stored in the phonological store.[21] The second subsidiary part is called the visuospatial sketch pad that stores visual spatial information, that is thought to set up and manipulate mental images. The visuospatial sketch pad also controls attention and processing and organizing information. Baddeley added a third component in 2000,[22] called the episodic buffer, responsible for integrating and manipulating material, which again has limited capacity and depends heavily on executive processing.

The relevance to this discussion is that working memory also known as short term memory has a limited capacity and is affected by a number of factors such as intelligence, age, lack of sleep, physical fitness, anxiety, emotions and stress and by an appreciation of how individuals are made up and react to external stimuli will inform error avoidance or prediction heuristics.

Professionalism and non-professionalism

Professionalism and non-professionalism are another layer that may impact on an individual's approach to their work and sense of responsibility and how they discharge their duties. A professional is someone who has trained for and is committed to a profession or a vocation and that profession is governed by a set of beliefs or a code of conduct. One basic difference between professional and non-professional is that the former may *down tools* promptly at the end of their shift whereas a professional would ensure that patients come first and would work longer to complete a task for the benefit of a patient.

Quality management systems (QMS) and visual prompts

Quality management systems are commonplace in industry but less so in healthcare, although this is changing as health services come to terms with the economy and the need to provide high quality services by means of productive and efficient organisation.[23] One of the strategies that industry have deployed to achieve greater productivity have been the implementation of QMS and the use of tools such as lean management, an improvement approach to improve flow and eliminate waste that was developed by Toyota.[24] Visual prompts have a long history in safety strategies in situations where operatives are required to manage complex systems that require multitasking in stressful situations.

The classic example is from the aviation industry where the first checklist was created and introduced in 1935 on October 30 at Wright field, Dayton Ohio, USA, where the United States army were conducting an evaluation of aircraft. One of the contenders was a Boeing 299 which took off, began a smooth climb and then stalled and crashed, both pilots later tragically died of their injuries. The pilots were unfamiliar with the aircraft and the aircraft was too complex for one person to remember everything that had to be done. Later it was determined that pilots needed some way of ensuring that they didn't overlook anything, and the checklist was born.[25] Checklists were introduced on a global scale by the work of Dr Atul Gwande and his team on behalf of the World Health Organization (WHO), Gwande In his book, "The Checklist Manifesto",[26] wrote that "failure in the modern world is really about errors of ineptitude, that is mistakes made because we don't make proper use of what we know. The routine tasks of surgeons, for example, have now become so incredibly complicated that mistakes of one kind or another are virtually inevitable".

The elements that have been incorporated into this study therefore are an appreciation of error causation, more specifically relevant basic risk factors (BRFs) that will give insight into the impact of the environment and processes on individuals. An understanding of the way in which individuals process information needs to be considered. Memory and memory capacity together with multi external stimuli capabilities of the brain form another element to this work. The impact of professionalism and non-professionalism is a factor and finally the benefit of quality management systems, lean management and the use of visual prompts complete the research strategy for this thesis and ought to show that it is feasible to identify latent errors in a particular working environment.

Literature review

The literature review searched for studies that had examined the impact of introducing risk and error reducing systems that were particularly mindful of known error causation factors such as stress and interruptions. The literature review also took into account papers that examined or referenced medicines management policies from paediatric hospitals that considered approaches to safety including error reduction strategies including utilizing teamwork and considering human factors, deploying visual aids and error reducing protocols, all within the given context of paediatric hospital pharmacies within the National Health Service in England. In other words, research that demonstrated learning from either their own errors or those that occurred within a similar environment and or other high-risk industries and have as a consequence implemented quality management systems to reduce risk learning from the errors that were made.

1   The impact of introducing risk and error reducing systems that were particularly mindful of known error causation factors such as stress and interruptions.

The search initially found 15,072 citations that were reduced to 11,446 after removal of duplicates. Further refinement based upon the exclusion/inclusion criteria reduced the citations to 746 potentially relevant articles that were healthcare safety systems interventions of which 16 articles matched both the topic and a paediatric setting.

2   Using workload as a proxy measure for stress.

Articles that discussed stress in the context of a paediatric hospital pharmacy and the impact, if any, that this may have had on efficiency, safety and professionalism.
    This produced 69 results of which four were considered to have some relevance.

3   Interruptions to the dispensing process.

This produced 53 articles of which two articles were considered to be relevant and one was of interest to review.

4   Medicines management policies from paediatric hospitals.

This section reviewed medicines management policies for error reduction strategies that included human factors and or quality management underpinning.
    This search yielded 52 citations of which three were deemed to be relevant, although none of the resulting articles referred to terminology or principles of human factors, teamwork or flattened hierarchies.

5   Quality systems and checklists

The impact on safety and efficiency of introducing a quality systems approach into one process on a ward area and this was the preparation and administration of IV injections. No relevant articles were found.

Gaps in the literature

Articles described the benefits of safety cultures and encouraging error reporting and lean management systems. None explicitly identified error causation at a root cause level. Working leaner improves efficiency but doesn't necessarily improve safety.

Of the articles describing stress related research, none were directly relevant to the research questions of this investigation. Business was identified as a causative factor in stress causation but not how to mitigate this.

One article described the impact of distraction on the prescribing not the dispensing process.

Medicines management policy literature research yielded articles that identified the importance of communication in reducing errors. Adherence to standard operating procedures (SoPs) by nurses was also examined, but interestingly the quality, relevance and usefulness of the SoPs were not. Another team looked at the benefit of introducing quiet zones supported by visual aids such as signs and instructions. The introduction of do not disturb red tabards were also investigated by one team in which interruptions were accepted as a factor in error causation but current practice was *adjusted* rather than being *rethought.*

Other teams examined the introduction of robotic dispensing units and another nursing systems and the impact that they may have on error causation.

In total 11896 citations were found and after further review and refinement this number was reduced to 34 citations for discussion.

Discussion

The rationale for the search strategy adopted, was to discover what the strategic approach to learning from errors in the healthcare system (UK). Then to develop a proactive rather than reactive approach to reducing the likelihood of the same type of error from reoccurring, that is predicting errors before they took place. Included in this initial literature search were sub topics, for example communication failures as a particular incident grouping; work on incident reporting itself; the benefit of standardizing procedures, protocols and medicine strengths; and introducing and assessing a safety culture.

The investigation began by examining the impact that the environment had on operatives working in the defined setting. Were systems such as workflows inefficiently designed? Might an operative's emotional state affect their perception of the environment to reduce their operating capacity as much as if the conditions were actually suboptimal.

How might operatives function under stress inducing conditions? Workload capacity is often discussed quite reasonably in many working environments and under many systems and an optimal workload capacity is defined. Not so in pharmacy, where workload is generally managed reactively rather than on the basis of capacity, safety and efficiency. The question asked was how increasing an individual's stress levels affected their approach to their tasks. Did this in turn increase or decrease the potential for errors to occur?

Error causation and error prevention strategies are well developed in many high risk industries, the nuclear industry, the oil industry, shipping, pharmaceuticals and classically amongst airlines, but less so within healthcare. These industries have developed sophisticated Quality Management Systems (QMS), processes are developed that are well designed, efficient and safe and continually performance monitored. SoPs are well written and training is appropriate together with continuous improvement embedded into their cultures. Teamwork and flattened hierarchies are commonplace and the use of visual prompts and other safety strategies ubiquitous. The literature review indicated that there is a gap in understanding in being able to predict error causation in the context of a healthcare environment.

Investigation of the impact of workload (proxy measure for stress) impacts on work performance

The aim of this study was to investigate how environmental factors such as workload affect the performance of pharmacists and technicians carrying out the final check of prescriptions. An insight into factors that may affect the accuracy of this process could help to prospectively identify the causes of errors in this population, and thus reduce the frequency of their occurrence. We investigated how dispensary workload affects: the time taken for pharmacists and technicians to perform the final accuracy check; the number of process steps carried out during checking; the number of purely safety checks carried out during checking.

Method

A non-participant, direct observation study of the final accuracy checking stage of the dispensing process was undertaken. Pharmacists (n=9) and technicians (n=9) were each observed for a 30-minute period, and their individual actions during the checking process were recorded and timed. Workload data was obtained from dispensary computer records, based on the number of items dispensed in a 30-minute period. The data was tabulated, linear regression analysis was carried out, and the standard error of the mean (SEM) was calculated for each indicator. A number of interviews, questionnaires and focus groups were then used to ascertain: the number of process steps (e.g. printing the label) and safety checks (e.g. checking the patient's name) staff members thought should be carried out to check a simple prescription (using a basic system analysis technique to compile a process map); staff members' opinions on the dispensary environment and workload; any comments on errors made during the checking process.

Separate focus groups were held for pharmacists and technicians to encourage uninhibited responses.

Results

Three data sets were obtained from the study, showing the impact of increasing workload on: the *time* taken to check items; the number of *process steps* used; and the number of *safety checks* carried out Table 20.2.

Table 20.2 The average time taken to check an item (pharmacist and technician groups) and the total work activity (number of items) for each 30-minute observation period

Technician	items per half hour	*Time per Prescription (mins)	Pharmacist	items per half hour	*Time per Prescription (mins)
1	36	3.8	10	38	4.4
2	24	5.0	11	29	3.8
3	22	6.7	12	27	2.3
4	20	6.0	13	22	3.2
5	19	3.8	14	21	3.4
6	15	10.0	15	20	3.3
7	12	7.7	16	19	3.1
8	11	7.5	17	17	3
9	11	4.3	18	14	2.9

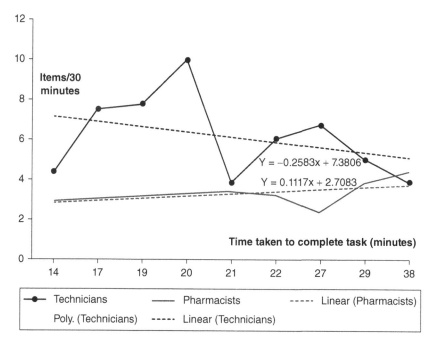

Graph 20.1 Shows the average time each participant took to accuracy check an item and how this varied as work activity increases within the dispensary (data is in Table 20.2)

In the technician group, the time spent checking an item decreased as workload increased. See Graph 20.1.

The mean time taken to check one prescription was 6.09 minutes (SEM ±1.4). In the pharmacist group the time taken to check items increased as workload increased. The mean time taken to check one prescription for the pharmacist group was 3.27 (SEM ± 0.4). As workload increased, the number

Table 20.3 The average number of steps* a technician checker and pharmacist checker makes per item checked and items passing through the dispensary in a half hour period as a measure of work activity (data for graph 20.2)

Pharmacist	average steps	items per half hour	Technician	items per half hour	average steps
11	35.8	38	1	36	21.0
12	35.8	29	2	24	7.0
13	22.1	27	3	22	26.0
14	23.3	22	4	20	24.4
15	22.2	21	5	19	21.7
16	42.5	20	6	15	90.0
17	26.9	19	7	12	11.0
18	28.7	17	9	11	23.7
19	28.9	14	8	11	26.0

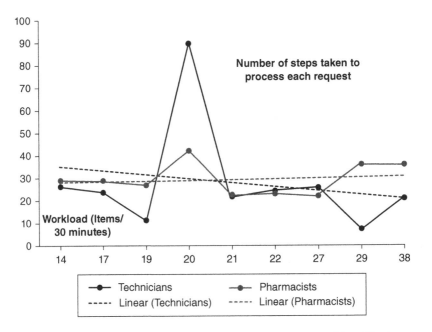

Graph 20.2 Workload of dispensary (number of items processed in a 30 minute period) vs steps taken to process each request (data is in Table 20.3)

of process steps taken to check an item reduced in the technician group but increased in the pharmacist group.

The mean number of steps was 27.9 (SEM ± 16.1) for the technician group and 29.6 (SEM ± 4.72) for the pharmacist group.

Safety steps as workload increased, the number of safety checks carried out by the technicians increased. The reverse was seen in the pharmacist

Table 20.4 Safety checks carried out and items passing through the dispensary per half hour. The mean for the Pharmacist group is 63.7 +/- 10.5 (SEM) and for the Technician group is 45.9 +/- 17.1.(SEM) [data for graph 20.3]

Technicians	Safety checks	items per half hour	Pharmacists	Safety Checks	items per half hour
1	63	36	10	75	38
2	19	24	11	37	29
3	30	22	12	72	27
4	54	20	13	40	22
5	49	19	14	57	21
6	90	15	15	68	20
7	3	12	16	82	19
9	57	11	17	69	17
8	48	11	18	73	14

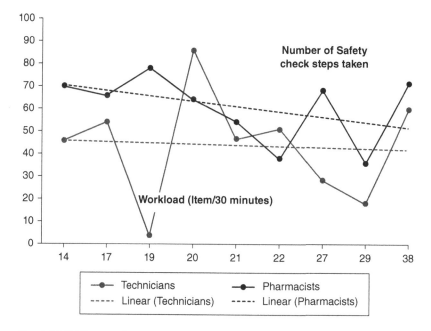

Graph 20.3 The number of steps that could be said to have safety implications* vs workload (the number of items passing through the dispensary in a 30 minute period) (data is in table 20.4)

group. The mean number of safety checks was 45.9 (SEM ± 17.1) for the technician group and 63.7 (SEM ± 10.5) for the pharmacist group.

Other environmental factors the focus groups and questionnaires revealed were that pharmacists and technicians felt more affected by the working environment (e.g the dispensary temperature) when workload was higher. Respondents reported feeling stressed in one in three dispensary sessions.

Those who reported making checking errors were more likely to report feeling stressed because of the working environment. Other environmental issues that staff reported as having a negative effect on their work included: pressure to work faster, noise, distractions, poor workplace design and poor paperwork design. All of these are recognised risk factors for errors.

Discussion

Time The results of the impact of workload on checking time were unexpected. The observation that as workload increased, technicians spent less time checking prescriptions, but pharmacists spent more time, could imply that technicians are more task-orientated, while pharmacists are concentrating on other priorities and responsibilities, such as safety. Future work could examine the link, if any, between education, training and professional culture and the approach taken to dispensary work. **Process steps** The above trend was repeated in the data set relating to the number of steps in the checking process. As workload increased, pharmacists tended not only to slow down, but to carry out more steps in the checking process. An interesting observation is that many of these steps were repeat steps, for example when a pharmacist double-checked an element of the prescription following an interruption. During the focus sessions, both pharmacists and technicians said that the average number of steps it would take to check a simple prescription was five or six. In practice, the number of steps taken averaged 27, because of double and triple checks.

Safety checks The number of safety checks increased for the technician group when workload increased. Reasons for this are unclear, although it may be related to recovery from interruptions. Although the number of safety checks undertaken by pharmacists decreased when workload increased, the total number of safety checks remained higher than that for the technician group, as seen in Table 20.2.

Conclusion

Although this is a small-scale study, we have identified that increased workload increases the time taken for pharmacists to check prescriptions, and the number of steps in the checking process, but decreases the time and number of steps taken by technicians. The reasons for this warrant further investigation. The number of safety checks carried out by pharmacists decreased as workload increased, and the reverse was true for technicians. However, it should be noted that the number of safety checks carried out by

Table 20.5 Summary of questionnaire results, of how an individual perceived the dispensary environment, that is, how it was impacting on them during a particular dispensary session

Number of completed forms* n=34	Did they feel challenged or out of their depth in that session	Noise — Were noise levels distracting	Temperature — Did the dispensary feel too cold or too hot	Lighting — Was the lighting harsh	Did the work place appear to be cluttered	Did they feel stressed in that session	Did they make an error during that session
Busy							
20	13 / 65%	8 / 40%	9 / 45%	5 / 25%	14 / 70%	10 / 50%	4 / 20%
Not busy							
14	3 / 21%	2 / 14%	5 / 36%	1 / 7%	2 / 14%	2 / 14%	1 / 7%
Stressed							
12	7 / 58%	6 / 50%	7 / 58%	2 / 17%	9 / 75%	-	5 / 42%
Not Stressed							
22	9 / 41%	4 / 18%	7 / 32%	3 / 14%	7 / 32%	-	0 / 0%

pharmacists was already at a high level. Environmental factors including dispensary noise, interruptions and design clearly affect staff performance.[27]

The impact of interruptions on individuals in error causation

Interruptions adversely affect task performance and other factors including stress and professionalism also would appear to adversely affects work performance.[28]

This research set out to measure the effect that interruptions have on dispensary accuracy checkers. In the UK Pharmacists are as a minimum masters level educated and are regulated by the General Pharmaceutical Council (GPhC). Pharmacy technicians became regulated by the GPhC on 30 June 2011 but were not so at the time of this observation. A technician requires an NVQ level 3, obtained through a two year day release programme. Qualified pharmacy technicians can then go on to train to become accuracy checkers.

Materials and methods

The study instrument was non-participant, direct observation of a discrete, clearly identifiable step within the dispensing process. A prescription requiring two containers of medicines to be dispensed was created. The medicines were labelled and placed in a tray together with the required paperwork, additional spoons or oral syringes and a dispensing bag. The operatives, both pharmacists and pharmacy technicians, volunteered to participate and were told that they would be timed accuracy checking a prescription and that the object of the observations was to measure the effect of the environment on their work only and that they were not being assessed. The prescriptions were non-complex, and did not contain any deliberate errors. Each observation (n=34) consisted of two arms as determined by a Latin square.

The observations were undertaken in the dispensary and in the quiet environment of an office. A designed interruption was introduced into some of the variants. Operatives were asked to complete a brief questionnaire relating to their work environment, workloads, personal focus, workspace, available equipment, and references and knowledge base for the work assigned.

Results

A statistical analysis of variables was carried out using Minitab. A calculation of least squared means for time showed that individuals were 28.41% less efficient when interrupted in the dispensary. The mean time taken to accuracy check the standard prescription increased from 121.20 seconds to 155.63.

The questionnaires were completed during the four-month observational period. The results showed that 20 questionnaires (n=34) were completed after a *busy* checking and 14 after a *quiet* dispensary session. Respondents were more likely to feel challenged in a busy session (81%) than in a less busy session (19%). They were more likely to find noise levels distracting in a busy session (40%) than in a non-busy session (14%). The dispensary temperature was more likely to be perceived as too hot or too cold in a busy session (45%) than a non-busy session (36%); the lighting to be too harsh (25%) compared with (7%); the dispensary appeared to be cluttered (70%) to (14%) and to feel stressed (50%) as compared with (14%). Participants recorded that they felt stressed during 12 of the 34 occasions that a form was completed (35.3%) and errors were made and noted on five questionnaires (41.7%). No errors were noted on questionnaires completed when the participants did not feel stressed.

Conclusion

Interruptions adversely affected dispensary accuracy checkers. Stress is an error forcing condition and interestingly a clear difference between professional and non-professional worker attitude was observable.

The time differences were calculated net of any interruption that was made.

Statistics

The data was analysed in Minitab using a general linear model that analyses balanced or unbalanced ANOVA (analysis of variance) models with crossed or nested and fixed or random factors. Variance was tested against length of interruption, location, individuals and order in which events took place. There were 44 timed events, carried out by 12 individuals in two locations. Although all the variants had an impact the two that were most significant were the length of the interruption (p=0.012) and the individual (p=0.001).

An analysis of variance for time, using adjusted sum of the squares (SS) for tests was carried out. Adjusted SS is used in general linear models without orthogonality that is where the factors are not completely independent of each other. The total variation was 69044.5 and the variable that caused the greatest variation was the individual (p=0.001). The order of location, that is the sequence in which the tests were carried out, whether in the dispensary first and then the office and so on, was shown to be insignificant (p=0.140).

A plot of least square means for time showed that it takes 22% longer (155.63-121.2) when interrupted in the dispensary than in the office.

Discussion

The most significant outcome of the observations was that between the two professional groups. The only task that was required of them was to accuracy check the dispensed prescription.

Table 20.6 Observed times for each variant by individual subject; the date on which the observation took place and, which variant was being observed. Key: Disp= Dispensary; Di = Dispensary with a controlled interruption; Oi = Office with a controlled interruption; Office=Office

Ref	Operative	Date	Time	Variant	Disp	Disp_i	Office	Oi	Interruption	net	% change	Disp	Di	Office	Oi
1	R	30/04/2008		9	**215.73**	–	128.16	119.64	8.52		93.35%	1		2	3
2	KM	07/05/2008		9	**129.92**	–	106.07	125.67	19.6		118.48%	1	–	2	3
3	MM	05/05/2010	12:12	9	**166.73**	–	113.95	126.2	12.25		110.75%	1	–	2	3
4	SM	10/05/2010		5	–	168.01	113.48	–	6.18		148.05%	–	1&2	3	–
5	R	12/05/2010		5	–	212.92	155.39	–	8.44		137.02%	1&2	–	3	–
6	KM	17/05/2010	10:30	5	–	122.8	118.51	–	4.56		103.62%	–	2	1	–
7	SO	20/05/2010		8	–	200.36	153.07	–	24.12		130.89%	–	2	1	–
8	DB	20/05/2010		8	–	120.06	106.7	–	14.94		112.52%	–	2	1	–
9	R	03/09/2010	12:08	12	–	–	130.2	164.41	34.21	145.82	112.00%	–	–	2	1
10	RS	03/09/2010	12:25	12	–	–	97.22	139.73	10.38	129.35	133.05%	–	–	2	1
11	SM	03/09/2010	12:36	12	–	–	105.32	123.54	19.18	104.36	99.09%	–	–	2	1
12	MM	03/09/2010	12:50	5	–	120.08	98.2	–	22.33	107.75	109.73%	–	1	2	–
13	DR	03/09/2010	13:00	4	96.92	127.07	–	–	17.88	109.19	112.66%	2	1	–	–
14	CA	04/11/2010	12:15	4	225	264.67	–	–	15	249.67	110.96%	2	1	–	–
15	DR	04/11/2010	12:45	3	73.01	–	–	97.67	16.86	80.81	110.68%	1	–	–	2
16	SO	04/11/2010	13:15	4	119.45	195.67	–	–	40.95	154.72	129.53%	2	1	–	–
17	SO	22/11/2010	11:05	3	112.95	–	–	99.73	23.24	76.49	67.72%	1	–	–	2

#		Date	Time												
18	AH	22/11/2010	11:16	3	134.82	–	–	138.8	20.01	118.79	88.11%	1	–	–	2
19	IP	22/11/2010	11:28	2	75.86	–	65.61	–	–	–	115.62%	1	–	2	–
20	DB	22/11/2010	11:35	2	94.7	–	81.09	–	30	121.61	116.78%	1	–	2	–
21	HU	22/11/2010	11:45	2	151.61	–	74.09	–	21.7	111.07	164.14%	1	–	2	–
22	JA	22/11/2010	11:55	1	143.67	132.77	–	–	16.35	117.04	77.31%	1	2	–	–
23	SC	22/11/2010	12:10	1	111.98	133.39	–	–	16.35	117.04	104.52%	1	2	–	–
24	TG	22/11/2010	12:30	1	107.77	127.2	–	–	25	130.77	108.42%	1	2	–	–
25	JW	22/11/2010	12:40	7	127.98	–	155.77	–	–	–	97.87%	2	–	1	–
26	DvR	22/11/2010	12:55	7	128.16	–	125.22	–	–	–	102.35%	2	–	1	–
27	CH	25/11/2010	11:00	7	93.22	–	111.34	–	–	–	83.73%	2	–	1	–
28	KM	25/11/2010	11:58	6	–	126.63	–	125.33	–	–	101.04%	–	1	–	2
29	LP	25/11/2010		6	–	235.36	–	132.18	–	–	178.06%	–	1	–	2
30	NR	25/11/2010		6	–	163.18	–	138.86	–	–	117.51%	–	1	–	2
31	DB	25/11/2010	11:46	10	74.07	–	–	108.92	11.98	96.94	130.88%	2	–	–	1
32	SO	25/11/2010	11:50	10	109.45	–	–	110.73	11.15	99.58	90.98%	2	–	–	1
33	MM	25/11/2010	11:52	10	62.89	–	–	102.32	17.01	85.31	135.65%	2	–	–	1
34	TA	25/11/2010	11:15	11	–	53.01	–	79.1	–	79.1	67.02%	2	2	–	1
35	IP	25/11/2010	11:30	11	–	68.39	–	110.26	–	–	62.03%	2	2	–	1
36	JA	25/11/2010		11	–	103.93	–	132.45	14.46	117.99	78.47%	2	2	–	1

Table 20.7 Table of an analysis of variance for time using adjusted sum of the squares for tests Seq SS is sequential sum of squares, Adj Sum is adjusted sum of squares, Adj mean is adjusted mean, F stands for the F statistic and P is the probability.

Source	DF	Seq SS	Adj Sum	Adj Mean	F	P
Location	1	3502.2	1289.9	1289.9	3.44	0.073
Interruption	1	2530.2	2681.3	2681.3	7.16	**0.012**
Subject No	11	51771.4	51771.4	4706.5	12.56	**0.001**
Error	30	11240.6	11240.6	374.7		
Total	43	69044.5				

Table 20.8 Least Squares Means (LSM) for time

Location	Means (LSM)
Dispensary	138.42
Office	124.80
Interruption	123.44
0	139.78
1	
Location* interrupt	
Dispensary (no interruption)	121.20
Dispensary (interruption)	**155.63**
Office (no interruption)	125.67
Office (interruption)	123.92

Of the 25 usable results 20 (80.0%) showed that when interrupted an operative took **22%** longer to accuracy check a prescription in the dispensary as compared with a quiet office without interruptions.

When an operative was subject to a direct planned interruption or indirect interruptions as is experienced working in a noisy environment then in 32.0% of the observations it took an additional 10–18% to carry out a standard task and in 28% of observations it took an additional 29%–48% of time to carry out a standard task.

Conclusion

The mean additional time taken to complete a standard task in one phase step within the dispensing process, accuracy checking, was observed to be 18.75%. The additional time that one third of individuals took was observed to be between 30% and 48%. Interruptions, an accepted part of dispensary routine, directly reduce an individual's efficiency and have the potential

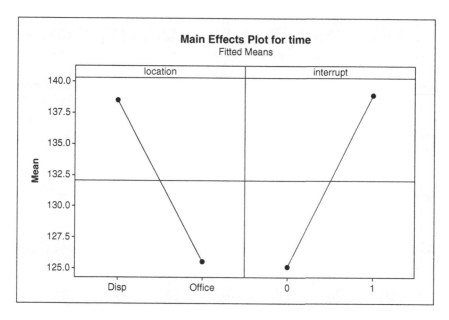

Figure 20.1 Showing graphically a main effects plot for time shows that it takes longer when interrupted in the dispensary than when not interrupted in the office.

consequently to therefore impact upon safety. Interruptions have been shown to be latent system failures.

The place of checklists, SoPs and multi-disciplinary skill mix in error prevention

The introduction of pharmacy technicians onto an acute ward, onto one medicine round a day, as part of the nursing team and partnered with a nurse, and tasking them with intravenous (IV) medicine preparation and patient administration duties has had a direct and measurable impact on reducing medicine related incidents and therefore on patient safety.

The pharmacy technician and nurse team were supported with a specially designed protocol, inspired by the WHO Safer Surgery checklist, and based on the five rights[29] of medicines administration that in addition has a calculation aide tool on the reverse. The pharmacy technicians received additional training in IV medicine administration together with an introduction to patient assessment skills. A senior medicines management nurse based within the pharmacy department oversaw the project. The pharmacy technician-nurse team was a genuine multidisciplinary skill mix team. The pharmacy technicians brought a protocol driven proactive medicine safety approach to the partnership, whilst the nurse added the necessary patient facing clinical skills.

Initial anecdotal data indicated that the addition of a pharmacy technician to the medicines administration process prevented one to three incidents (known as IRIs) a day. The pharmacy technicians proactively addressed errors or potential errors before proceeding with the IV preparation process, questioning ambiguity or identifying actual errors on the medicine administration charts. In paediatric hospitals, two nurses, registered with the Nursing and Midwifery Council (NMC), are required to check medicines before being administered.

Introduction

What could the health service learn from other high-risk industries with regard to the use of quality management Systems to reduce risks and increase efficiency? Would the introduction of a protocol-driven strategy into the medicines administration process impact positively?

Project overview

Pharmacy technicians,[30] part of the pharmacy team, working under the supervision of a pharmacist, were allocated to the project for a three-month period in the first instance.

They were assigned to the busiest medicine round particularly to IV medicine preparation and administration. They also recorded any interventions that they made or concerns that they had.

Context

In a paediatric hospital, nurses are tasked with the selection, preparation or more accurately reconstitution including if necessary dilution and administration of IV drugs to patients. IV monographs prepared by the pharmacy department give guidelines as to what options are available for reconstituting a particular drug, what precautions might need to be taken and what routes and in which way a particular drug may be administered. Depending upon the route and method of administration selected, the dilution calculations will vary and are usually complex. There is therefore no one correct way to approach these tasks and each nurse is likely to have determined a strategy based upon their knowledge and experience.

Treatment rooms where these medicines are prepared are seldom quiet. Nurses are often interrupted either directly by colleagues asking questions or indirectly by a range of activity and noises around them.

A variety of techniques have been deployed previously to make medicine related critical processes safer with varying degrees of success. Examples of these strategies include reducing the likelihood of direct interruptions by staff engaged in medicines administration wearing red tabards, these indicate that the operative must not be interrupted. Indirect interruption strategies include having designated quiet work areas. Interruptions have been shown to reduce work efficiency by 18.75%.[31]

The pharmacy technicians brought with them a cultural work practice of strict adherence to protocols and a proactive stance with regard to errors or incipient errors.

Method

The study design was a retrospective analysis of reported intervention data over a four-week period on an acute oncology ward at a tertiary NHS (UK) Hospital.

The Kotter[32] model of change management, usually used as a tool for achieving meaningful change and accelerating practice change, was deployed.[33] The pharmacy technicians were integrated into the nursing team.

A Quality Management System strategy was introduced, and this included a strictly protocol-driven approach to the process, which included the introduction of a specially designed checklist, on the reverse of which was a calculation aide. The preparation area taken over by the project became a 'quiet area' and was used in preference to the usually used noisy bedside areas for complex flow rate calculations.

Two staff surveys were conducted of nursing staff involved in the project. The first an unstructured survey collecting general views of the project from 15 (25.9%; n= 58) ward nursing staff who were present on the ward on the day shift over three days in July 2015. The second was a structured interview in the form of a questionnaire.

Results

Pharmacy technicians recorded 15 near misses (trapped errors) during a designated error-recording week in the project.

Survey – staff nursing satisfaction

Fifteen nurses were interviewed, which represented 25.9% (n=58) of the total possible ward nursing compliment. Regarding the nurses interviewed, 100% (n=15/15) were satisfied with the introduction of a pharmacy technician as part of their team.

Table 20.9 Summary of the near misses recorded. The addition of a pharmacy technician to the medicines administration process appeared to prevent 1–3 medicine related incidents (IR1s) a day from occurring.

Near misses	[Date] – Description
Wrong route	[08/06/15] – The IV Tazocin as indicated when the last IV dose was to be administered. The patient had been written up for oral Flucloxacillin and it was here that the instructions for the IVshad also been written on the prescription. The instruction said to stop all IV antibiotics and commence orals from 12 noon. The Doctor hadn't verbally communicated this to the nurse and she hadn't noticed the new instructions so had continued to proceed with the preparation of the Tazocin. I (technician) pointed out the new instructions and asked the Doctor to clarify what was required.
Wrong time (Duration of administration)	[09/06/15] – The intention was to give as a bolus over 3–5 minutes – advised that the dose must be infused as 21mg/kg over a minimum of 15 minutes
Wrong time (Duration of administration)	[09/06/15] – It was planned to administer the dose as a bolus over 3–5 minutes however as it is prescribed as 40mg/kg I advised that it needed to be given over 15–30 minutes as a bolus.
Wrong time (Duration of administration)	[11/06/15] – Drug was going to be given as a bolus – but should be administered as an infusion because of the dose.
Wrong time (Duration of administration)	[17/06/15] – Concern – pushed in one movement and not given over 5 minutes.
Wrong time (Duration of administration)	[17/06/15] – Concern – pushed in one movement and not given over 5 minutes.
Wrong time (Duration of administration)	[23/06/15] – Concerns over speed of bolus prep
Wrong time	[24/06/15] – Query Rasburicase 3mg should have been given at 5am – nurse chasing (due at 5am given at 2:40pm). Supply sorted by tech as IP needed it as soon as possible.

DATE	Comment if not checked/ prepared – Reason why or interventions
08/06/15	The IV Tazocin.or indicated when the last IV dose wan to be administered. The patient had been written up for oral Flucloxacillin and it was here that the instructions for the IVs were had alto been written on the prescription. The Instruction said to slop all IV antibiotics and commence orals from 12 noon. The Doctor hadn't verbally communicated this to the nurse and she hadn't noticed the new instructions so had continued to proceed with the preparation of the Tazocin. I (technician) pointed out the new instructions and asked the Doctor to clarify what was required.

09/06/15	Wrong volume of water drawn up – 19.1ml instead of 19ml. It was planned to bolus the dose but I advised that it must be infused after 15–30 minutes as it is 40mg/kg
09/06/15	The intention was to give as a bolus over 3–5 minutes. Advised that the dose must be infused as 21mg/kg over a minimum of 15 minutes
09/06/15	It was planned to administer the dose as a bolus over 3–5 minutes however as it is prescribed as 40mg/kg I advised that it needed to be given over 15–30 minutes as a bolus. Also advised her to change maintenance fluid to saline as glucose not compatible with Meropenem.
10/06/15	Vancomycin was prepared in error – left on side for 30 minutes while Meropenem was being infused.
11/06/15	Drug was going to be given as a bolus – but should be administered as an infusion because of the dose.
17/06/15	Flow rate wrong at bedside – needed help
17/06/15	Concern – pushed in one movement and not given over 5 minutes
17/06/15	Concern – pushed in one movement and not given over 5 minutes
18/06/15	1ml drawn up instead of 2ml (intervention)
19/06/15	Intention was to dilute to 20ml – changed to 30ml
22/06/15	Wrong dose prescribed – Doctor amended
23/06/15	**Concerns over speed of bolus preps**
24/06/15	Query Rasburicase 3mg should have been given at 5am. Nurse chasing (due at 5am given at 2:40pm) Supply sorted by tech as IP needed it as soon as possible
30/06/15	Level due at 6pm – rate needs to be at least 90minutes – discussion: talked through monograph and explained how to work out rate over 90 minutes rather than 60 minutes.

Discussion

Although slips are vastly more common than mistakes, healthcare has typically responded to all errors as if they were mistakes, resorting to remedial education and/or added layers of supervision. Such an approach may have an impact on the behaviour of an individual who committed an error but does nothing to prevent other frontline workers from committing the same error, leaving patients at risk of continued harm unless broader, more systemic, solutions are implemented.[34]

An axiom of human factors engineering is that processes should be standardized whenever possible,[35] whether to improve safety or improve efficiency. This approach is at the heart of all quality systems strategies such as Lean Six Sigma, Total Quality management (TQM), ISO9000 and other such quality systems.

Working to standardised protocols and visual cues are particularly beneficial when working in busy time-pressured, high-risk situations and where staff are subject to direct or indirect distractions. Checklists, evidence-based protocols are particularly helpful in ensuring that each necessary step in a process has been completed. In the project the use of a standardised protocol in the form of a checklist together with a calculation tool was particularly useful in trapping errors and giving assurance to those involved that they hadn't made an error.

Incorrect awareness of the time over which a bolus injection had actually been given was a function of multiple contributory causes not least of which was the impracticality of attempting to give a bolus over a prolonged period, two minutes has been suggested as a practical threshold beyond which compliance becomes difficult. The lack of a readily available means of monitoring the time, however even with such a means available, interruptions and workload pressure are likely to have reduced situational awareness as the operative would need to manage competing multiple pulls on their attention. This becomes an example of the normalisation of deviance from accepted practice. The risk of non-adherence to a recommended bolus times is a condition known as Speed Shock which occurs when a foreign substance usually a medication is rapidly introduced into the circulation.[36]

Errors in flow rate calculations could be attributed to attempting these complex data manipulations at a stage in the process that is subject to most interruptions whether direct, being asked a question by a colleague or a patient's parent or indirect that is locational, working within a noisy workspace at the patient bedside. A simple solution would be to complete the complex decisions at the start of the process in a controlled environment such as a treatment room rather than at the bedside.

This concept was proposed to the nursing team and emulated in that the task of performing complex infusion flow rate calculations, typically carried out at the patient's bedside where the environment can be noisy and distractions abound were transferred to the quiet of the treatment room and performed at the outset of the preparation sequence. In this way the nurses were less hurried and less distracted.

In paediatrics, IV administration is known to be a high-risk procedure due to the additional complexities associated with administration and dose calculations that children require.

Human factors have it that many accidents are blamed on the actions or omissions of an individual who was directly involved in operational or maintenance work. This typical but short-sighted response ignores the fundamental failures which led to the accident. These are usually rooted deeper in the organisation's design, management and decision-making function.

It is accepted that hospital wards are stressful, from the sense of workload pressure, environments in which distractions are common place and coping mechanisms may well include deviation from protocol as an acceptable approach.

Concluding remarks

Error causation theory was discussed at length and the two approaches that theorists use: the human approach, that is errors arise due to human weakness due to the fact that individuals can be careless, inattentive, perverse or simply incompetent. The second approach is that of flawed systems, flawed in the sense that error enforcing conditions, emanating from basic risk factors, lie dormant in systems but the systems are poorly designed and fail to trap these flaws from making themselves known as errors.

An appreciation of how memory works and how the brain responds to multiple stimuli formed another arm of the foundational theory required for this research.

A comprehensive literature review revealed that to date little consideration had been given to either an appreciation of how individual operatives react in a stressful healthcare environment containing multiple stimuli impacting continually on these individuals or the implication of professionalism for how operatives approach their work.

This first study showed that there was a measurable connection between workload, used as a proxy measure for stress and performance. It was noted that the pharmacist cohort and the pharmacy technician cohort behaved in explicitly different ways to each other, raising the question of professional versus non-professional behaviour. Interruptions appeared to cause both sets of operatives to become distracted and lose awareness of what they had completed and what task remained to be completed and this led to unnecessary task repetition.

It was widely accepted in the literature, that, a priori, interruptions lead to errors rather than post priori, that is from observation. It was therefore determined to establish whether interruptions in reality actually impacted adversely on operatives leading to errors occurring or whether this was merely an assumption but in reality something else was happening.

If the sole cause of errors in healthcare task related scenarios were due to external interruptions, then introducing a "do not interrupt zone" would work effectively. On the other hand, if the causes were multiple then the solution would arguably be more complex.

This reasoning led to the next project in this research and that was to establish the impact that interruptions whether direct or indirect actually had on an operative performing a designated controlled task. This post priori approach was unique in the literature in a healthcare setting. It was found that interruptions did in fact impact adversely on efficiency and output performance.

A project was devised in which the emergent hypothesis was tested. Pharmacy technicians partnered ward-nursing staff on specific medicine administration ward rounds. The medicine administration process was completely redesigned taking into account data collected during the course of this research and lessons learnt from error causation theory and error reports from in particular the airline industry.

Notes

1 Reason, James. (1990). *Human Error.* Cambridge: Cambridge University Press.
2 "Skills, Rules and Knowledge; Signals, Signs and Symbols and other distinctions in Human performance Models." *IEEE TRansaction Systems, Man and Cybernetics* 13(3): 257–266.
3 Rasmussen, J. (1986). *Information Processing and Human-Machine Interaction: An Approach to Cognitive Engineering.* New York: Elsevier Science Inc.
4 Embrey, D. (1990). *Understanding Human Error and Behaviour.* Wigan: Human Reliability Associates Ltd.
5 Groeneweg, J. (2002). *Controlling the Controllable: Preventing Business Upsets.* Leiden: DSWO Press.
6 http://innovbfa.viabloga.com/files/Herbert Simon theories of bounded rationality 1972.pdf (accessed August 8, 2016).
7 Groeneweg, J. (2002) *Controlling the uncontrollable: Preventing Business Upsets.* Leiden: DWSO Press.
8 Boehm-Davis, D. A. and R. Remington (2009). "Reducing the disruptive effects of interruption: A cognitive framework for analysing the costs and benefits of intervention strategies." *Accident Analysis & Prevention*, 41(5), 1124–1129.
9 Speier, C., Valacich, J. S. and Vessey, I. (1999) "Influence of task interruption on individual decision making: An information overload perspective", *The Decision Sciences*, 30(2): 337–360.
10 Trafton, J. G. and Monk, C. M. (2008) "Task interruptions". *Reviews of Human Factors and Ergonomics*, 3(4): 111–126.
11 Hicks R. W., Becker S. C. and Cousins, D. D., eds. (2008). MEDMARX data report. A report on the relationship of drug names and medication errors in response to the Institute of Medicine's call for action. Rockville, MD: Center for the Advancement of Patient Safety, US Pharmacopeia; www.usp.org/pdf/EN/medmarx/2008MEDMARXReport.pdf (accessed February 2009).
12 Clifton-Koeppel, R. (2008). "What Nurses Can Do Right Now to Reduce Medication Errors in the Neonatal Intensive Care Unit" *Newborn and Infant Nursing Reviews*, 8(2): 72–82.
13 www.ecfr.gov/cgi-bin/text-idx?SID=51e6ec25df70113f148981a89ed2d944&mc=true&node=se14.3.135_1100&rgn=div8 (accessed August 2009).
14 Oulasvirta, A. and Saarilouma, P. (2004). "Long-term working memory and interrupting messages in human-computer interaction". *Behaviour and Information Technology*, 64: 941–961.
15 Trafton, J. G. and Monk, C. M. (2008). "Task Interruptions". *Reviews of Human Factors and Ergonomics*, 3(4): 111–126.
16 Cades, D. M., Trafton, J. and Boehm-Davies, D. (2006). "Mitigating disruptions: Can resuming an interrupted task be trained?" Proceedings of the Human Factors and Ergonomics Society 50th annual meeting (368–371). Santa Monica CA: Human Factors & Ergonomics Society.
17 Pashler, H. (1994). "Dual-task interference in simple tasks: data and theory". *Psychol Bull*, 116(2), 220–244.
18 Speier, C. S. and Vessey, I. (1999/3). "The influence of task interruption on individual decision making."www.scopus.com/inward/citedby.url?scp=0038930 657&partnerID=8YFLogxK 30 (accessed August 8, 2016).
19 Rubinstein, J. S., et al. (2001). "Executive control of cognitive processes in task switching." *Journal of Experimental Psychology: Human Perception and Performance*, 27(4), 763–797.
20 Baddeley, A. D. and Hitch, G. (1974). "Working memory". In G. H. Bower (ed), *The psychology of learning and motivation: Advances in research and theory* (Vol. 8, 47–89). New York: Academic Press.

21 http://aspsychologyblackpoolsixth.weebly.com/working-memory-model.html (accessed August 8, 2016)

22 Baddeley, A. D. (2000). "The episodic buffer: a new component of working memory?" *Trends in Cognitive Science*, 4, 417–423. doi:10.1016/S1364-6613(00)01538-2. PMID 11058819.

23 www.gov.uk/government/uploads/system/uploads/attachment_data/file/499229/Operational_productivity_A.pdf (accessed February 2016).

24 www.institute.nhs.uk/building_capability/general/lean_thinking.html (accessed August 9 2016)

25 Schamel, J. (2012 updated); www.atchistory.org/History/checklst.htm (accessed August 6, 2016)

26 Gwande, A. (2011). *The CheckList Manifesto*. London: Profile.

27 Groeneweg, J. (1998) *Controlling the controllable: the management of safety*. 4th edition. Leiden: DSWO Press.

28 Sinclair, A., Terry, D. and Slimm M. (2012) "To investigate how disruptive interruptions are on paediatric dispensary accuracy checkers." *European Journal of Hospital Pharmacy: Science and Practice*, 19, 90.

29 www.nursingtimes.net/clinical-archive/medicine-management/five-rights-of-medication-administration (accessed 23 February 2019).

30 www.pharmacyregulation.org/education/pharmacy-technician (accessed August 8, 2016).

31 Sinclair, A.G. (2012). "The impact of interruptions on individuals carrying out the accuracy-checking step of the dispensing process". *European Journal of Hospital Pharmacy: Science and Practice*, 19: 90.

32 Kotter, J. (2015). "The 8-Step process for leading change." from www.kotter international.com/our-principles/changesteps/changesteps (accessed August 10, 2018).

33 Guérin, A., Hall, K., Lebel, D., et al. (2015). "Change management in pharmacy: a simulation exercise and identification of change barriers by pharmacy leaders." *Int J Pharm Pract*, 189–193.

34 http://psnet.ahrq.gov/primer.aspx?primerID=21, accessed July 2015.

35 http://psnet.ahrq.gov/primer.aspx?primerID=20, accessed July 2015.

36 Phillips, L. et al. (2014). *Manual of I.V. Therapeutics: Evidence-Based Practice for Infusion Therapy*. 6th edn. Philadelphia: F.A. Davis.

21 The risks and benefits of climate change policy on business operation
A system dynamic analysis

*Thomas A. Tsalis, Foteini Konstandakopoulou,
Konstantinos I. Evangelinos and
Ioannis E. Nikolaou*

Introduction

Today, Stern (2008) points out that climate change is one of the most significant risks for economies since it could threaten the steady progress of the global economy. The severity of climate change impacts on the business operation is varied according to the business sector such as the sport, tourism and the agricultural sector. The reduction of the ski season and the weakness of agribusiness to produce specific products could be some significant impacts of climate change. It is known that ski operators and their investors have lately identified the vulnerability of the ski industry from the growing negative impacts of climate change and the higher costs for businesses to face such problems (Scott and McBoyle 2007). Similarly, the agricultural sector has many options to face the impacts of climate change such as technological solutions, various governmental programs and production and financial management methods (Smit and Skinner 2002).

The impacts of climate change could have also financial risks for a number of business sectors. Some types of business risks have been classified by CERES (2010) into four categories such as physical risk, regulatory risk, reputation risk and litigation risk. The physical risks (e.g. droughts and floods) for the business sector could cause problems for the daily operation and production processes, discontinuing of supply of raw materials, the replacement of their business units and transportation problems. Blyth et al. (2007) consider that regulatory risks might go hand in hand with additional costs of business to adopt climate change mitigation and adaptation strategies. Similarly, a bad reputation for businesses could be associated with climate change risks and through obsolete technical equipment with high environmental emissions, the rising of energy consumption, and the high level of greenhouse gas (GHG) emissions of businesses. Finally, risks could be caused to businesses from litigation failures which are accompanied by heavy fines for businesses as a result of their reluctant to fulfil the requirements of regulations regarding climate change protection.

However, some of these risks are not always obstacles for business operation but sometimes are considered incentives for businesses to make new innovations and exploit new opportunities. Many large businesses have achieved new innovations in their attempt to comply with climate change requirements such as technological innovations, complementary capabilities and socio-technical innovations (Pinkse and Kolk 2010). The integration of such risks into their strategic management of their businesses in an attempt to exploit some innovative resources and capabilities to avoid the possible organizational failures is considered vital issue for their viability (Linnenluecke and Griffiths 2010).

This chapter aims to design a system dynamic model to assist scholars and practitioners in improving their understanding regarding the relationship between business operation and climate change impacts. The emphasis of the suggested model is to identify the influences of the climate change policy on business operation. The design of the model is based on some basic propositions which are developed from the current literature of corporate climate change.

The rest of the chapter is consisted of five sections. The first section includes the methodology of the chapter. The second section analyzes the current literature on potential climate change impacts on business operation. The third section describes a causal model diagram through the STELLA software program. The next section tests some scenarios for the strategies undertaken by businesses to face climate change risks and the final section analyzes the conclusions.

Methodology

The suggested methodology includes two parts. The former analyzes some certain propositions and the latter describes the structure of the system dynamic model. The propositions are based on the theory of case study research mainly to identify the basic variables of the suggested model through the relevant literature review. One of the gurus of case study research, Yin (1994) has pointed out that significant information could arise from various case studies since the experimental narrative of a meticulous subject arises from various data sources. This research strategy assists scholars in examining an evolution of a subject in a real situation and in providing suitable information in building a general theory. Another significant author in the field of the case study research, Eisenhardt and Graebner (2007) has considered that information from studies and relative literature play a critical role in building appropriate propositions. According to Yin (1994) the appropriate number of cases for a research study could range from three to 12.

This chapter is based on seven case studies which have been carried out in the agribusiness sector. Particularly, the perception of agribusiness' managers regarding the direct relationship of the climate change risks on their business operation is examined. Additional information was drawn from corporate climate change literature.

Following this, some propositions are developed in order to identify the basic variables of the model and their feedback. A system dynamic thinking model is developed to employ some certain examples in order to create correct archetypes (Wolstenholme 2003).

Figure 21.1 shows the structure of the methodology of this book chapter. It is constituted by three basic steps. The first step indicates the most important sources for drawing the available information for developing the key variable and propositions for their relations. The second step provides the relationship between case study research and system dynamic model. Finally, a number of scenarios are tested to identify the significance of various basic variables to the overall system.

Propositions development

Some scholars have examined the manners in which climate change might affect the costs and the benefits of businesses and market (Tol 2002). Indeed, many potential positive and negative aspects of businesses have identified. The relative climate change literature shows financial costs, production and the operation risks, the worry of stakeholders for the viability

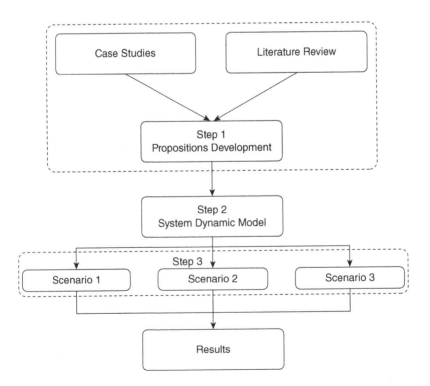

Figure 21.1 Research structure

of businesses as well as observations for increase of share market and new innovations.

Actually, there are two main schools of thoughts which examine the relationship of businesses with the natural environment such as environmental economics which face environmental degradation as externalities which affect negatively business operation and corporate environmental management which identifies these externalities as opportunities for new innovations and entrepreneurships (Reinhardt 1999). According to environmental economics, climate change may cause many barriers for businesses owing to the requirements of further financial resources to eliminate the consequences of the likely risks or to adapt their operation with the requirements of the climate change legislation. The corporate environmental management points out that, under specific circumstances, the requirements for climate change legislation are likely to offer a competitive advantage to businesses and new opportunities for benefits. The next analysis is classified in two general sub-sections such as risks and benefits of businesses from climate change impacts.

Business risks from climate change policy

Climate change policy could cause many risks for businesses. There is an ongoing debate regarding the relationship between climate change policy and business operation. The climate change policy risks could be classified into a number of categories. According to the Coalition for Environmentally Responsible Economies (CERES 2010) there are four classical types of climate change risks for businesses including physical, regulatory, reputation and litigation risks. The physical risk causes various negative impacts on business operation. Many scholars have suggested frameworks to assist the most vulnerable business sectors in integrating certain issues into their strategic management to organize businesses to face the expected large-scale impacts of climate change (Linnenluecke et al. 2011; Nikolaou et al. 2015). Many scholars have pointed out that the protection of environment quality is a considerable factor (among other external factors: economic, social, political and technological) which plays a vital role in the continual operation of businesses. Focusing on the skiing industry, Scott and McBoyle (2007) argued that many threats they could face as results of climate change include the decreasing of the regularity of snowfalls and limiting of the skiing period. Much evidence has been identified to facilitate businesses to incorporate strategies into the supply chain management in order to face climate change effects (Griffiths et al. 2007). The relative literature has shown that floods and droughts affect the production capabilities of agriculture and other economic sectors such as the food industry. This risk is presented here only for completeness reasons. It is not utilized in the suggested model.

The regulatory risks include the financial costs of businesses to invest in mitigate and adaptive strategies in order to fulfill the requirements of climate change policy (Evangelinos et al. 2015). The climate change policy is classified in three general categories of instruments. The first category encompasses the *"command and control"* instruments which enforce businesses to implement certain practices to face climate change (Reid and Toffel 2009). The second category focuses on *"market-based"* instruments (e.g. energy taxes, tradable permits) which pressure businesses to design strategies for facing climate change (Roughgarden and Schneider 1999). The final category refers to voluntary instruments adopted by businesses to face climate change challenges as well as to identify economic opportunities (Boiral 2006; Kolk and Pinkse 2007). The conducted case studies and relative literature conclude with the following propositions:

P1: The "command and control" climate change instruments force businesses to adopt climate change mitigation and adaptation practices positively related with business total costs.

P2. The "market-based" climate change instruments force businesses to adopt climate change mitigation and adaptation strategies with positive relationship with business total costs

P3: The stakeholders' views might affect businesses to adopt climate change adaptation and mitigation strategies with positive effects on their total costs

The third category of climate change risk includes damages to business reputation as a result of their poor performance on climate change, business accidents affecting climate change and lack of appropriate business strategies to face climate change impacts. There are some business sectors which are considered as the main ones responsible for climate change because of their higher carbon footprint than other business sectors (Röös et al. 2010). Additionally, the reputational profile of some business sectors seems to be more vulnerable to climate change risks. Many businesses respond to such risks by designing certain climate change strategies mainly for changing the negative public opinion for the current status of business climate change performance (Kolk and Pinkse 2007). Business adaptation strategies are mainly adopted in order to protect their reputation (Arnell and Delaney 2006). Through case study research and relative literature emerged the following rational proposition:

P4: The lack of business climate change strategies might affect negatively the business reputation.

The litigation risk is caused in the case where businesses fail to meet the requirements of legislation and they are forced to pay the corresponding fines. These risks might cause sudden expenditures for businesses and generate requirements for extra financial capitals. A rational proposition is:

> P5: *The non-compliance of businesses with climate change policy requirements might create additional costs.*

Business benefits from climate change policy

However, the policies for climate change could create conditions for businesses to adopt adaptation and mitigation strategies to face climate change risks and to advance their climate change performance. Aside from the benefits for the natural environment, these strategies have been linked many times to new businesses innovations and positive results for their reputation and total costs over a long-term period (Berkhout et al. 2006). Moreover, many business mitigation strategies (e.g. attempts to offset their carbon footprint) might offer a clear indication to consumers who are looking to buyproducts from low carbon footprint businesses and then assist businesses in increasing their share market (Iribarren et al. 2010). This analysis and relative literature could conclude with the following propositions:

> P6: *The businesses' mitigation and adaptation strategies to face climate change might cause innovations with positive influences on the total business costs.*

> P7: *The business strategies to eliminate carbon footprint might increase the consumption of goods produced by such businesses.*

Causal loop diagrams

The first step includes the design of the causal loop diagram which makes clear the structure of the explored system. In relation to Sterman (2000), this happens since the casual loop diagram provides manageable representations of the key variables and their feedback regarding the examined system. The acceptance of casual loop diagrams is associated with the aid that they provide to researchers in order to make clear the exact construction and function of explored models. It is also a very simple tool that uses stock and flows to represent the fundamental relationships of the key variables of the explored systems. These variables are represented either as module or as flows which illustrates the features that play a crucial role in the behavior of a system and their feedbacks. The arrows are assigned with polarity signs (+ or −) which demonstrate the likely or unlikely tendency of variables.

Figure 21.2 illustrates the casual loop diagram which is based on the above propositions. A number of crucial variables are provided for the business performance, the climate change performance and the three types of

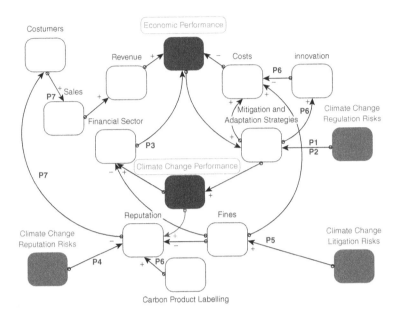

Figure 21.2 A causal loop

climate change risks. It denotes a common context to identify the essential variables of both sides: business and climate change risks. It worth noting that it could be modified according to the necessities of different business sectors to be closer to the actual situations. Some certain colors are selected to point out the crucial elements of the proposed model, such as the dark blue for economic performance and climate change performance, while the green color is assigned for the three types of climate change risks.

In this model, the climate change risks are considered external factors that affect the operation of businesses. The variable of *Climate Change Regulation Risks* represents risks from new regulations for climate change which enforce managers (as explained in the propositions 1 and 2 – P1, P2) to adopt from a bundle of various types of *Mitigation and Adaptation Strategies* (Tsalis and Nikolaou. 2017). The final decision of businesses for appropriate strategy to face climate change risks is associated with the ability of business to invest in this strategy (*Economic Performance*). Additionally, the effective application of the mitigation and adaptation strategies is expected to make new *Innovations* (according to the proposition 6 – P6) and decrease the total *Costs*. Additionally, the higher level of the *Mitigation and Adaptation Strategies* could lead to better *Climate Change Performance* for businesses. This is likely to have a positive influence on business *Reputation* in the case where climate sensitive consumers and the financial sector are willing to invest in such businesses.

The reputational risk (*Climate Change Reputation Risks*) affects the *Reputation* of businesses as shown in the proposition 4 (P4). The bad and good *Reputation* of businesses could be associated with the decision of *Consumers* and *Financial Sector*. On the one hand, the good reputation could encourage *Consumers* to buy products from businesses with better climate change performance and consequently to increase their *Revenues*, while on the other hand the bad reputation (bad *Climate Reputation Risks*) could negatively influence the preference of consumers for the products from businesses with bad *Climate Change Performance*. Similarly, the good *Reputation* could influence the decision of the *Financial Sector* (e.g. investors, banking sector and insurance companies) and business *Economic Performance* since the *Financial Sector* has asked for warranties for their investments. The businesses' bad reputation (as a result of climate change) is likely to make the *Financial Sector* reluctant to invest in those businesses.

The litigation risk is associated with the lack of compliance of businesses with the requirements of environmental legislation, a fact that might increase their costs from potential fines and penalties of non compliance. Additionally, the potential fines might have a negative influence on the reputation of businesses and possibly on the decisions of the financial sector and consumers in relation to such businesses.

A dynamic climate change business model – a scenario-based approach

A simple system dynamic model is designed in the STELA software environment. It aims to capture the "*what if*" relationships between climate change risks and business performance in the case where some of the above-mentioned climate change risks will have happened. In this section, the same colors that are assigned (red, green and blue) for variables in the casual loop section continue to be used.

The suggested model is developed in according to the basic parameters of the casual loop diagram. Two parameters are the basis for the suggested model such as the business financial performance (e.g. *Economic Performance*) and the business operation changes as result of climate change (e.g. corporate climate change performance). The measurement of *Corporate Economic Performance* variable is made with the standard mathematical formula of the revenue minus the costs (π=TR–TC).

The Total *Revenue* (TR) is estimated as the product between *Price* (P) and *Sales* (Quantity). Following the classical economic books, the total *Costs* (TC) is the sum of two parts: Fixed Costs (FC) and Variable Costs (VC). For the purposes of the suggested model, the total *Cost* contains the expenditures of businesses for the *Climate Change Adaptation and Mitigation Strategies* and relative *Rate of Fines*. The expenditures of businesses are also affected by the regulatory costs and the compliance costs

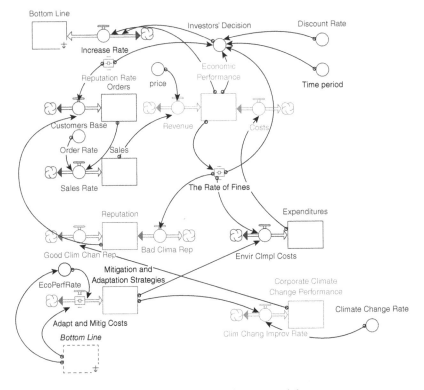

Figure 21.3 The dynamic business climate change model

through *The Mitigation and Adaptation Strategies Costs.* These types of costs could arise through requirements of legislation (propositions 1 and 2).

The adoption of climate change strategies from businesses affects their *Climate Change Performance* through lower carbon footprint of products. Some scholars have identified that large businesses which have adopted GHG strategies have achieved limited effects on their carbon performance due to weak legislative framework (Jones and Levy 2007). The better *Climate Change Performance* might positively influence the *Reputation* of businesses and the preference of *Customers* to increase their *Orders.* This also affects *Sales* and *Revenues* (Chakrabarty and Wang 2013). The improvement of reputation might have an effect on the decisions of the financial sector as it is illustrated in the casual loop diagram. At this juncture, only the effect of preference investors to invest in the businesses with better climate change performance is examined (Demertzidis et al. 2015). Some crucial parameters that affect the *Investors' Decision* regarding climate change reputation are *The Economic Performance, The Potential Costs of Physical Risks,* and *The Rate of Fines.* It is also calculated as the *Net Present Value of Economic Performance* added to *The Rate of Fines* for *Time Period* and *Discount Rate* (Blyth et al. 2007). Finally, the decision of investors affects the *Bottom Line* of businesses and their *Economic Performance.*

Regulatory climate change risk

The first scenario assumes that businesses invest in mitigation and adaptation strategies in order to face the climate command and control instruments and economic instruments. Essentially, businesses have many climate change strategies to comply with the requirements of the public policy such as carbon footprint strategies, GHG technologies, and tradable permits. The implementation of climate change strategies will be positively associated with *Economic Performance, Climate Change Performance* and *Reputation* in the case where businesses implement well such strategies (Figure 21.3). On the one side, the adoption of the *Mitigation and Adaptation Strategies* might mean progress on the business climate change performance. On the other side, the progress of *Climate Change Performance* is likely to increase the *Economic Performance* due to the improvement of businesses' *Reputation* and the preferences of customers to buy lower carbon footprint products.

Business climate change risk and reputation

The second scenario emphasizes the climate change failures of business operations and their reputation. A bad reputation for businesses regarding climate change could influence negatively their financial performance. A strategy for environmental quality might stimulate consumers to select to buy products from businesses with good environmental performance and better environmental image (Skjærseth and Skodvin 2001). Literature also shows that unexpected environmental accidents affect negatively the business reputation and their product demand, the decisions of investors in relation to the businesses and the views of suppliers.

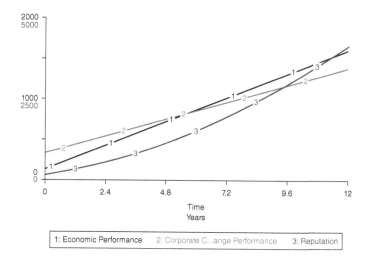

Figure 21.4 The trends of main variables of first scenario

Litigation from climate change risk

The third scenario assumes that businesses might pay fines as result of non-compliance with the requirements of relative legislation. Figure 21.5 illustrates after the fine a displacement of the economic performance curve, while the reputation and environmental performance of business are improved.

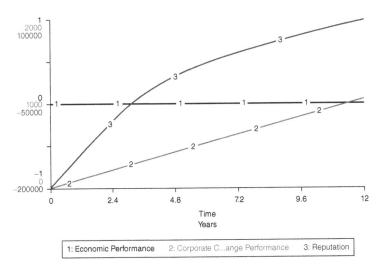

Figure 21.5 The trends of main variables of second scenario

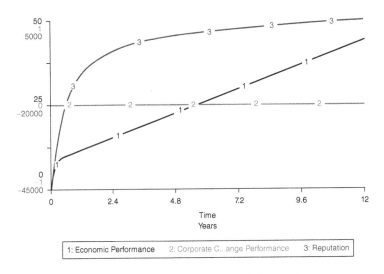

Figure 21.6 The trends of main variables of third scenario

Discussion and Conclusions

This chapter suggests a model for investigating the dynamic relationships between climate change risks and business operation. It aims at identifying the key parameters of the complex system of business climate change and examining various scenarios regarding the potential impacts of climate change on the business operation. It assists in examining the significance of climate change policy on the business environment. The suggested model improves scholars' understanding and develops a mental reproduction to help businesses to design a plan to avoid climate change risks.

The findings indicate that the businesses which adopt strategies in response to the climate change policy possibly increase business total costs. However, in the case where these strategies are followed by innovations it seems to have positive effects on the business reputation. This scenario could offer higher benefits for businesses which make a voluntary move to implement climate change strategies seeing that they gain a competitive advantage.

It is identified also that litigation and reputation risks could be associated with climate change legislation and the inability of businesses to achieve the requirements of such legislation. The model explains that the litigation risk could imply great fines which affect the business financial performance. The strictness of fines is likely to make investors hesitant to invest in these businesses so as to protect their capital from a future deteriorate economic performance. The reputational risks could occur as a result of the quick growing of the requirements of climate change legislation, the permanent nature or the height of sudden litigation risks and bad environmental performance.

In spite of the effort to advance the understanding of scholars concerning climate change and business operation, this chapter also adds to current literature by defining dynamic relationships between climate changes and business economic performance. A methodology is suggested which links proposition development and system dynamic methodologies. This assists in overcoming the subjective nature of the choosing of the crucial variables of a system and their linkages.

References

Arnell N.W. and Delaney E.K., 2006. Adapting to climate change: public water supply in England and Wales, *Climatic Change*, 78, 227–255.

Berkhout F., Hertin J. and Gann D.M., 2006. Learning to adopt: organizational adaptation to climate change impacts, *Climate Change*, 78, 135–156.

Blyth W., Bradely R., Bunn D., Clarke C., Wilson T. and Yang M., 2007. Investment risks under uncertain climate change policy, *Energy Policy*, 35, 5766–5773.

Boiral O., 2006. Global warming: should companies adopt a proactive strategy?, *Long Range Planning*, 39, 315–330.

Coalition for Environmentally Responsible Economies (CERES), 2010. Climate change risk perception and management: a survey of risk managers.

Chakrabarty S. and Wang L., 2013. Climate change mitigation and internalization: the competitiveness of multinationals, *Thunderbird International Business Review*, 55(6), 673–688.

Cohen B. and Winn M.I., 2007. Market imperfections, opportunity and sustainable entrepreneurship, *Journal of Business Venturing*, 22, 29–49.

Demertzidis, N., Tsalis, T.A., Loupa, G. and Nikolaou, I.E. (2015). A benchmarking framework to evaluate business climate change risks: A practical tool suitable for investors decision-making process. *Climate Risk Management*, 10, 95–105.

Eisenhardt K.M. and Graebner M.E., 2007. Theory building from cases: opportunities and challenges, *Academy of Management Journal*, 50(1), 25–32.

Evangelinos, K., Nikolaou, I. and Leal Filho, W. (2015). The effects of climate change policy on the business community: a corporate environmental accounting perspective. *Corporate Social Responsibility and Environmental Management*, 22(5), 257–270.

Griffiths A., Haigh N. and Rassias J., 2007. A framework for understanding institutional governance systems and climate change: the case of Australia, *European Management Journal*, 25 (6), 415–427.

Iribarren D., Hospido A., Moreira M.T. and Feijoo G., 2010. Carbon footprint of canned mussels form a business-to-consumer approach. A starting point for mussel processors and policy makers, *Environmental Science & Policy*, 13, 509–521.

Jaffe A.B., Newell R.G. and Stavins R.N., 2005. A tale of two market failures: technology and environmental policy, *Ecological Economics*, 54, 167–174.

Jones C.A. and Levy D.L., 2007. North American business strategies towards climate change, *European Management Journal*, 25 (6), 428–440.

Kolk A. and Pinkse J., 2007. Multinationals' political activities on climate change, *Business & Society*, 46 (2), 201–228.

Linnenluecke, M.K. and Griffiths, A., 2010. Corporate sustainability and organizational culture. *Journal of World Business*, 45(4), 357–366.

Linnenluecke M.K., Griffiths A. and Winn M., 2012. Extreme weather events and the critical importance of anticipatory adaptation and organizational resilience in responding to impacts, *Business Strategy and the Environment*, 21, 17–39.

Linnenluecke M.K., Stathakis A. and Griffits A., 2011. Firm relocation as adaptive response to climate change and weather extremes, *Global Environmental Change*, 21, 123–133.

Nikolaou, I., Evangelinos, K. and Leal Filho, W. (2015). A system dynamic approach for exploring the effects of climate change risks on firms' economic performance. *Journal of Cleaner Production*, 103, 499–506.

Pinkse J. and Kolk A., 2010. Challenges and trade-offs in corporate innovation for climate change, *Business Strategy and the Environment*, 19, 261–271.

Reid E.M. and Toffel M.W., 2009. Responding to public and private politics: corporate disclosure of climate change strategies, *Strategic Management Journal*, 30, 1157–1178.

Reinhardt F., 1999. Market failure and the environmental policies of firms: economic rationales for "Beyond Compliance" behavior, *Journal of Industrial Ecology*, 3 (1), 9–21.

Röös E., Sunberg C. and Hansson P.-A., 2010. Uncertainties in the carbon footprint of food products: a case study on table potatoes, *International Journal of Life Cycle Assessment*, 15, 478–488.

Roughgarden T. and Schneider S.H., 1999. Climate change policy: quantifying uncertainties for damages and optimal carbon taxes, *Energy Policy*, 27, 415–429.

Scott D. and McBoyle G., 2007. Climate change adaptation in the ski industry, *Mitigation and Adaptation Strategy for Global Change*, 12, 1411–1431.

Skjærseth J.B. and Skodvin T., 2001. Climate change and the oil industry: common problems, different strategies, *Global Environmental Politics*, 1(4), 43–64.

Smit B. and Skinner M.W., 2002. Adaptation options in agriculture to climate change: a typology, *Mitigation and Adaptation Strategies for Global Change*, 7, 85–114.

Sterman, J. (2000) *Business Dynamics: Systems Thinking and Modeling for a Complex World*. Boston: Irwin/McGraw-Hill.

Stern N., 2008. The economics of climate change, *American Economic Review: Paper & Proceedings*, 98(2), 1–37.

Tol R.S.J., 2002. Estimates of the damage costs of climate change, *Environmental and Resource Economics*, 21, 135–160.

Tsalis, T.A. and Nikolaou, I.E. (2017). Assessing the effects of climate change regulations on the business community: a system dynamic approach. *Business Strategy and the Environment*, 26(6), 826–843.

Wolstenholme E.F., 2003. Towards the definition and the use of a croe set of archetypal structures in system dynamics, *System Dynamic Review*, 19(1), 7–26.

Yin, R.K. 1994. *Case study research: design and methods* (2nd ed.). Newbury Park, CA: Sage.

22 Political risk management during instability

The experience of readymade garment manufacturing firms in Bangladesh

Krish Saha

Introduction

Political instability at the country level is generally assessed as the susceptibility of constitutional or unconstitutional change in the government (Alesina et al. 1996). Political instability literature, therefore, investigates the nature of change in government and its general impact on the socio-economic condition. The political risk literature, on the other hand, is concerned about the potential implications (mainly negative) of such instability on firms. A further categorisation of this stream of literature divides the published works into two main categories, (i) the nature of political risk that covers the classification of risks and their impacts on firms (Clark 1997; Quer et al. 2012), and (ii) the risk assessment methodology applied to assess political risk (Xiaopeng and Pheng 2013; Rios-Morales et al. 2009).

This chapter explores the nature of political risks and their management strategies adopted by the readymade garment (RMG) manufacturing firms in Bangladesh and contributes to the first stream of political risk literature. Moreover, exploration of the assessment methodology has also been conducted to make a contribution to the political risk management discipline. Therefore, we seek to determine: (i) the nature of the political risks; (ii) the risk assessment procedures adopted by the RMG firms; and (iii) the risk management strategies adopted by the RMG firms in Bangladesh. Bangladeshi RMG firms are good examples of manufacturing businesses that are the backbone to the economic growth of a developing country that frequently experiences political instability. The RMG sector is the highest earning industry in Bangladesh, contributing 18% of GDP and 76% of export earnings. More than 4300 RMG firms currently employ 4.4 million people (BGMEA 2018). As a developing country with the highest population density, weak governance and vulnerability to natural disaster Bangladesh is prone to a significant level of political instability the extent of which is summarised in Table 22.1. The RMG firms in Bangladesh thrived within such a politically volatile market over the past decades and provide an adequate scope to investigate the political risk approach in a developing market context. Such contextualisation is a proven approach in the risk literature (Jakobsen 2010; Jiménez and Delgado-García 2012; Jiménez et al.

2014; Han et al. 2018). Moreover, a series of papers (e.g. Huq et al. 2014; 2016; Huq and Stevenson 2018; Taplin 2014) that investigated various aspects of the RMG industry in Bangladesh adopted the case study method to explore nuances in the understanding of the industry's dynamics vis-à-vis various socio-political and economic challenges. The global value chain, economic significance and exposure to policy dialogue at the national and international level makes the RMG industry suitable for political risk analysis. Despite its exposure to political risk and suitability of analysis there hasn't been any study, to the best of our knowledge, conducted scoping the RMG industry in Bangladesh.

The risk literature has explored the perception of risk primarily in areas of security of individual/property and natural environment mainly through the Likert scale-based risk rating by managers (Al Khattab et al. 2007; Graham et al. 2016). The predominance of risk literature within the international business discipline focused mainly on the impact of political risk on market entry decision (Zarkada-Fraser and Fraser 2002), or on general business activities (Keillor et al. 2005). The institutionalisation of political risk assessment procedure has received extensive coverage as well (Al Khattab et al. 2008a). The general critique of such literature, however, is their narrow theoretical scope, oversight of the influence of countries political characteristics on risk perception, and its broader economic impact.

In this chapter, we draw on the political economy of Bangladesh to explore how this may have affected the perception of risk within the business community; and also the wider economic impact that pervades beyond the stakeholder of the RMG industry. The chapter is organised in several sections that review the extant literature to develop the theoretical framework (section two); outlines the methodology of the study (section three); presents and discusses the results with its contributions and limitations in sections four and five.

Theoretical framework

Impact of political risk-country level

Political risks mainly arise from political instabilities and susceptibilities of regime change. Published literature indicates four categories of political instabilities: (i) moderate-regular (within regime), (ii) major-regular (within regime), (iii) irregular (regime level change), and (iv) severe changes (revolutions, coup d'états, civil wars and political assassinations) (Campos and Nugent 2002; Feng 1997). The regular instabilities present lower risks to firms. But the irregular and severe ones are far riskier since they cause social chaos. The irregular and severe instabilities, e.g. coup d'état pose greater risks to firms and at times even reverse economic growth (Table 22.1). However, in many cases, regular instabilities within a regime may deliver a positive outcome for firms as the instability is perhaps caused by the incumbent regime's poor performance in achieving a promised political and economic outcome. For example, in the aftermath of the Brexit referendum, we have seen several cabinet and executive changes within the Conservative government.

In a democracy, voters' motivation to re-elect or change the ruling regime is mainly guided by the economic performances of the government. The democratic system is far more flexible to adjust such within-government changes and arising instabilities. Although such changes generally bring some short-term uncertainties, they facilitate greater accountability in favour of growth within the existing political system.

Not surprising that the economic growth literature sees the relationship between political risk and economic growth as a bi-directional one (Alesina et al. 1996; Darby et al. 2004; Jong-a-Pin 2009). Political risk affects a country by decreasing the total factor productivity (TFP) and physical and human capital accumulation (Aisen and Veiga 2013). Even uncertainty related to an unpredictable cabinet reshuffle may lead to a 2.39% decrease in the real annual GDP per capita.

On the other hand, political risk affects human capital accumulation by reducing investments in skill development similarly as it inhibits firms to

Table 22.1 Regime type, and instability (1972–2018)

Regime	Time Horizon	Regime Type	Category of Instability	Dominant Group
Awami League	1972–75	Parliamentary Democracy	Severe	Politician
Ziaur Rahman	1976–79	Military	Severe	Military
Ziaur Rahman (BNP)	1980–82	Presidential	Severe	Military-Civil bureaucracy
H.M. Ershad	1983–86	Military	Irregular	Military-Civil bureaucracy
H.M. Ershad (Jatiya Dal)	1987–91	Presidential	Major regular	Military-Civil bureaucracy
BNP	1992–96	Parliamentary Democracy	Major regular	Politician
Awami League	1997–01	Parliamentary Democracy	Moderate regular	Politician
BNP	2002–06	Parliamentary Democracy	Major regular	Politician
Caretaker	2007–08	Civil-Military	Irregular	Military-Civil bureaucracy
Awami League	2009–14	Parliamentary Democracy	Moderate regular	Politician
Awami League	2014–18	Parliamentary Democracy	Moderate regular	Politician

Source: *Adopted from Saha (2018)*

invest in R&D works. Political risk disqualifies economic agents from bene-fiting from investment in human capital. Therefore, they become reluctant to invest in human capital as they are sceptical of the return of that invest-ment when political uncertainty continues for a more extended period. Heightened political risk shifts resource allocation from developing human capital towards national security and induces brain-drain in developing countries (Gyimah-Brempong and De Camacho 1998; Hodge et al. 2011; Pellegrini 2011; Pellegrini and Gerlagh 2004).

The effect of the political risk on international trade is another stream of literature, although quite narrow in size and scope, that contributes to the country level impact assessments. Works of Blomberg and Hess (2006); Blomberg et al. (2004); Mirza and Verdier (2008) and Muhammad et al. (2013) report structural changes in sectoral, bilateral and commodity trade around the world as a result of terrorism, internal and external con-flicts. They classified political risk as a hidden tax that could range from 18% to 24%.

Impact of political risk-firm level

Political risk impedes productive economic decisions since a higher suscep-tibility to political instability signals an uncertain policy environment which makes risk-averse firms hesitant towards investment. There is a plethora of literature in international business that indicates how political instabil-ity dictates managerial decision choice (Al Khattab et al. 2007; 2008a; b; Busse and Hefeker 2007; Jakobsen 2010). The main risks faced by firms are the breach of contract and property right enforcement (Acemoglu et al. 2001; Levchenko et al. 2010) increase of institutional corruption, regual-tory shifts in a politically unstable country.

Radical political change often influenced by ideologically motivated government's arbitrary expropriation of property rights also enhances the level of insecurity. For example, we have seen phenomenon such as whole-sale nationalisation of private properties in Zimbabwe, Venezuela and Bolivia, forced contract renegotiations in the mining industry in Guinea (see Jakobsen 2010 for more examples in the mining sector). Keefer and Knack (2002) link these forms of arbitrary expropriation with the regime's credibility and time horizon. They argue that governments with relatively longer time horizons are less likely to engage in arbitrary confiscation since it reduces their ability to extract future economic rents due to resulting large scale divestments in the private sector, while regimes with shorter time horizons care far less about future rents. However, the regime's time horizons are determined, to a large extent, by political events such as wars, revolutions and ideological shift. Political uncertainty severely restricts the efficacy of the bureaucracy that institutionalises and enforces economic transaction conducted by firms as well. Firms that trade internally require a higher level of efficacy from the incumbent bureaucracy to stay competitive in the global market (Chan 2002).

Regimes mired in political turmoil aggressively use the monopoly of coercive power for self-aggrandisement. As a result, political competition becomes powerless to limit rent-seeking behaviours of the regimes. Such political environment is the breeding ground for institutional corruption. Zafarullah et al. (2001) outlined how coercive regimes have limited the efficacy of the state bureaucracy in Bangladesh and endorsed institutional corruption during the military rule. A corrupt and inefficient administration increases red tape and time delays for firms. Cudmore and Whalley (2003) and De Jong and Bogmans (2011) assessed the time value of delays to find that bribing government officials is the only practical way to fast track processing of the necessary administrative documentation. As a result, firms incur a higher cost of capital by spending management time negotiating the form and amount of bribe with corrupt bureaucrats in addition to the amount of bribe they are paying (Meon and Sekkat 2008).

It is evident that corruption increases during political unrest since loyalty shifts towards socio-political ties from the nation state (Fukuyama 2014; Robbins 2000). This shift in loyalty endorses discretionary behaviour within the administrative system in a way that undermines the rule of law and accountability. Myrdal's (1968; 1970) exploration of corruption and poverty in postcolonial South Asia provided substantial evidence to support such loyalty-based discretion within the administrative system. The South Asian countries struggle for independence from Britain, the separation of the Indian sub-continent and the resulting socio-political upheaval in the immediate aftermath of independence incentivised discretion and loyalty towards the elite groups. Firms that belonged to the elite groups gained immensely from various trade incentives designed for them to monopolise economic transaction. The license raj in India (1947–90) is evidence of the elite capture of economic transaction as a result of bureaucratic loyalty towards particular groups within the Indian state (Aghion et al. 2008). Similar elite capture has also been evidenced in Pakistan and Bangladesh due to the lack of institutionalisation of the political system and failure of checks and balance on the bureaucratic autonomy (Robbins 2000).

However, political risks are a lot lower in countries where frequent political unrest is the norm since firms are experienced in dealing with the consequences of political turmoil. Firms are generally aware of the likelihood of the occurrence of such events, but they do not necessarily know the timing and means of the unrest. Therefore, we see how a higher propensity of political instability can reduce the perception of political risk.

Not only the multi-national corporations, but political risks also motivate the domestic firms to exit from the home country and invest abroad. The nature of this exit is contested in earlier work on political risk driven capital flights by Alesina and Tabellini (1989). Using the examples of Latin American and Asian developing countries, they demonstrate that accumulation of massive external debts by the public sector initiates significant foreign asset accumulation by private sectors.

The severity of political risk varies according to the country's economic status. Firms from developed countries generally receive knowledge of ensuing policy and regulatory changes comparatively earlier than firms hosted by developing countries (Hood and Nawaz 2004). Firms risk of exposure somewhat depends on their country of origin. Developed countries' firms generally face import restriction, exchange control, labour interference, price controls, profit repatriation and contract problems as part of their political risk exposure, whereas socio-political disorder is the most prominent risk faced by firms in developing countries (Jiménez and Delgado-García 2012; Pahud de Mortanges and Allers 1996; Rice and Mahmoud 1990). For developing countries' firms, it is necessary to maintain close ties with the host government in order to favourably condition the impact of ensuing policy changes (Shen et al. 2001).

Political risk assessment process

Despite recognising political instability as one of the most significant risk factors for business (Shapiro 2006), the assessment process of such risk is generally subjective, unsystematic and industry specific (Pahud de Mortanges and Allers 1996; Hood and Nawaz 2004). This is, perhaps, due to the lack of awareness of political risks or the systematic process of analysing political risk (Burmester 2000). However, financial, mining and construction firms are ahead of institutionalising the political risk assessment and management process, primarily due to their higher exposure to government's policies on expropriation and spending budget, compared to manufacturing firms as they are far less influential to the economy of developed countries (Minor 2003). Even though the manufacturing firms are the backbone to the economies of developing countries institutionalisation of the risk assessment process is absent or at best at its infancy due to reasons mentioned earlier. On the other hand, high tech industries are considerably less vulnerable to political risk than low tech industries in which large scale standardisation has already taken place (Kobrin 1982).

The level of institutionalisation is determined by the propensity to start and the intensity of carrying out a political risk assessment (Burmester 2000), and also by the quality of the evaluation (Blank et al. 1980). The risk assessment process generally starts by assigning responsibility to employees or a dedicated team within a firm or commissioning an external consultancy. Assigning risk assessment to employees is often quite informal for most firms in the developing world as managers may already conduct risk assessment without having the task explicitly outlined in their job description (Al Khattab et al. 2008a). Whereas, the intensity or regularity of evaluation in most developing country firms is generally reactive and often initiated by political events (e.g. sudden regime change, socio-political unrest, mass protest and so on) that are threatening to firms' economic outcomes (Oetzel 2005). The crisis-oriented risk assessment process is commonly based on the senior managers' judgement on the significance and magnitude of the ensuing political risks and their perceived impact on firm's performance. Senior managers' judgement of political risk may often be informed by political insiders. Notwithstanding, the existing literature recognises that, although firms are exposed to the same political risk, the degree of impact affecting them varies according to few typical characteristics: geographical spread (Al Khattab et al. 2007; Howell 2001), type of industry (Al Khattab et al. 2007; Zarkada-Fraser and Fraser 2002); years of experience (Al Khattab et al. 2008a; Oetzel 2005); the size of the firm (Oetzel 2005; Al Khattab et al. 2007). These characteristics determine the level of institutionalisation of political risk within those firms as well.

Research method

A multiple case study method is the adopted method for this study. Three RMG firms and one trade association from Bangladesh supplied the data for the case study approach. The embryonic nature of the political risk research in the developing country/RMG firm context calls for an inductive exploratory study (Eisenhardt 1989; 1991; Yin 2009). The multiple case study method is robust to explore new areas. The method enhances external validity, reduce/eliminate observer bias, augment triangulation and improve generality as well (Voss 2008; Yin 2009).

Case sample

The case selection process adheres to several criteria: (i) they are general members of the Bangladesh Garments Manufacturing and Exporters Association (BGMEA) or Bangladesh Knitwear Manufacturers and Exporters Association (BKMEA); (ii) they are a first tier supplier of multinational fashion brands; (ii) they have been in the business for ten years or more to have gone through some degree of political instability; and (iv) they employ more than 500 people (Table 22.2). Smaller manufacturers, mainly order manufacturers, that did not fulfil these criteria are excluded since they sit in the second or third tier in the local supply chain. Their exposure to political risk is primarily through their buyers. It was also identified at the sample selection stage that their understanding and awareness of the political risk would not add significant value to the study. The qualitative evidence gathered from the trade association facilitates the triangulation exercise as they hold significant collective bargaining power in the political decision-making process in Bangladesh. Beside semi-structured interviews of 20 participants from four organisations conducted during 2015–17, the author has visited several factories and attended the Bangladesh Apparel & Textile Exposition (BATEXPO) in 2017 to collect a combination of rich and thick primary and secondary data.

Table 22.2 Summary of the data sample

Organisation	Type	Interviewee	Nos. of Interviewees	Age	Size
F1	Public Limited Company	Managing Director; Chief Executive Officer; Director of Quality; Executive Director; Head of Merchandising	5	20	5000 employees
F2	Private Limited Company	Managing Director; Chief Executive Officer; General Manager	3	10	750 employees
F3	Public Limited Company	Executive Director; Director of Export; Director of Quality	3	35	7300 employees
TA	Trade association	Assistant General Secretary (AGS); Chairman, Members	4	39	4369 members

Analytical procedure

The analytical procedure treated the collected data as representing phenomena and developed a more meaningful structure through construct coding and categorisation (Ritchie et al. 2013). The coding structure is systemised by keywords searching determined by the number of times a specific keyword appears in the text, the adjectives and adverbs used to qualify those words, expressions and gestures used by the interviewees in saying those words during the time of interviews.

It is noteworthy that Bangladeshi people are quite politically engaged and are not shy about expressing their views on political matters in a safe environment. Thus, there was a particular risk of exaggerating political risks and their impacts. Although political enthusiasm and engagement were very much noticed among the interviewees, the extent of exaggerating was limited. In addition, the transcripts were sent for review by the interviewees for further validation. Thus, the quality of the transcript is ensured to capture information disseminated during the interviews. Then, the interview transcripts from each case were analysed individually before cross-case analysis to select common themes (Creswell 2009). Such steps of reliability and validity were undertaken throughout the analytical process.

The thematic construction of the generated codes followed the two-stage model of Corbin and Strauss (2008) to unbundle the raw data at the first stage of open coding. Later, the axial coding method is applied to establish connections among the codes to develop broad themes. The axial coding disaggregates core themes according to categories and concept. The logical sorting and consistent ordering of the data took place through the selective coding method for main variables identification at the final stage of qualitative analysis.

Findings

The findings from the analysis, political risk associate to the RMG industry, assessments methods and management strategies are discussed in this section. The section is structured according to the research questions set out at the beginning of this chapter.

The nature of the political risks in Bangladesh

To a large extent, political risk is determined by the evolution and nature of politics of a country. Since its independence Bangladeshi politics has gone through several major periods of instabilities. Coup d'état, counter coup d'état, military rule, mass movement for democracy and election rigging are all quite common in Bangladeshi political history. The severity of the instabilities was so grave that they restricted the time horizon of the regimes and crippled the capacity of the state by hollowing out the administrative system with patronisation. The patronisation of the civil and military bureaucracy reached a level where the regimes relied on their support for survival, and as result accountability within the administrative system suffered significantly. The serious lack of accountability gave way to corruption and eventually institutionalised it. Table 22.1 summarises the various regime types, their time horizons, nature of political instability and dominant groups in Bangladesh. However, a form of political settlement has given democracy a foundation within the governance mechanism, although the nature of such democracy is not beyond question. Nevertheless, the political instability in Bangladesh poses at least three risk factors at the macro level: (i) diminishing the administrative capacity of the bureaucracy that offer primary services to people and businesses; (ii) institutionalised and systematic corruption; and (iii) severely affecting the integrity in politics.

Administrative capacity is linked with the socio-political and economic conditions; and the level of accountability present within the governance mechanism (Fukuyama 2013; Zafarullah and Rahman 2008). In addition, a conducive institutional environment that brings about necessary political support, intra-group cooperation and required resources to ensure those supports and collaboration are also needed. Unfortunately, the conducive institutional environment struggles to flourish in Bangladesh; and therefore, the existing administrative capacity is not transforming into proactive and responsive policy guidelines in Bangladesh. Moreover, the resourcefulness of the administrative services in Bangladesh is impaired by the extreme politicising of the bureaucracy and its exposure to the malign influence of unaccountable coercive power and corruption, our interviews suggest. Data indicates that politicising bureaucracy, lack of institutional coordination, corruption, lack of transparency and accountability, and incorrect implementation of reform initiatives are to blame for a diminishing administrative capacity in Bangladesh.

Corruption within the administrative system is systematic and highly institutionalised. Our interviewees believe that the scope of corruption is deeply rooted in the system of governance. Our data on corruption is also supposed by the corruption perception index (CPI) in which Bangladesh appears as one of the corrupt countries (Transparency International 2016). The pervasive presence of corruption has a profound influence on the national psyche of Bangladesh. Politician, businessmen, professionals and military personnel are all involved in corruption as our interviewees claimed.

Our data also indicate the existence of both cash-based and non-cash-based corruption within the administrative system. The cash-based corruption includes institutional and personal bribery, embezzlement; whereas non-cash-based bribery involves cartel, power abuse, harassment, extortion and so on. A non-cash-based bribery is often called '*tadbir*' (excessive lobbying in the Bengali language) (Zafarullah and Siddiquee 2001).

Obscurity and lack of information regarding service provision and complaint procedures deter transparency and fuel administrative corruption. Generally, public demand for information disclosure is watered down with the reference of the Officials Secrets Act 1923 and the Government Service Rules 1979 in Bangladesh. Therefore, citizens have no means to monitor the quality of government services. A member of the trade association (TA) believes that the root cause of administrative corruption:

> . . . is mainly due to the obscure accountability procedure and the 'job for life' arrangement for the civil servant. There is hardly a case of disciplinary measure applied on civil servants after a complaint has been lodged by the general public other than locational transfers. The transfer is part of the civil service, and it is hard to understand how locational transfer can be regarded as a punitive measure to address inappropriate practice by the civil servants. The absence of the practice of administrative laws allows public organisations to treat our needs with indifference.

'Integrity is a rare commodity in Bangladeshi politics' candidly said the founding CEO of firm F2. Politicians who uphold integrity are gradually marginalised from the centre of politics. Rewarding candidatures in general elections to big donors has become a trend for all the mainstream political parties in Bangladesh. The Daily Star (2019) quoted the Prime

Minister accusing the main opposition of such practice in a public meeting: 'They [BNP] put their nomination in auction and the highest bidder got nomination. Many of them met with me and spoke of their anguish'. It is an alarming indicator of non-integrity in Bangladeshi politics and resulted in a decreasing number of career politicians in the parliament as MPs (Figure 22.1).

The questionable motivation of those non-career politicians turned MPs to join politics exposes them in a manner that allows corrupt bureaucrats to exert their malign influence. Their lack of political experience makes them vulnerable to the domination of the more politically experienced and professionally qualified bureaucrats. The CEO of firm F2 also suggested a tense relation between the politician and senior bureaucrats that fuels corruption: 'One of the reasons for corruption is the tension between the bureaucrats and the politicians. Politicians are corrupt because they live in constant paranoia of what would happen when the loss power, so they do corruption as much as possible to gain wealth to secure future, thus loss moral authority to control corruption which facilitates bureaucratic corruption'.

Senior bureaucrats are routinely rewarded for their political loyalty or personal relationship with politicians. The rules of performance measurement

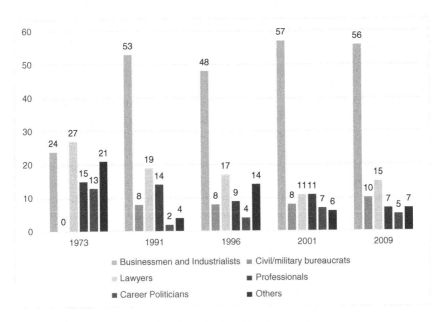

Figure 22.1 Parliament members' professional background

Note: *Data for the 2014 and 2018 general elections is not included due to a widespread boycott of the election by the main opposition parties.*

Source: *Adopted from Saha (2018)*

based on merit are often bypassed within the administrative system; and due to policy obscurity, the meritocracy falls victim of malpractice. A World Bank (2000) study highlights that frequent transfers, job rotation and increased numbers of ministerial posts in recent government have stretched the implementation capacity, complicated the existing coordination problem and further exacerbated the regulatory intrusiveness.

It is worthy of mentioning also that politics in Bangladesh revolves around personalities, dynasty and political faction[1] (Rahman 2014). Unfortunately, the democratisation in post-1991 politics has not produced much difference in the patrimonial nature of Bangladeshi politics. Zafarullah (2005) observes the eventfulness of the so-called democratic periods in Bangladesh as an impulsive/abnormal process caused by the main political parties, their mutual distrust and disrespect. Those abnormalities have sporadically thrown the country into violent political chaos and jeopardised the socio-economic growth.

Risk assessment procedure within the RMG firms

Admittedly, the nature of political risk in Bangladesh is rooted in its legislative and administrative systems and the interaction between them. Therefore, it is necessary to determine whether the risk assessment procedure within our data sample recognises the distinctive nature of the political risk that pervades in Bangladesh.

All our interviewees unequivocally recognised the importance of political risk to the performance of their businesses. However, it is also identified in our interview that the assessment process is very informal and unsystematic although senior managers are aware of the existing risk assessment procedures. The institutionalisation process of risk assessment has not even started in most of the RMG firms, confirmed by the Chairman of the trade association (TA).

Similar to the financial or oil industry, the RMG industry in Bangladesh is highly exposed to government policies for its global competitiveness. At present, the Bangladeshi government has several policy incentives (e.g. tax rebate, back-to-back LC, start-up fund, capacity building funds, bonded warehouse) dedicated only to the RMG industry. There is growing criticism from other sectors due to the discriminatory policy incentives. Spokesmen representing other industries argue that the RMG industry has received incentives for over three decades which made the industry globally competitive. Therefore, the incentives should be diverted to other industries with high growth potency, i.e. ship building and light engineering.

Furthermore, the RMG firms are also exposed to the importing countries trade policies since their landing cost is based on the availability of the generalised system of preferences (GSP). Through the GSP, the developed countries offer partial or full exemption of tariffs on the RMG imports to scale up the industry hosted mainly by developing countries. According to Europa.com (2018) and USTR.Gov (2015) India, Vietnam and Bangladesh are the biggest beneficiaries of the GSP facility. Econometric analyses of the impact of the GSP facility also confirm its trade facilitating roles. Rose (2004) and Blomberg and Hess (2006) identify that the GSP nearly doubles the amount of bilateral trade and has a more positive impact on bilateral trade than the effect of the WTO membership. The US withdrawal of the GSP for Bangladesh in June 2013 after the collapse of the Rana Plaza (Brennan 2013) has led to a decline of Bangladeshi RMG export to the American market.

The propensity and intensity of political risk assessment for the RMG firms are mostly reactive in nature. The executive director of firm F1 openly acknowledged the absence of systematic risk assessment procedure although the firm is listed in the Dhaka stock exchange. He is aware that the share market reacts to political events regularly, and a negative reaction can mean millions being wiped off from the company's share holdings. He explained that almost all RMG firms started as family owned SMEs and grew in an unregulated environment over the years. 'Nothing was done in a professional way in the RMG factories', he suggested.

However, the extensive growth in recent years and competitive pressures from competitors abroad forced the RMG firms to professionalise their approach to management. Most employees in the industry came with very little or no education and moved their way up into management as they gained required technical skills. But, 'things have changed now', the executive director of firm F1 gleefully commented, as most management trainees in his business are business school graduates; and quite a few senior managers joined the firm with foreign education. The new breed of managers has a global outlook and familiar with state-of-the art of risk assessment process. Pointing at the Head of Merchandising (HoM) (F1), he suggested that the HoM volunteered to lead the in-house risk assessment process alongside his merchandising role. The HoM explained that the current practice of assessing domestic political risk is to approach political insiders for insights and scenario planning based on insider's foresights. He regularly browses the mainstream news portals of all the client countries to keep himself up-to-date with unfolding political events. His anxiety about Brexit and related risks was quite evident in the interview as a few British high street retailers are his clients.

However, the HoM acknowledged that the process is, to a large extent, dependant on his political judgement. Yet, he ruled out the use of quantitative risk analysis saying that the data is often narrow; and the analytical outcome generally exaggerates the potential impact of political risk. He was not willing to be too risk averse and gave away business opportunities based on such analysis. Our findings correspond with Al Khattab et al. (2008b) as they found that, in the Middle East, managers were assessing political risk informally as part of their job.

Managers from other firms in our interviews acknowledged similar types of informal and judgement-based risk assessment practice. The Executive Director (F3) presented us with an example of investment in sewing technology in relation to political unrest. He withheld the investment decision for several months in the run-up to the 2018 general election. Politics becomes quite violent during the election period. As soon as the regime loosens its political control, street violence starts – this used to be the scene during the election period before.

> Many RMG workers are part of the labour wings of the major political parties, and they participated in violence advertently or inadvertently. If they knew that the owner of the firm has an opposite political inclination they did not hesitate to vandalise the factories. Thus, it caused huge financial loss for the employer and risk of lay-off for themselves. It is callous, but people act out of whim sometimes.

Therefore, risking significant investment in machineries, didn't seem reasonable to him.

There have been 122 workers action undertaken in 2016 which often turned into moderate to severe violence. Typical workers action in the RMG sector involves mass gathering, highway blockade, strike, and factory seize and vandalism as the Bangladesh Institute of Labour Studies (BILS) (2018) reports. However, these mass protests often serve as the social justice to sanction accountability in a country with minimal basis of accountability.

The BGMEA works as the central association to assess and mitigate political risks that have the potency to affect the RMG industry. Therefore, our interviews with the trade association were of great value as we could conduct a probability assessment of potential risk (Table 22.3) according to the risk bearer (country/industry/firm). Although not exhaustive this is perhaps the stepping stone of institutionalising the risk assessment procedure in the RMG sector. We have consulted risk literature, the literature on Bangladeshi politics, content from Bangladeshi press and the International Country Risk Guide (ICRG) alongside the interview data.

Political risk management strategies for the RMG sector

The political risk management strategy within the RMG firms is primarily guided by the firms' political connections. The Chairman of the TA indicated that running a successful business is seemingly impossible without the right political connections in Bangladesh. Proximity to political power provides firm owners access to policy decisions well in advance and thus reduces the impact of policy shock. We discovered that the BGMEA as the largest trade association in Bangladesh holds significant lobbying power. It was formed in 1982 as a trade association to serve the collective needs of the factory owners and exporters (BGMEA 2018). The BGMEA is organised to

Table 22.3 Probability assessment of risk

Risk bearer	Risk sources	Probability assessment	
		Bangladesh	Export destination
Country/ Industry	Regime change (resulting to change in preferential trade agreements)	N	M
	Cabinet change (resulting to change in preferential trade agreements)	N	M
	New overarching trade agreements	N	M
	Policy change relating to foreign investors/ buyers	M	N
	Coup d'états, civil wars, revolution, political assassinations	L	N
	Military invasion/ war	N	N
	Terrorist attacks	L	M
	Organised crime syndicates	H	L
	Arbitrary expropriation	L	N
	Non-convertible currency	N	N
	Predatory taxation	N	N
	Import and/or export restrictions	L	L
	Economic sanctions	N	N
	Mass demonstrations	H	M
	Sectarian violence	M	N
Firm level	Abrupt regulatory/rule change (export/ import licences, export/import duties, rules on profit repatriation)	M	L
	Subsidies/incentives/protection for selective companies	H	L
	Targeted attacks/vandalism	H	L
	Firm specific sabotage, extortion	H	L
	General breach of property rights	M	N
	Corruption	H	L
	Bureaucratic harassment	M	L
	Personal safety/security threat	M	L
	Shipping risk/piracy	M	M
	Money transfer risk due to policy change	L	L

Note: *H, M, L and N denotes probability of risk occurrences as high, moderate, low and none.*

Source: *Based on Al Khattab et al. 2007; Alon and Martin 1998; Alon and Herbert 2009; Robock 1971; Rugman and Collinson 2009; Saha 2018.*

support factory owners in getting new orders and liaising with foreign buyers. The organisation plays a significant role in export quota negotiations and in the administration of the quotas.

The influence of the organisation on the Bangladeshi power circle is evidenced in its success of making political leaders pledge that the apparel sector would be outside the purview of political strikes (Yunus and Yamagata 2012: 19). It is deemed to be a remarkable declaration due to the frequent use of strike in Bangladesh as the most effective resort of political protest. The following quote from the general manager of firm F2 signifies the damages strikes have done to the RMG firms: 'Long strikes, vandalism, arson attacks were common before. One of my HGV trucks were attacked and on fire. The whole container was burnt. Very upsetting situation and we lost a considerable amount of money due to violent politics'. He continued, 'Political violence is bad for us; we lost so many clients during the unrest a few years ago. A group of Russian businessman were coming to visit us, but they went back from the airport after finding out how violent political activist have become. It would have been excellent if we got that order, but what could we do. We need to reduce risk and uncertainty'.

Our data also reveals that the equilibrium of power is maintained by a complicated but delicate political settlement agreed by the political and industrial elites. The political settlement is negotiated through the distribution of informal rents to ensure clients' loyalty to political patrons (Roniger 2004). The patron delivers the rents to the chosen client group by allowing access to strategic and operational resources, vital trade information, exclusive government contracts and licences (Ahmed et al. 2014). By limiting the non-elite industrial groups' access rights to these rents the patron manages the balance of power and enforces cooperation among the industrial elites. The balancing act of collaboration is based on a mutual understanding that any conflict between the client and the patron reduces the rents on offer. We found evidence of such arrangement in much of the global south in the neo-institutional literature pioneered by Douglas North and his colleagues. Their limited access order theory of development argues the necessity of such an arrangement in the developing world (North et al. 2007).

In circumstances of unavoidable political shocks, negotiations between the patron and client for the rents take place. These negotiations have the potency to replace one political patron by another, and with that, the privileged industrial group also changes. However, the overreliance of the Bangladeshi economy on the RMG industry makes it the inevitable client of the political patron. The AGS of the TA reiterated the importance of the apparel industry to the Bangladeshi government: 'The apparel industry is government's favourite sector due to its contribution in GDP.'

In return for the favourable state support, the industry provides the government with the highest proportion of the export income and employ the largest number of the unskilled labour force in the country. Although political governance has become weaker in the process of constant shifts and balance of power over the years, stable privatisation and export oriented macroeconomic policy environment existed throughout. Therefore, the political settlement in Bangladesh maintains an equilibrium of weak government but a stable economic policy with a consensus for limited market intervention, privatisation, liberalisation, and export-led growth among all the dominant political actors. Thus, the informal settlement smoothed the negative impact of the volatile political environment in Bangladesh.

Beyond using the political lobby power, the RMG sector has been creative in managing political risks as our interviewees indicated. Managers revealed that they started the shift at 5.30 am and finished at 6.30 pm to avoid picketing during political strikes which are generally 6 am–6 pm long. During the political unrest in the run up to the general election in 2014 strikes continued for several weeks and took violent turns. In order to maintain a healthy working environment, during the long shifts, employees were provided three meals and entertainment breaks. Female employees with infants were given nursery facilities so that they could bring their infants at work and see/feed them throughout the shifts. These measures helped the RMG firms maintain their lead time and reduce political risk related losses.

The second example of risk management is also unique. Although the restriction on trade unionism limited the risk of labour unrest to a large extent, it suffocated the worker's voice and significantly lowered their bargaining power. Many firms felt the need to create a platform for the collective voice of the workforce. They part funded the recent labour association election that took place on the 25 August 2016. The initiative indicated improving the labour-owner relationship and reduced the likelihood of violent labour actions seen many times in the past.

Discussion and closing thoughts

From our evaluation of the political risk, its assessment and management strategies in the RMG sector in Bangladesh a few major themes emerged: (i) political risk is embedded in the nature of politics of a country and how it has evolved over the years; (ii) although mired in political uncertainties, firms from countries such as Bangladesh do not institutionalise the risk assessment process; (iii) political risks are managed by the political connections of firms or industry's lobby capacity; (iv) experience of doing business in politically volatile environments provides managers with unique risk management strategies; and finally, (v) firms habituated to political volatility generally have high risk appetite as the trade-off between risk and gain is significant. However, it is noteworthy that the size and scale of the firms determined, to a large extent, the quality of in-house risk assessment and survival when faced by severe political risks. There is also a higher preference for political insider/ judgement-based risk assessment within the RMG industry, and no indication of hiring a risk consultant is spotted in our data.

In terms of the political risk assessment of the exporting countries, the RMG firms are overreliant on their buyer. There is a significant level of trust in the political system of the western buyer countries, although events such as Brexit made Bangladeshi suppliers jittery about the weaknesses in the seemingly robust governing system.

It is also worth noting that the RMG sector faces more firm/industry level political risk, e.g. labour unrest, protest for higher wages and better working conditions, compared to the country level at present. While in the past the macro level risks, e.g. political assassination, terrorism, revolution, coup d'états, were more eminent. The shift in the nature of risk is evidence of general progress within the governance system in Bangladesh. Although the pace is slow, it provides assurance to the investors of a business friendly future. It is expected that significant investment will be dedicated to automation and skill enhancement.

We also discovered that significant improvement in the social sustainability dimension and moderate improvement in lowering gender-based discrimination had taken place in recent years, but the improvement in the process and value-addition for the economic sustainability of the sector is still a work in progress. We understand the social sustainability improvement has the potency to reduce many firm/industry level political risks that arise from social-political injustice and inequality. However, the risk of large scale labour lay-off due to automation in the advance manufacturing pioneered by China can have a severe effect on the risk exposure of the Bangladeshi RMG firms. For a robust, risk management approach these areas need consideration.

This research will benefit the RMG firms in Bangladesh as well as the FDI within the sector. The connection between national political traditions, political risks at various levels and their impacts inform managers well for a robust risk strategy suitable for the business. The development of the probability assessment of risk is undoubtedly going to offer managers a stepping stone for further bespoke risk assessment procedures. Similarly, recognising the risk return and the risk-gain trade off vis-à-vis the RMG sector may inspire reluctant investor to change their minds.

This study was constrained by the lack of sector specific literature and therefore had to develop the conceptual framework from a synthesis of political economics, political history and risk literature. In that sense the study is multi-disciplinary. On the other hand, the sectoral and country specific contextualising of the research scope may reduce the generalisability of the findings. Nonetheless, the relevance of the study can be claimed without much disagreement due to its appropriate scope. Bangladesh, as a developing country, the second largest contributor to the global RMG trade, and home to thousands of RMG business portrays a much generalisable landscape for political risk studies.

This study is the first academic work, to the best of our knowledge, scoping the political risk in the RMG sector in Bangladesh. Therefore, the work is inclined to claim thought leadership in the field as it is undeniable that this study set the path for future academic work. Nonetheless, we do not claim that the discoveries are written on stone. They are, of course, subject to further scrutiny and refinement.

Note

1 Approximately 17% of the members of the current parliament and almost the entire top political leadership are dynastic (Rahman 2014).

References

Acemoglu, D., Johnson, S. and Robinson, J.A. (2001) Colonial origins of comparative development: An empirical investigation. *American Economic Review*, 91, 1369–1401.

Aghion, P., Burgess, R., Redding, S.J. and Zilibotti, F. (2008) The unequal effects of liberalization: Evidence from dismantling the License Raj in India. *American Economic Review*, 98(4), 1397–1412.

Ahmed, F.Z., Greenleaf, A. and Sacks, A. (2014) The paradox of export growth in areas of weak governance: The case of the Ready Made Garment Sector in Bangladesh. *World Development*, 56, 258–271.

Aisen, A. and Veiga, F.J. (2013) How does political instability affect economic growth? *European Journal of Political Economy*, 29, 151–167.

Alesina, A. and Tabellini, G. (1989) External debt, capital flight and political risk. *Journal of International Economics*, 27(3–4), 199–220.

Alesina, A., Ozler, S., Roubini, N. and Swagel, P. (1996) Political instability and economic growth. *Journal of Economic Growth*, 1(2), 189–211.

Al Khattab, A., Anchor, J. and Davies, E. (2007) Managerial perceptions of political risk in international projects. *International Journal of Project Management*, 25(7), 734–743.

Al Khattab, A., Anchor, J. and Davies, E. (2008a) The institutionalisation of political risk assessment (IPRA) in Jordanian international firms. *International Business Review*, 17(6), 688–702.

Al Khattab, A., Anchor, J. and Davies, E. (2008b) Managerial practices of political risk assessment in Jordanian international business. *Risk Management*, 10(2), 135–152.

Alon, I. and Martin, M. (1998). A normative model of macro political risk assessment. *Multinational Business Review*, 6(2), 10–19.

Alon, I. and Herbert, T. (2009). A stranger in a strange land: Micro political risk and the multinational firm. *Business Horizons*, 52(2), 127–137.

BGMEA (2018) *Bangladesh Garment Manufacturers and Exporters Association*. Available at: www.bgmea.com.bd/ [Accessed 19 March 2018].

Blank, S., Basek, J., Kobrin, S. and Lapalombara, J. (1980) *Assessing the Political Environment: An Emerging Function in International Companies* . New York, Conference Board Inc.

Blomberg, S.B. and Hess, G.D. (2006) How much does violence tax trade? *Review of Economics and Statistics*, 88(4), 599–612.

Blomberg, S.B., Hess, G.D. and Orphanides, A. (2004) The macroeconomic consequences of terrorism. *Journal of Monetary Economics*, 51(5), 1007–1032.

Brennan, C. (2013) 'Bangladesh factory collapse toll passes 1,000' *BBC*, 10 May. Available at: www.bbc.co.uk/news/world-asia-22476774 [Accessed: 19 June 2016].

Burmester, B. (2000) Political risk in international business. In Tayeb, M. (Eds), *International Business: Theories, Policies and Practices*. Harlow: Pearson Education Limited, 247–272.

Busse, M. and Hefeker, C. (2007) Political risk, institutions and foreign direct investment. *European Journal of Political Economy*, 23(2), 397–415.

Campos, N.F. and Nugent, J.B. (2002) Who is afraid of political instability? *Journal of Development Economics*, 67(1), 157–172.

Chan, K.S. (2002) Trade and bureaucratic efficiency. *Economic Development and Cultural Change*, 50(3), 735–754.

Clark, E. (1997) Valuing political risk. *Journal of International Money and Finance*, 16(3), 477–490.

Corbin, J. and Strauss, A. (2008) *Basics of Qualitative Research: Techniques and Procedures for Developing Grounded Theory*. London: Sage.

Creswell, J.W. (2009) *Research Design: Qualitative, Quantitative, and Mixed Methods Approaches*. Thousand Oaks: Sage Publications.

Cudmore, E. and Whalley, J. (2003) *Border Delays and Trade Liberalization*. [pdf] Cambridge, MA: National Bureau of Economic Research (NBER Working Paper W9485). Available from: www.nber.org/papers/w9485.pdf [Accessed 7 June 2015].

Darby, J., Li, C. and Muscatelli, A. (2004) Political uncertainty, public expenditure and growth. *European Journal of Political Economy*, 20(1), 153–179.

De Jong, E. and Bogmans, C. (2011) Does corruption discourage international trade? *European Journal of Political Economy*, 27(2), 385–398.

Daily Start (2019) *Find out reasons for defeat*, January 11, 2019. Online www.thedailystar.net/bangladesh-national-election-2018/bnp-itself-responsible-december-30-national-election-debacle-sheikh-hasina-1685617.

Eisenhardt, K.M. (1991) Better stories and better constructs: The case for rigor and comparative logic. *Academy of Management review*, 16(3), 620–627.

Eisenhardt, K.M. (1989) Building theories from case study research. *Academy of Management Review*, 14(4), 532–550.

Fukuyama, F. (2014) *Political Order and Political Decay: From the Industrial Revolution to the Globalization of Democracy*. New York: Farrar, Straus, & Giroux.

Fukuyama, F. (2013) What is governance? *Governance*, 26(3), 347–368. Available at: www.dphu.org/uploads/attachements/books/books_4962_0.pdf [Accessed 2 July 2016].

Graham, B.A., Johnston, N.P. and Kingsley, A.F. (2016) A unified model of political risk. In *Strategy Beyond Markets* (119–160). Bingley: Emerald Group Publishing Limited.

Gyimah-Brempong, K. and De Camacho, S.M. (1998) Political instability, human capital, and economic growth in Latin America. *The Journal of Developing Areas*, 32(4), 449–466.

Han, X., Liu, X., Xia, T. and Gao, L. (2018). Home-country government support, interstate relations and the subsidiary performance of emerging market multinational enterprises. *Journal of Business Research*, 93, 160–172.

Hodge, A., Shankar, S., Rao, D.S. and Duhs, A. (2011) Exploring the links between corruption and growth. *Review of Development Economics*, 15(3), 474–490.

Huq, F.A. and Stevenson, M. (2018) Implementing socially sustainable practices in challenging institutional contexts: Building theory from seven developing country supplier cases. *Journal of Business Ethics*, 1–28.

Huq, F.A., Chowdhury, I.N. and Klassen, R.D. (2016) Social management capabilities of multinational buying firms and their emerging market suppliers: An exploratory study of the clothing industry. *Journal of Operations Management*, 46, 19–37.

Huq, A.F., Stevenson, M. and Zorzini, M. (2014) Social sustainability in developing country suppliers: An exploratory study in the ready made garments industry of Bangladesh. *International Journal of Operations & Production Management*, 34(5), 610–638.

Hood, J. and Nawaz, M.S. (2004) Political risk exposure and management in multi-national companies: is there a role for the corporate risk manager? *Risk Management*, 6(1), 7–18.

Howell, L.D. (2001) *The Handbook of Country and Political Risk Analysis* 3rd ed., USA, Political Risk Services Group.

Jakobsen, J. (2010) Old problems remain, new ones crop up: Political risk in the 21st century. *Business Horizons*, 53(5), 481–490.

Jiménez, A. and Delgado-García, J.B. (2012) Proactive management of political risk and corporate performance: The case of Spanish multinational enterprises. *International Business Review*, 21(6), 1029–1040.

Jiménez, A., Luis-Rico, I. and Benito-Osorio, D. (2014) The influence of political risk on the scope of internationalization of regulated companies: Insights from a Spanish sample. *Journal of World Business*, 49(3), 301–311.

Jong-a-Pin, R. (2009) On the measurement of political instability and its impact on economic growth. *European Journal of Political Economy*, 25(1), 15–29.

Keefer, P. and Knack, S. (2002) Polarization, politics and property rights: Links between inequality and growth. *Public Choice*, 111(1–2), 127–154.

Keillor, B.D., Wilkinson, T.J. and Owens, D. (2005) Threats to international operations: Dealing with political risk at the firm level. *Journal of Business Research*, 58(5), 629–635.

Kobrin, S.J. (1982) *Managing Political Risk Assessment: Strategic Response to Environmental Change*. (Vol. 8). California: University of California Press.

Levchenko, A.A., Lewis, L.T. and Tesar, L.L. (2010) The collapse of international trade during the 2008–09 crisis: in search of the smoking gun. *IMF Economic review*, 58(2), 214–253.

Meon, P.G. and Sekkat, K. (2008) Institutional quality and trade: which institutions? Which trade? *Economic Inquiry*, 46(2), 227–240.

Minor, J. (2003) Mapping the new political risk. *Risk Management*, 50 (3), 16–21.

Mirza, D. and Verdier, T. (2008) International trade, security and transnational terrorism: Theory and a survey of empirics. *Journal of Comparative Economics*, 36(2), 179–194.

Muhammad, A., D'Souza, A. and Amponsah, W. (2013) Violence, instability, and trade: evidence from Kenya's cut flower sector. *World Development*, 51, 20–31.

Myrdal, G. (1968) *Asian drama: An enquiry into the poverty of nations*, vol 2. New York: The Twentieth Century Fund. Reprint. In: A.J. Heidenheimer, M. Johnston and V.T. LeVine (eds.) (1989) *Political Corruption: A Handbook*. Oxford: Transaction Books, 953–961.

Myrdal, G. (1970) *The Challenge of World Poverty*. London: Allen Lane.

North, D.C., Wallis, J.J., Webb, S.B., and Weingast, B.R. (2007) Limited access orders in the developing world: A new approach to the problem of development. [pdf] Washington D.C.: The World Bank (Policy research paper No. 4359). Available at: https://openknowledge.worldbank.org/bitstream/handle/10986/7341/WPS4359.pdf?sequence=1&isAllowed=y [Accessed 12 October 2015].

Oetzel, J. (2005) Smaller may be beautiful but is it more risky? Assessing and managing political and economic risk in Costa Rica. *International Business Review*, 14(6), 765–790.

Pahud de Mortanges, C.P. and Allers, V. (1996) Political risk assessment: Theory and the experience of Dutch firms. *International Business Review,* 5(3), 303–318.

Pellegrini, L. (2011) The effect of corruption on growth and its transmission channels. In Pellegrini, L., ed. *Corruption, Development and the Environment.* Dordrecht: Springer, 53–74.

Pellegrini, L. and Gerlagh, R. (2004) Corruption's effect on growth and its transmission channels. *Kyklos,* 57(3), 429–456.

Quer, D., Claver, E. and Rienda, L. (2012) Political risk, cultural distance, and outward foreign direct investment: Empirical evidence from large Chinese firms. *Asia Pacific Journal of Management,* 29(4), 1089–1104.

Rahim, M.M. (2017) Improving social responsibility in RMG industries through a new governance approach in laws. *Journal of Business Ethics,* 143(4), 807–826.

Rahman, A. (2014) Traits of politicians in Bangladesh – Do they matter? *The Daily Star ,* 24 December. Available at: www.Thedailystar.Net/Traits-Of-Politicians-In-Bangladesh-56733 [Accessed 24 July 2017].

Rice, G. and Mahmoud, E. (1990) Political risk forecasting by Canadian firms. *International Journal of Forecasting,* 6(1), 89–102.

Rios-Morales, R., Gamberger, D., Šmuc, T. and Azuaje, F. (2009) Innovative methods in assessing political risk for business internationalization. *Research in International Business and Finance,* 23(2), 144–156.

Ritchie, J., Lewis, J., Nicholls, C.M. and Ormston, R. (2013) *Qualitative Research Practice: A Guide for Social Science Students and Researchers.* London: Sage.

Robbins, P. (2000) The rotten institution: corruption in natural resource management. *Political Geography,* 19(4), 423–443.

Robock, S. (1971). Political risk: identification and assessment. *Columbia Journal of World Business,* 6(4), 6–20.

Roniger, L. (2004) Political clientelism, democracy, and market economy. *Comparative Politics,* 36(3), 353–375.

Rugman, A. and Collinson, S. (2009). *International Business,* 2nd edn. Harlow: FT Prentice Hall.

Saha, K. (2018) *Governance and Trade: A Mixed-Method Analysis of Governance Impact on Bangladeshi Apparel Trade Performance,* PhD thesis, Birmingham City University, Birmingham.

Shapiro, A.C. (2006) *Multinational Financial Management,* 8th ed. New Jersey: Wiley.

Shen, L.Y., Wu, G.W. and Ng, C.S. (2001) Risk assessment for construction joint ventures in China. *Journal of Construction Engineering and Management,* 127(1), 76–81.

Taplin, M.I. (2014) Who is to blame? A re-examination of fast fashion after the 2013 factory disaster in Bangladesh. *Critical Perspectives on International Business,* 10(2), 72–83.

Voss, C. (2008) Case research in operations management. In: Karlsson, C. ed. *Researching Operations Management.* Abingdon: Routledge, 162–195.

World Bank (2000) *Anticorruption in Transition. A contribution to the policy debate.* [pdf] Washington, D.C.: The World Bank. Available at: https://siteresources.worldbank.org/INTWBIGOVANTCOR/Resources/contribution.pdf [Accessed 4 April 2015].

Xiaopeng, D. and Pheng, L.S. (2013) Understanding the critical variables affecting the level of political risks in international construction projects. *KSCE Journal of Civil Engineering,* 17(5), 895–907.

Yin, R. (2009) *Case Study Research: Design and Methods.* Thousand Oaks: Sage.

Yunus, M. and Yamagata, T. (2012) The garment industry in Bangladesh. In: Fukunishi, T. ed. *Dynamics of the Garment Industry in Low-Income Countries: Experience of Asia and Africa (Interim Report).* [pdf] *Chousakenkyu Houkokusho, IDE-JETRO.* Ch 6, 1–28. Available at:www.ide.go.jp/library/English/Publish/Download/Report/2011/pdf/410_ch6.pdf [Accessed 27 May 2014].

Zafarullah, H. and Rahman, R. (2008) The impaired state: assessing state capacity and governance in Bangladesh. *International Journal of Public Sector Management,* 21(7), 739–752.

Zafarullah, H. (2005) Surplus politics, democratic deficit: impediments to democratic rule in Bangladesh. *South Asian Survey,* 12(2), 267–285.

Zafarullah, H., Khan, M.M. and Rahman, M.H. (2001) The civil service system of Bangladesh. In: Burns, J.P. and Bowornwathana, B. eds. *Civil Service Systems in Asia.* London: Edward Elgar, 24–78.

Zafarullah, H. and Siddiquee, N.A. (2001) Dissecting public sector corruption in Bangladesh: Issues and problems of control. *Public Organization Review: A Global Journal,* 1(4), 465–486.

Zarkada-Fraser, A., and Fraser, C. (2002) Risk perception by UK firms towards the Russian market. *International Journal of Project Management,* 20(2), 99–105.

Index

Locators in *italics* refer to figures and those in **bold** to tables.

Printed in the United States
by Baker & Taylor Publisher Services